Gigantism and Acromegaly

Gigantism and Acromegaly

Edited by

Constantine A. Stratakis*

Former Scientific Director and Chief Section on Endocrinology and Genetics, Eunice Kennedy Shriver National Institute of Child Health and Human Development (NICHD), Bethesda, MD, United States of America

ACADEMIC PRESS

An imprint of Elsevier

*Dr. Stratakis is currently the Chief Scientific Officer, ELPEN, Inc. in Athens, Greece, and Senior Investigator at FORTH (www.forth.gr) in Heraklion, Crete, Greece.

Academic Press is an imprint of Elsevier
125 London Wall, London EC2Y 5AS, United Kingdom
525 B Street, Suite 1650, San Diego, CA 92101, United States
50 Hampshire Street, 5th Floor, Cambridge, MA 02139, United States
The Boulevard, Langford Lane, Kidlington, Oxford OX5 1GB, United Kingdom

Notices
Knowledge and best practice in this field are constantly changing. As new research and experience broaden our
understanding, changes in research methods, professional practices, or medical treatment may become necessary.

Practitioners and researchers must always rely on their own experience and knowledge in evaluating and using any
information, methods, compounds, or experiments described herein. In using such information or methods they
should be mindful of their own safety and the safety of others, including parties for whom they have a professional
responsibility.

To the fullest extent of the law, neither the Publisher nor the authors, contributors, or editors, assume any liability
for any injury and/or damage to persons or property as a matter of products liability, negligence or otherwise, or
from any use or operation of any methods, products, instructions, or ideas contained in the material herein.

British Library Cataloguing-in-Publication Data
A catalogue record for this book is available from the British Library

Library of Congress Cataloging-in-Publication Data
A catalog record for this book is available from the Library of Congress

ISBN: 978-0-12-814537-1

For Information on all Academic Press publications
visit our website at https://www.elsevier.com/books-and-journals

Publisher: Stacy Masucci
Acquisitions Editor: Ana Claudia A. Garcia
Editorial Project Manager: Grace Lander
Production Project Manager: Swapna Srinivasan
Cover Designer: Miles Hitchen

Typeset by MPS Limited, Chennai, India

Working together
to grow libraries in
developing countries

www.elsevier.com • www.bookaid.org

Contents

CHAPTER 7   The 3PAs syndrome and succinate dehydrogenase deficiency in pituitary tumors **127**

Paraskevi Xekouki, Vasiliki Daraki, Grigoria Betsi, Maria Chrysoulaki, Maria Sfakiotaki, Maria Mytilinaiou and Constantine A. Stratakis

CHAPTER 8   *CDKN1B* (p27) defects leading to pituitary tumors **157**

Sebastian Gulde and Natalia S. Pellegata

List of contributors

Sylvia L. Asa
Department of Pathology, University Hospitals Cleveland Medical Center and Case Western Reserve University

Albert Beckers
Service d'Endocrinologie, Domaine Universitaire du Sart Tilman, Université de Liège, Liège, Belgium

Grigoria Betsi
Department of Endocrinology, Diabetes and Metabolism, University Hospital of Heraklion, School of Medicine, University of Crete, Heraklion, Greece

Rosa Catalano
Department of Clinical Sciences and Community Health, University of Milan, Milan, Italy; PhD Program in Endocrinological Sciences, Sapienza University of Rome, Rome, Italy

Prashant Chittiboina
Neuroendocrine Surgery Unit, National Institute of Neurological Diseases and Stroke, Bethesda, MD, United States; Surgical Neurology Branch, National Institute of Neurological Diseases and Stroke, Bethesda, MD, United States

Maria Chrysoulaki
Department of Endocrinology, Diabetes and Metabolism, University Hospital of Heraklion, School of Medicine, University of Crete, Heraklion, Greece

Adrian F. Daly
Department of Endocrinology, Centre Hospitalier Universitaire de Liège, Liège Université, Liège, Belgium

Vasiliki Daraki
Department of Endocrinology, Diabetes and Metabolism, University Hospital of Heraklion, School of Medicine, University of Crete, Heraklion, Greece

Shereen Ezzat
Department of Medicine, University Health Network and University of Toronto

Sebastian Gulde
Institute for Diabetes and Cancer, Helmholtz Zentrum München, Neuherberg, Germany

Fady Hannah-Shmouni
Section on Endocrinology and Genetics, *Eunice Kennedy Shriver* National Institute of Child Health and Human Development (NICHD), Bethesda, MD, United States

Laura C. Hernández-Ramírez
Section on Endocrinology and Genetics, *Eunice Kennedy Shriver* National Institute of Child Health and Human Development (NICHD), National Institutes of Health (NIH), Bethesda, MD, United States

Anjelica Hodgson
Department of Laboratory Medicine and Pathobiology, University of Toronto, Toronto, ON, Canada

Elizabeth Hogan
Neuroendocrine Surgery Unit, National Institute of Neurological Diseases and Stroke, Bethesda, MD, United States; Department of Neurosurgery, George Washington University, Washington, DC, United States

Federica Mangili
Department of Clinical Sciences and Community Health, University of Milan, Milan, Italy

Giovanna Mantovani
Department of Clinical Sciences and Community Health, University of Milan, Milan, Italy; Endocrinology Unit, Fondazione IRCCS Ca' Granda Ospedale Maggiore Policlinico, Milan, Italy

Ozgur Mete
Department of Pathology, University Health Network, University of Toronto, Toronto, ON, Canada; Department of Laboratory Medicine and Pathobiology, University of Toronto, Toronto, ON, Canada

Maria Mytilinaiou
Department of Endocrinology, Diabetes and Metabolism, University Hospital of Heraklion, School of Medicine, University of Crete, Heraklion, Greece

Sara Pakbaz
Department of Pathology, University Health Network, University of Toronto, Toronto, ON, Canada

Natalia S. Pellegata
Institute for Diabetes and Cancer, Helmholtz Zentrum München, Neuherberg, Germany

Patrick Petrossians
Service d'Endocrinologie, Domaine Universitaire du Sart Tilman, Universitè de Liège, Liège, Belgium

Erika Peverelli
Department of Clinical Sciences and Community Health, University of Milan, Milan, Italy

Carolina R.C. Pieterman
Section of Surgical Endocrinology, Department of Surgical Oncology, The University of Texas MD Anderson Cancer Center, Houston, TX, United States

Iulia Potorac
Department of Endocrinology, Centre Hospitalier Universitaire de Liège, Liège Universite, Liège, Belgium

Liliya Rostomyan
Service d'Endocrinologie, Domaine Universitaire du Sart Tilman, Universitè de Liège, Liège, Belgium

Maria Sfakiotaki
Department of Endocrinology, Diabetes and Metabolism, University Hospital of Heraklion, School of Medicine, University of Crete, Heraklion, Greece

Constantine A. Stratakis
Section on Endocrinology and Genetics, *Eunice Kennedy Shriver* National Institute of Child Health and Human Development (NICHD), Bethesda, MD, United States

Christina Tatsi
Section on Endocrinology and Genetics, *Eunice Kennedy Shriver* National Institute of Child Health and Human Development (NICHD), National Institutes of Health, Bethesda, MD, United States

Donatella Treppiedi
Department of Clinical Sciences and Community Health, University of Milan, Milan, Italy

Steven G. Waguespack
Department of Endocrine Neoplasia & Hormonal Disorders, The University of Texas MD Anderson Cancer Center, Houston, TX, United States

Paraskevi Xekouki
Department of Endocrinology, Diabetes and Metabolism, University Hospital of Heraklion, School of Medicine, University of Crete, Heraklion, Greece

Kevin C.J. Yuen
Barrow Pituitary Center, Barrow Neurological Institute, Departments of Neuroendocrinology and Neurosurgery, University of Arizona College of Medicine and Creighton School of Medicine, Phoenix, AZ, United States

About the editor

Constantine A. Stratakis, MD, D(Med)Sci, PhD (hc)
Former Scientific Director and Chief Section on Endocrinology and Genetics,
Eunice Kennedy Shriver National Institute of Child Health and Human
Development (NICHD), Bethesda, MD, United States of America

Dr. Constantine A. Stratakis is internationally recognized for his research, mentorship, and leadership; he is one of the best-known endocrine investigators in the world.

Dr. Stratakis has been extremely successful in employing genetic linkage and other genome-wide tools to identify most of the genes known to be responsible for adrenocortical tumor (ACT) and pituitary adenoma (PA) formation. His work led to the identification of more than 35 genes for diseases affecting the endocrine glands. His laboratory has identified the genetic defects responsible for Carney complex, Carney Triad, the dyad of paragangliomas and gastric stromal sarcomas (Carney-Stratakis syndrome) and described new entities such as X-LAG (X-linked acrogigantism) and 3-PAs (the 3 Ps association of paragangliomas pheochromocytomas and PA). It was his identification of the protein kinase A (PKA) regulatory subunit type 1A (PRKAR1A) as the gene responsible for Carney complex in 2000 that led to the identification of the *PRKACA* gene as the most commonly mutated gene in all benign ACTs in 2014 by his and other laboratories. Dr. Stratakis described novel genes, such as *GPR101* in gigantism, and succinate dehydrogenase (SDH) defects in a variety of tumors, including a disease that bears his name, and led the efforts to show that many gastric stromal sarcomas (GIST) form due to SDH deficiency.

Dr. Stratakis is the author of more than 800 publications, his work has been cited > 26,000 times, and he served in major Editorial roles of leading journals, including Deputy Editor, Journal of Clinical Endocrinology and Metabolism (2010-2015) and co-Editor, Hormone & Metabolic Research and Molecular & Cellular Endocrinology (2016-today). Dr. Stratakis received the 1999 Award for Excellence in Published Clinical Research and the 2009 Ernst Oppenheimer Award, both from the Endocrine Society, and the 2019 Dale Medal from the Society for Endocrinology and many other honors.

Dr. Stratakis has also been named Visiting Professor in academic centers around the world from Harvard University, Boston, MA to the University of Adelaide, Australia, Belgrade University, Serbia, Kayseri University, Turkey, National University of Singapore, Singapore, Chinese University of Hong-Kong, Hong-Kong to name a few. Dr. Stratakis has received two honorary degrees from the Universities of Liege and Athens, respectively.

As a mentor, Dr. Stratakis has trained more than 200 students, residents, pre-doctoral, post-doctoral and clinical fellows, and now nurtures the careers of junior faculty members and investigators. He has received numerous recognitions for his mentorship and teaching. Dr. Stratakis is widely recognized for his leadership, including in professional organizations. He just completed his year as the 2018-2019 President of the Society for Pediatric Research (SPR).

Dr. Stratakis served as NICHD Scientific Director from 2009 to 2020, where he oversaw one of the largest intramural research programs at NIH, with more than 70 investigators, more than 1,000 staff, and an annual budget now exceeding $200 M. During his time, more than 80% of NICHD's space was either constructed anew or completely renovated and a new administrative structure was established after the successful completion of NICHD IRP's first ever strategic plan that he oversaw and implemented. Dr. Stratakis was committed to increasing the diversity of NICHD's IRP staff. During his time, the Board of Scientific Councilors of NICHD acquired a majority of female scientist leaders who are now advising NICHD's research; Dr. Stratakis oversaw the hire of many investigators of diverse backgrounds as well as early-stage investigators (ESIs). Dr. Stratakis supported LGBTQ+ issues, as he was one of the leaders of the NIH LGBTQ+ research work team (2014-2016). Dr. Stratakis recently assumed a position as Chief Scientific Officer and Founding Director of a new Research Institute in Athens, Greece. His laboratory will be at the Foundation for the Research and Technology-Hellas, Heraklion, Crete where he was recently elected as Senior Investigator and Research Director.

An introduction to "Gigantism and Acromegaly: Genetics, Diagnosis, and Treatment"

Constantine A. Stratakis

*Section on Endocrinology and Genetics (SEGEN), Eunice Kennedy Shriver National Institute of Child Health &
Human Development (NICHD), National Institutes of Health (NIH), Bethesda, MD, United States*

Giants (from the Greek "Γίγαντες") are known from ancient times in all societies and cultures. In Greek mythology, the giants rebelled against the Olympian Gods in a supernatural fight that ended with their final defeat, resulting in Olympian sovereignty on the Earth. They initially appeared in Hesiod's Theogony as children of Gaia (Earth) and Uranus (Sky) and were described as powerful, savage, and fearless beings with a single circular eye on their forehead. The giants challenged the Olympian Gods for supremacy in the world in the battle known as Gigantomachy [1].

In Hebrew the Nephilim were giants that existed before the flooding that occurred at Noah's time. Goliath was a giant Philistine warrior that was defeated by David, the man who was to become the king of Israel. It is said that Goliath's height was 9 ft. 6½ in., almost 3 m [2]. Akhenaten, the reformer Egyptian pharaoh who moved Egypt's capital and established a new religion around the Sun, may also have had a form of gigantism that was passed on to his family. Both *gigantes* and nephilim were related to the divine but seen as different, even vile; they had to be eliminated: in fact, one of the reasons for Noah's flooding was to get rid of the nephilim. To this day, Enceladus (Ἐγκέλαδος), a giant who was buried under Etna, is one of the Greek words for "earthquake," adding to the connotation of large size as disastrous. Even at war, where one might think being larger might confer an advantage, giants were understood to be not effective soldiers (probably due to their comorbidities); that is why Frederic the Great disbanded the regiment of "Potsdam Giants" created by his father, Frederick William I of Prussia.

But there is no question that giants always attracted attention [1,3,4]. The Irish giants (Fig. 1) in the 17th and 18th century made a living by participating in exhibitions throughout Europe; James Byrne, whose skeleton was kept at the Royal College of Surgeons of England, was 7 ft. 8 in. (2.34 m) tall. The Alton giant (Robert Pershing Wadlow, 1918—40) was the tallest man of his time and of international fame; just like recently every move of a 2.51-m-tall Turkish farmer, Sultan Kösen, was being recorded as "the world's tallest living man."

Overgrowth was recognized as an abnormality (although not necessarily a medical condition) as early as the CE 7th century: St. Isidore, Archbishop of Seville, included large size of the body in his list of growth defects. However, the credit for the first medical description of growth hormone (GH)-caused overgrowth as a disease should be given to Pierre Marie, who also coined the term "acromegaly" in 1886. Today we know, as Prof. Wouter W. de Herder has told us, that others like

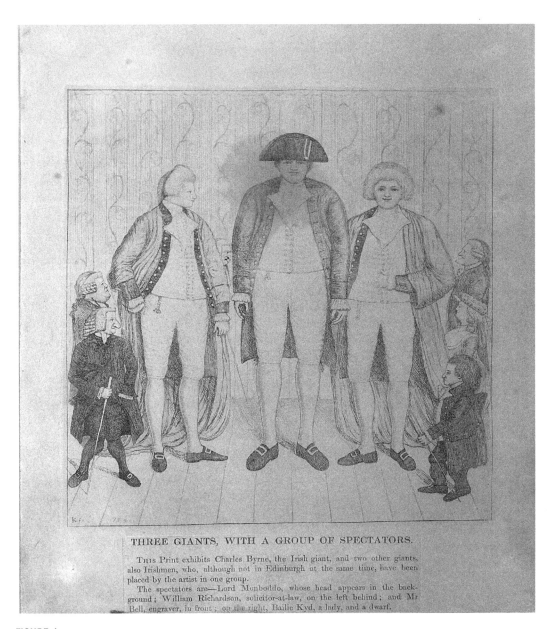

THREE GIANTS, WITH A GROUP OF SPECTATORS.

This Print exhibits Charles Byrne, the Irish giant, and two other giants, also Irishmen, who, although not in Edinburgh at the same time, have been placed by the artist in one group.

The spectators are—Lord Monboddo, whose head appears in the background; William Richardson, solicitor-at-law, on the left behind; and Mr Bell, engraver, in front; on the right, Bailie Kyd, a lady, and a dwarf.

FIGURE 1

The Irish giants (including Charles Byrne) with a group of spectators.

Courtesy Dr. Alan Guttmacher, USA.

the Dutch physician Johannes Wier had described the disease earlier [4]. Minkowski and Cushing linked acromegaly to pituitary tumors, and the discoveries of radiation therapy and transsphenoidal surgery followed [3].

The last 30 years has produced amazing discoveries in the genetics of gigantism and acromegaly [5]: *GNAS* defects (activating mutations of the G-protein stimulatory subunit-alpha) lead to gigantism in McCune−Albright syndrome, and in acromegaly when confined to sporadic GH-producing adenomas (GHPAs) [6]. *PRKAR1A* mutations that lead to increased cAMP (cyclic adenosine monophosphate) signaling, just like *GNAS* mutations, are responsible for gigantism or acromegaly in the context of Carney complex but have not been found in sporadic GHPAs [7]. *MEN1* (menin) gene mutations lead to gigantism and/or acromegaly in the context of multiple endocrine neoplasia type 1 (MEN1) but only rarely in sporadic acromegaly [8]. Mutations in the cyclin-dependent kinase (CDKN) 1B (*CDKN1B*) are found in MEN type 4 (MEN4) [1]; other CDKNs, all essential molecules in the regulation of cell cycle, growth, and proliferation, are mutated, rarely, in MEN1-/MEN4-like syndromic gigantism and/or acromegaly but not in sporadic GHPAs [1,5,8]. Most of the patients with gigantism before puberty have defects on the X-chromosome in the context of a condition that is named X-linked acrogigantism or X-LAG [9]; their presentation is rather dramatic and does not mimic the usually insidious onset of sporadic acromegaly [10]. Patients with familial isolated GHPAs (familial isolated pituitary adenoma) also exist and may present with only gigantism or acromegaly; their genetic defect is in the aryl hydrocarbon receptor-interacting protein or AIP [11]. Finally, syndromic gigantism or acromegaly can occur in the context of succinate dehydrogenase mutations, where it is found in association with paragangliomas and pheochromocytomas [12] and/or other states, associated with multiple tumors and caused by yet unknown genetic defects [13].

Although genetics has unraveled the mysteries of gigantism and/or acromegaly in the last two decades, there have been fewer advances in the treatment of these two conditions. And despite excellent reviews on their history or genetics, there have been fewer books or other comprehensive collections on all their aspects, including diagnosis and treatment [14].

In this book, we have been fortunate to get some of the best clinicians and investigators in the field of GH research to write absolutely wonderful and totally up-to-date chapters on all aspects of gigantism and acromegaly, from their history, to diagnosis, treatment, genetics, surgical management, and the overlap with other genetic syndromes or GH-producing conditions.

Indeed, gigantism and acromegaly today are not what they were a few years ago: for example, who would have thought two decades ago that, by 2021, most children with gigantism would have an identifiable genetic cause? Although molecularly unexplained cases remain, they account for a relatively small percentage of GH excess in pediatrics and young adults. In addition to genetics, transsphenoidal surgery, molecular-targeted medical therapies, and novel modes of irradiation make 2021 an exciting time to complete this book, which I hope that you will all enjoy and use for teaching our trainees, as well as for improving the care of our patients and their families.

Acknowledgments

This work was supported by the Intramural Research Program of the *Eunice Kennedy Shriver* National Institute of Child Health and Human Development.

References

[1]   Stratakis CA. A giant? Think of genetics: growth hormone-producing adenomas in the young are almost always the result of genetic defects. Endocrine 2015;50(2):272−5.

[2]   Markantes GK, Theodoropoulou A, Armeni AK, Vasileiou V, Stratakis CA, Georgopoulos NA. Cyclopes and giants: from Homer's Odyssey to contemporary genetic diagnosis. Hormones (Athens) 2016;15(3):459−63.

[3]   Sheaves R. A history of acromegaly. Pituitary 1999;2(1):7−28.

[4]   De Herder WW. The history of acromegaly. Neuroendocrinology 2016;103(1):7−17.

[5]   Barry S, Korbonits M. Update on the genetics of pituitary tumors. Endocrinol Metab Clin North Am 2020;49(3):433−52.

[6]   Vasilev V, Daly AF, Zacharieva S, Beckers A. Clinical and molecular update on genetic causes of pituitary adenomas. Horm Metab Res 2020;52(8):553−61.

[7]   Bertherat J, Horvath A, Groussin L, Grabar S, Boikos S, Cazabat L, et al. Mutations in regulatory subunit type 1A of cyclic adenosine 5'-monophosphate-dependent protein kinase (PRKAR1A): phenotype analysis in 353 patients and 80 different genotypes. J Clin Endocrinol Metab 2009;94(6):2085−91.

[8]   Agarwal SK, Mateo CM, Marx SJ. Rare germline mutations in cyclin-dependent kinase inhibitor genes in multiple endocrine neoplasia type 1 and related states. J Clin Endocrinol Metab 2009;94(5):1826−34.

[9]   Trivellin G, Daly AF, Faucz FR, Yuan B, Rostomyan L, Larco DO, et al. Gigantism and acromegaly due to Xq26 microduplications and GPR101 mutation. N Engl J Med 2014;371(25):2363−74.

[10] Beckers A, Lodish M, Trivellin G, Rostomyan L, Lee M, Faucz FR, et al. X-linked acrogigantism (X-LAG) syndrome: clinical profile and therapeutic responses. Endocr Relat Cancer 2015;22(3):353−67.

[11] Daly AF, Beckers A. Familial isolated pituitary adenomas (FIPA) and mutations in the aryl hydrocarbon receptor interacting protein (AIP) gene. Endocrinol Metab Clin North Am 2015;44(1):19−25.

[12] Xekouki P, Szarek E, Bullova P, Giubellino A, Quezado M, Mastroyannis SA, et al. Pituitary adenoma with paraganglioma/pheochromocytoma (3PAs) and succinate dehydrogenase defects in human and mice. J Clin Endocrinol Metab 2015;100(5):E10−19.

[13] Mai PL, Korde L, Kramer J, Peters J, Mueller CM, Pfeiffer S, et al. A possible new syndrome with growth-hormone secreting pituitary adenoma, colonic polyposis, lipomatosis, lentigines and renal carcinoma in association with familial testicular germ cell malignancy: a case report. J Med Case Rep 2007;1:9.

[14] Neggers S. Acromegaly: 130 years later. Neuroendocrinology 2016;103(1):5−6.

History of the identification of gigantism and acromegaly

Liliya Rostomyan, Albert Beckers and Patrick Petrossians

Service d'Endocrinologie, Domaine Universitaire du Sart Tilman, Université de Liège, Liège, Belgium

1.1 Historic times

Heroes with large body size and strong features appear in the myths and legends of almost all cultures. One of the oldest mentions of "giants" can be found in the Old Testament. The biblical Goliath of Gat (the legendary hero of the Philistines) was supposed to measure about 3 m (Fig. 1.1):

> ... And there came out from the camp of the Philistines a champion named Goliath of Gath, whose height was six cubits and a span ...
>
> **Sam 17:4, English standard translation**

> A biblical cubit corresponds to 45–57 cm, and a span, 22–28 cm, according to different interpretations. The Jewish scriptures (Torah) in Deuteronomy, the fifth book of the Hebrew Bible, describe King Og the ruler of Bashan, whose height can be estimated as 3.5–4.0 m according to the description of his iron bed:
> 9 cubits long and 4 cubits wide
>
> **Deut. 3:11, English standard translation**

Greek mythology describes in detail Giant characters, assigning them different roles. These include the children of Uranus and Gaia (Earth and Sky), predecessors of the Olympian gods, Titans, Cyclops, as well as mortal and deterrent Giants. Similar figures appeared later in many other cosmologies and mythologies like Gog and Magog in British legends, the Irish Fionn mac Cumhaill and the Norse Ymir.

The fossil findings, historical documents, and visual artworks in the museums hint about real historical personages with gigantism and acromegaly in past centuries. The images of the Egyptian pharaoh Akhenaten (Amenhotep IV) from the XVIII dynasty depict signs that could be related to the disease with typical acromegalic facial features, elongated mandible, enlarged lips, and thickened eyebrow arches (Fig. 1.2). However, one could not exclude that these features were the manifestation of a kind of artistic interpretation. Historical facts about the Roman emperor Gaius Julia Vera Maximinus, who ruled in Rome between CE 235 and CE 238, claim that his height was

Gigantism and Acromegaly. DOI: https://doi.org/10.1016/B978-0-12-814537-1.00014-2

FIGURE 1.1

David und Goliath.

Source: Color lithography by Osmar Schindler, 1888.

2.60 m, that he had a large nose and chin, thick eyebrows, and a prominent forehead. Indeed, these features stand out in his images on ancient coins.

Historical cases depicted in the visual arts have also been described. In the University Museum of Cultural History in Marburg in the Landgrave's Castle, a portrait, dated 1583, features a 26-year-old woman with typical signs of acromegaly, who was probably the chambermaid of the second wife of Louis IV, Landgrave of Hesse-Marburg, Barbara Orth (Fig. 1.3). She was named "Die große Barb" (The big Barb) due to her height reaching 4 ells (a German length measure), the equivalent of 224 cm [1].

There are notable giants whose size and feats of strength gained them the attention of kings and brought them to court. For example, the giant William Evans whose height was 2.29 m became a porter to King Charles I. A portrait of Cornish Giant Anthony Payne, who measured 2.24 m and

FIGURE 1.2

Relief of Head of Akhenaten by unknown artist, 18th Dynasty, c. 1350 BCE, Neues Museum, Staatliche Museen, Berlin.

Source: Photo © bpk - Photo Agency / Ägyptisches Museum und Papyrussammlung, Staatliche Museen, Berlin.

FIGURE 1.3

"Die große Barb".

Source: Oil on canvas by unknown painter, 16th century. The University Museum of Cultural History in Marburg, Germany.

was Sir Bevill Grenville's bodyguard, was commissioned by King Charles II for his loyalty and bravery at the battle of Lansdowne in which Sir Bevill was killed. King of Prussia Frederick William I, known for his military approach to government, created his own army of giants (Giants from Potsdam) from which he recruited only men with a height exceeding 2 m.

At all times, individuals with outstanding body size and disproportionate features were distinguished by their unusual appearance and attracted the attention of the general public.

For centuries, many people with gigantism caught the spotlight of their contemporaries, becoming objects of popular contemplation, in particular as circus and theater attractions (Fig. 1.4). They wandered around Europe, earning money in performances. Among those were the Irish Cornelius Magrath (2.36 m), Charles Byrne (2.31 m), and Patrick Cotter O'Brien (2.44 m); the Russian Feodor Machnow (2.38 m); and the Chinese Chang Yu Sing (2.44 m). In 1880 the marriage of a couple of two famous circus giants Anna Swan (2.41 m) and Captain Bates (2.36 m) became an almost sensational event in the whole of Europe.

FIGURE 1.4

Charles Byrne, also known as the Irish Giant (1761—83), presented with the Knipe Brothers who also had gigantism and dwarfs. Three Giants with A Group of Spectators by John Kay in National Portrait Gallery Collection.

Source: From "Kay's Originals Volume 1: A series of original portraits and caricature etchings by the late John Kay, with biographical sketches and illustrative anecdotes." Hugh Patton (1837).

1.2 **First medical reports**

In the past, no one suspected that what they were impressed by was actually a pathological condition that is associated with increased morbidity and mortality. The medical professionals, however, have long had a particular interest in the nature of the excessive stature of those individuals with gigantism. Early description of clinical symptoms and signs of acromegaly including cases with gigantism appeared since the 18th century in medical reports and in notes written by some patients [2,3].

The first medical case report on postmortem examination was done by Andrea Verga in 1864 [4]. He described a woman with what he called prosopectasia (Greek: *prosopon*—face, *ektasis*—enlargement), who had dysmorphic facial features. At postmortem examination, he found a destroyed sphenoid bone and displaced optic chiasm by a walnut-sized sellar tumor. A few years later, in 1877 the Italian Vincenzo Brigidi described similar changes, but he supposed that bone disease was the main cause of the disease [5].

The first medically credible recognition as a disease and the use of the term "acromegaly" (Greek: *acro*—limb, *mega*—great) comes from the French neurologist Pierre Marie in 1886 [6] (Fig. 1.5). In his report, Marie presented two cases of acromegaly and their autopsy descriptions.

FIGURE 1.5

Pierre Marie (1853–1940).

Source: Photograph by Eugéne Pirou, Collection of portraits in Bibliothéque interuniversitaire de santé, Paris, France.

Pierre Marie was the first to question whether the observed pituitary enlargement is primary or secondary to the enlargement of other organs (visceromegaly) and skeletal deformation. In 1887 Oskar Minkowski, describing a case of acromegaly, first suspected the relationship of two elements: enlargement of the pituitary and acromegaly [7], but there was no experimental evidence to support this thesis. The results of autopsy of one of Maries's cases by Broca demonstrated the enlarged sella [8]. In 1890 José Dantas de Souza-Leite included this case and several other available autopsy results in his thesis, where he definitively demonstrated pituitary enlargement in acromegaly [9]. With the advances in roentgenographic visualization of the bone structures inside the skull, Oppenheim demonstrated in 1901 an enlarged sella on X-ray in a patient with acromegaly [10].

Over the next years, there were several more reports of similar cases. In light of earlier studies and his own observations, the Italian doctor Roberto Massalongo suggested that acromegaly and gigantism are part of the same pathological process caused by pituitary hypertrophy [11]. Finally, Brissaud and Meige claimed that gigantism and acromegaly are the same diseases, and "gigantism is the acromegaly of the young" [12]. Several detailed reports of acromegaly and gigantism cases ensued in the monograph by Launois and Roy, describing large pituitary lesions [13].

Further evidence for the pituitary etiology of the disease was obtained with a thorough examination of the skull of the giant Charles Byrne by Harvey Cushing showing the deformation and destruction of the *sella turcica* [14]. In 1909 Harvey Cushing performed "partial hypophysectomy" in a patient with acromegaly and concluded that pituitary surgery leads to remission of disease symptoms [15].

1.3 The discovery of the role of growth hormone

Many histopathological studies reported pathological changes in the pituitary, like the description of eosinophilic cells' accumulation in the pituitary by Benda in 1900 [16]. In 1927 Cushing and Davidoff clearly established that acromegaly was the result of an eosinophil pituitary adenoma [17].

In 1921 Evans and Long demonstrated the presence of a growth-stimulating factor in the pituitary. In their experiment on rats, intraperitoneal administration of an extract of the anterior pituitary to animals caused changes similar to acromegaly [18]. These findings supported the idea that the cause of acromegaly is pituitary hyperfunction. Over the next 35 years, many studies followed, giving indirect evidence on the substance which stimulated growth. Thus the administration of an extract from the anterior pituitary gland of acromegaly patients was able to outline its metabolic activity and effects, in particular the induction of bone growth [19,20]. It was not until 1950 that Li and Papkoff, in California, and Raben, in Massachusetts, isolated human growth hormone (GH) from pituitaries collected at autopsy [21,22]. Five years later, with the advent of biochemical analysis with radioimmunoassay techniques, the GH serum concentrations were measured, and elevated GH levels in acromegaly patients was confirmed [23,24].

Along with the isolation of GH, there have been reports of a secondary substance that mediated the effects of GH. In their studies, Salmon and Daughaday showed that GH requires another compound for its action on skeletal tissue, in particular for the GH-dependent incorporation of sulfate into chondroitin sulfate [25]. This substance was initially called the "sulfation factor." Furthermore, the elevated activity of this compound was also observed in serums of patients with active acromegaly, whereas its level generally came back to the normal ranges when the disease became inactive [26]. Subsequent studies expanded the effects of this GH-dependent plasma factor beyond cartilage sulfation and other metabolic actions were discovered. Due to its intrinsic insulin-like biological effects that could not be inhibited by antiinsulin antibodies, it was termed "nonsuppressible insulin-like activity" [27]. In 1972 it was considered a key mediator of GH action and the term "somatomedin" was proposed [28]. Purification and measurements of serum concentration of the basic peptide (somatomedin C) by radioimmunoassay followed, permitting a more accurate assessment of its activity in different clinical conditions, in particular acromegaly [29]. In 1978 when Rinderknecht and Humbel demonstrated that active somatomedins from human plasma display structural homology with proinsulin, they were finally renamed to insulin-like growth factors.

1.4 The discovery of hypothalamic regulation

In the late 1940s British physiologist and neuroendocrinologist Geoffrey Harris' early research led to the concept of the "neurohumoral" regulation of the anterior pituitary by hypothalamus [30]. Further research included identification of the "hypothalamic releasing factors" released into the hypophyseal portal system by two groups, those of Drs. Andrew Schally and Roger Guillemin. Somatostatin was isolated, characterized, and sequenced in 1973 [31], while growth hormone–releasing hormone (GHRH) which was one of the first hypothalamic hormones to be sought [32,33] had to wait until 1982 to be discovered in a pancreatic tumor from an acromegalic patient [34–38]. Since then, exceptional cases of acromegaly have been reported due to hypothalamic or ectopic GHRH secreting tumors (gangliogliomas, pancreatic, and lung tumors), which constitute approximately less than 2% of all acromegaly cases [39].

The discovery of hypothalamic GHRH and somatostatin, as key components of central regulation of GH production, led to discussions on whether acromegaly is an exclusively pituitary disorder or whether it is the consequence of primary hypothalamic dysregulation. Most of the evidence favored the hypothesis that both conditions, acromegaly and gigantism, have an exclusive pituitary-born nature. Interestingly, recent advances in research related to the neuroendocrinology of the growth and the GH axis argued for a GHRH dysregulation and hypothalamic involvement in a particular form of early-onset acrogigantism due to *GPR101* abnormality [40].

1.5 Syndromic forms of acromegaly

Various syndromic forms of GH-secreting pituitary adenomas have been discovered. After Jacob Erdheim's autopsy report of a pituitary tumor and enlarged parathyroid glands in an acromegalic patient in 1903 [41], Cushing and Davidoff described in 1927 an acromegalic patient with multiple tumors of parathyroids and pancreatic islet cells [17]. In 1954 Wermer found that the condition featuring multiple adenomas of endocrine glands can be transmitted as a dominant trait [42]. The term multiple endocrine neoplasia (MEN) was coined later, in 1968, by Steiner et al. [43] and refers to syndromes characterized by combinations of endocrine tumors; Wermer's syndrome subsequently was renamed as MEN1.

Acromegaly associated with another multiorgan genetic syndrome was described by Carney in 1985 as an association of myxomas, skin hyperpigmentation, and endocrine overactivity [44]. Interestingly, Cushing appears to be the first to describe a patient, on whom he had operated between 1913 and 1932 (the precise date is unknown) and in whom clinical symptoms and post-mortem findings corresponded to the diagnostic criteria of Carney complex. This case was later confirmed genetically [17,45,46].

In historical records, Thomas Hasler, called the giant from Lake Tegernsee (in the Bavarian Alps, Germany), was described as having severe dysmorphic transformations of his face due to bone growth and a height of 2.35 m at the time of his death in 1876 when he was 25 years old (Fig. 1.6). Examination of his skeleton, which is housed in the Institute of Pathology at Munich University, revealed an enlargement of the sella turcica, indicated a large tumor of the pituitary, and fibrous dysplasia of the skull and multiple bones [47,48]. This case is referred to as probably

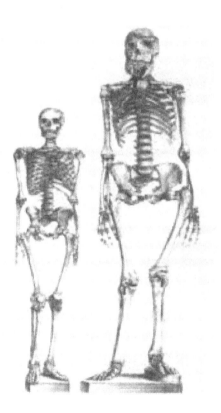

FIGURE 1.6

Skeleton of Thomas Hasler or the Tegernsee giant (1851—76) exposed next to the skeleton of an average man.

Source: Photograph from Launois and Roy, 1904 [13].

the oldest description of the association of acrogigantism with fibrous dysplasia, which is consistent with a rare congenital developmental disorder first described as a new multiorgan syndrome in 1937 by American doctors Donovan McCune and Fuller Albright [49,50].

Besides the complex syndromic presentations, the descriptions of some of the most famous cases in the history of acrogigantism, such as the Hugo brothers (Fig. 1.7) or the *Kentucky giant*, could have suggested the existence of familial forms. Another example of acromegaly is the portraits of Charles II and Philip IV, who were close relatives (son and father) (Fig. 1.8). The occurrence of pituitary tumors in a familial setting and their etiology have evoked continued interest among medical professionals. In the late 1990s a hereditary condition exclusively limited to pituitary tumors was described in Liége as familial isolated pituitary adenoma [51−55], including frequent cases of acromegaly and gigantism [53].

FIGURE 1.7

The Hugo brothers Battista (1876−1916) and Paolo Antonio (1887−1914) also known as "Géants des Alpes" depicted with their family on a postcard.

Source: Postcard from the collection of Pr Beckers.

FIGURE 1.8

An illustration of the difficulties of historic endocrinology: King of Spain Philip IV (1621−65) (left), painted here by Diego Vel´azquez and his son Charles II (1661−1700) (right), painted by Juan Carreño de Miranda. Father and son present with facial features that could suggest familial acromegaly. However, it is also well known that one of the Habsburgs family characteristics was prognatism (called the "Habsburg Jaw").

Source: Left: The National Gallery, London, United Kingdom. Right: Kunsthistorisches Museum Wien, Vienna, Austria.

1.6 The contributions of genetics

Molecular genetics uncovered the mystery of familial and syndromic forms of acromegaly−gigantism by identification of the causative role of the particular genes involved in the GH-axis disorders and pituitary tumorigenesis. Abnormalities in pituitary adenoma predisposition genes *MEN1* and *PRKAR1A* were discovered to set up a genetic background for acromegaly in the context of multiorgan syndromes, MEN1 and Carney complex [56,57]. Later, mutations in other genes (*CDKN1B*, *SDHx*, *PRKACB*, *MAX*, and *AIP*) were also unraveled to be involved in the development of GH-secreting adenoma.

Recently, the application of technological advances in studying the specific somatotropinoma populations permitted to reveal abnormalities in *GPR101* as being responsible for newly described X-LAG (X-linked acrogigantism) syndrome with a consistent clinical presentation of a distinct acrogigantism occurring in very young children [58−60]. Many of the tallest individuals described in human history had characteristics suggestive of X-LAG. Robert Pershing Wadlow (the Alton giant), the tallest man of all time, reached the height of 2.72 m at age of 22 when he died. According to historical information and medical records, he was born with normal anthropometric parameters but began to grow impressively fast from early infancy [61]. Such early disease onset and severe overgrowth refer to his case retrospectively being considered a strong candidate for the X-LAG diagnosis.

FIGURE 1.9

Julius Koch or the "Géant Constantin" (1873–1902).

Source: Postcard from the collection of Pr Beckers.

The refinement of pathophysiological concepts directed further research toward molecular confirmation in historical cases, whose skeletons were available for paleopathological examination. Julius Koch (2.59 m) who traveled the world under the name the Giant Constantin died in Mons in the early 1900s (Fig. 1.9). Historical materials, medical records, and postmortem autopsy findings evidenced the obvious presentation of early-onset overgrowth due to pituitary disorder. Recently, genetic analysis from a century-old specimen permitted to discover him as with X-LAG syndrome, thus making him the tallest case in history in whom this genetic diagnosis of gigantism has been established [62].

1.7 Going toward a new face of acromegaly

In the era of highly developed communications, the accumulation and synthesis of experience in diagnostics and management of rare diseases have become possible thanks to multicenter collaborations. The analysis of data collected from large populations reveals in detail the "true face" of the disease. New tools for creating registers and databases became helpful for collecting and analyzing the large amounts of accumulated data on acromegaly. Thus the Liege Acromegaly Survey—the largest international database of more than 3000 acromegaly patients shows interrelation in disease characteristics and trends in diagnosis and treatment of the disease over time [63].

1.8 Conclusion

Thus, while the presence of medical signs in historical records and artworks, depicting individuals with acromegaly and gigantism, continues to fascinate the general public, the historical medical literature suggests relevant background to trace the development of events in the scientific aspect. Accurate case observations have resulted in a precise description of the medical conditions, long before the discovery of their pathophysiological mechanisms. Then, more than a century-long period since the first recognition of acromegaly as a new disease entity was remarkable by the abundance of research into its etiology, pathophysiology, and epidemiology. Furthermore, along with the technological and scientific progress, returning to the historical medical records opens new research challenges in the face of a presence of disease signs suspicious for the acromegaly and gigantism in the historical cases. This discussion observes the evolution in our understanding of the nature of the disease, its clinical presentation, diagnostic challenges, and treatment development, enhancing an appreciation of comprehensive glance in the history of medicine in general and of particular medical conditions such as gigantism and acromegaly. Further inexhaustible interest in this rare pathology pursues the path to new discoveries in the field.

References

[1] Krause W, Rassner G, Happle R. "die grosse barb" in the museum of the University of Marburg. An early documentation of acromegaly. Hautarzt 2009;60(6):502−4. Available from: https://doi.org/10.1007/s00105-009-1762-9. Available from: https://www.ncbi.nlm.nih.gov/pubmed/19543868.

[2] de Herder WW. Acromegaly and gigantism in the medical literature. Case descriptions in the era before and the early years after the initial publication of Pierre Marie (1886). Pituitary 2009;12(3):236−44. Available from: https://doi.org/10.1007/s11102-008-0138-y. Available from: https://www.ncbi.nlm.nih.gov/pubmed/18683056.

[3] Pearce JM. Nicolas Saucerotte: acromegaly before Pierre Marie. J Hist Neurosci 2006;15(3):269−75. Available from: https://doi.org/10.1080/09647040500471764. Available from: https://www.ncbi.nlm.nih.gov/pubmed/16887764.

[4] Verga A. Caso singolare de prosopectasia. R Ist Lombardo Sci Lettere Bendiconti Cl Sci Mat Naturali 1864;1:111−17.

[5] Brigidi V. Studii anatomopatologica sopra un uomo divenuto stranamente deforme per chronica infirmita. Societe Medico-Fisica Fiorentina; 1877.

[6] Marie P. Sur deux cas d'acromegalie hypertrophie singuliere, non congénitale des extremites supérieures, inférieures, et céphalique. Rev de méd 1886;6:297−333.

[7] Minkowski O. Ueber einen fall von akromegalie. Berl Klin Wochenschr 1887;24:371−4.

[8] Broca A. Un squelette d'acromégalie. Archives générales de médecine (Paris) 1888;8(2):656−74.

[9]   Souza-Leite JDD. De l'acromégalie, Maladie de Marie Thèse de Paris, 1890.

[10] Oppenheim H. Berliner gesellschoft für psychiatrie und nervenkrankheiten. Sitzung von 13 November 1899. Arch für Psychiatrie und Nervenkrankheiten 1901;34:303−4.

[11] Massalongo R. Hyperfonction de la glande pituitaire et acromégalie; gigantisme et acromégalie. Rev Neurol 1895;3:225.

[12] Brissaud É, Meige H. Gigantisme et acromégalie. Journal de médecine et de chirurgie pratiques 1895; 66(4):49−76.

[13] Launois PE, Roy P. Études biologiques sur les géants. Paris: Masson; 1904.

[14] Keith A. An inquiry into the nature of the skeletal changes in acromegaly. Lancet 1911;177 (4572):993−1002. Available from: https://doi.org/10.1016/S0140-6736(01)62497-8. Available from: http://www.sciencedirect.com/science/article/pii/S0140673601624978.

[15] Cushing H. The hypophysis cerebri clinical aspects of hyperpituitarism and of hypopituitarism. J Am Med Assoc 1909;53(4):249−55.

[16] Benda C. Beitrage zur normalen und pathologischen histologie der menschlichen hypophysis cerebri. Klin Wochenschr 1900;52:1205−10.

[17] Cushing H, Davidoff LM. The pathological findings in four autopsied cases of acromegaly with a discussion of their significance. Monographs of the Rockefeller Institute for Medical Research, 131. New York: The Rockefeller Institute for Medical Research; 1927. p. pp.

[18] Evans HM, Long JA. The effect of the anterior lobe administered intraperitoneally upon growth, maturity and estrous cycles of the rat. Anat Rec 1921;21(1):62−3.

[19] Freud J, Levie L, Kroon D. Observations on growth (chondrotrophic) hormone and localization of its point of attack. J Endocrinol 1939;1(1):56−64.

[20] Kinsell LW, Michaels GD, Li CH, Larsen WE. Studies in growth: I. Interrelationship between pituitary growth factor and growth-promoting androgens in acromegaly and gigantism: II. Quantitative evaluation of bone and soft tissue growth in acromegaly and gigantism. J Clin Endocrinol 1948; 8(12):1013−36.

[21] Li CH, Papkoff H. Preparation and properties of growth hormone from human and monkey pituitary glands. Science 1956;124(3235):1293−4. Available from: https://doi.org/10.1126/science.124.3235.1293. Available from: https://www.ncbi.nlm.nih.gov/pubmed/13390958.

[22] Raben MS. Chromatographic separation of growth hormone and other peptides. Proc Soc Exp Biol Med 1956;93(2):338−40. Available from: https://doi.org/10.3181/00379727-93-22749. Available from: https://www.ncbi.nlm.nih.gov/pubmed/13379504.

[23] Glick SM, Roth J, Yalow RS, Berson SA. Immunoassay of human growth hormone in plasma. Nature 1963;199(4895):784−7. Available from: https://doi.org/10.1038/199784a0. Available from: https://www.ncbi.nlm.nih.gov/pubmed/14071191.

[24] Utiger RD, Parker ML, Daughaday WH. Studies on human growth hormone. i. A radio-immunoassay for human growth hormone. J Clin Invest 1962;41(2):254−61. Available from: https://doi.org/10.1172/JCI104478. Available from: https://www.ncbi.nlm.nih.gov/pubmed/13924012.

[25] Salmon WD, Daughaday WH. A hormonally controlled serum factor which stimulates sulfate incorporation by cartilage in vitro. J Lab Clin Med 1957;49(6):825−36.

[26] Daughaday WH, Salmon WDJ, Alexander F. Sulfation factor activity of sera from patients with pituitary disorders. J Clin Endocrinol Metab 1959;19(7):743−58. Available from: https://doi.org/10.1210/jcem-19-7-743. Available from: https://www.ncbi.nlm.nih.gov/pubmed/13664761.

[27] Froesch ER, Muller WA, Burgi H, Waldvogel M, Labhart A. Non-suppressible insulin-like activity of human serum. ii. Biological properties of plasma extracts with non-suppressible insulin-like activity. Biochim Biophys Acta 1966;121(2):360−74. Available from: https://doi.org/10.1016/0304-4165(66)90125-5. Available from: https://www.ncbi.nlm.nih.gov/pubmed/5962523.

[28] Daughaday WH, Hall K, Raben MS, Salmon WDJ, van den Brande JL, van Wyk JJ. Somatomedin: proposed designation for sulphation factor. Nature 1972;235(5333):107. Available from: https://doi.org/10.1038/235107a0. Available from: https://www.ncbi.nlm.nih.gov/pubmed/4550398.

[29] Furlanetto RW, Underwood LE, Van Wyk JJ, D'Ercole AJ. Estimation of somatomedin-c levels in normals and patients with pituitary disease by radioimmunoassay. J Clin Invest 1977;60(3):648−57. Available from: https://doi.org/10.1172/JCI108816. Available from: https://www.ncbi.nlm.nih.gov/pubmed/893668.

[30] Raisman G. An urge to explain the incomprehensible: Geoffrey Harris and the discovery of the neural control of the pituitary gland. Annu Rev Neurosci 1997;20(1):533−66. Available from: https://doi.org/10.1146/annurev.neuro.20.1.533. Available from: https://www.ncbi.nlm.nih.gov/pubmed/9056724.

[31] Ling N, Burgus R, Rivier J, Vale W, Brazeau P. The use of mass spectrometry in deducing the sequence of somatostatin—a hypothalamic polypeptide that inhibits the secretion of growth hormone. Biochem Biophys Res Commun 1973;50(1):127−33.

[32] Frohman LA, Szabo M, Berelowitz M, Stachura ME. Partial purification and characterization of a peptide with growth hormone-releasing activity from extrapituitary tumors in patients with acromegaly. J Clin Invest 1980;65(1):43−54. Available from: https://doi.org/10.1172/JCI109658. Available from: https://www.ncbi.nlm.nih.gov/pubmed/6243140.

[33] Reichlin S. Growth hormone content of pituitaries from rats with hypothalamic lesions. Endocrinology 1961;69(2):225−30. Available from: https://doi.org/10.1210/endo-69-2-225. Available from: https://www.ncbi.nlm.nih.gov/pubmed/13740467.

[34] Cronin MJ, Rogol AD, MacLeod RM, Keefer DA, Login IS, Borges JL, et al. Biological activity of a growth hormone-releasing factor secreted by a human tumor. Am J Physiol 1983;244(4):E346−53. Available from: https://doi.org/10.1152/ajpendo.1983.244.4.E346. Available from: https://www.ncbi.nlm.nih.gov/pubmed/6404176.

[35] Esch FS, Bohlen P, Ling NC, Brazeau PE, Wehrenberg WB, Thorner MO, et al. Characterization of a 40 residue peptide from a human pancreatic tumor with growth hormone releasing activity. Biochem Biophys Res Commun 1982;109(1):152−8. Available from: https://doi.org/10.1016/0006-291x(82)91578-9. Available from: https://www.ncbi.nlm.nih.gov/pubmed/7159418.

[36] Guillemin R, Brazeau P, Bohlen P, Esch F, Ling N, Wehrenberg WB. Growth hormone-releasing factor from a human pancreatic tumor that caused acromegaly. Science 1982;218(4572):585−7.

[37] Rivier J, Spiess J, Thorner M, Vale W. Characterization of a growth hormone-releasing factor from a human pancreatic islet tumour. Nature 1982;300(5889):276−8. Available from: https://doi.org/10.1038/300276a0. Available from: https://www.ncbi.nlm.nih.gov/pubmed/6292724.

[38] Spiess J, Rivier J, Thorner M, Vale W. Sequence analysis of a growth hormone releasing factor from a human pancreatic islet tumor. Biochemistry 1982;21(24):6037−40. Available from: https://doi.org/10.1021/bi00267a002. Available from: https://www.ncbi.nlm.nih.gov/pubmed/6295453.

[39] Borson-Chazot F, Garby L, Raverot G, Claustrat F, Raverot V, Sassolas G, et al. Acromegaly induced by ectopic secretion of GHRH: a review 30 years after GHRH discovery. Ann Endocrinol (Paris) 2012;73(6):497−502. Available from: https://doi.org/10.1016/j.ando.2012.09.004. Available from: https://www.ncbi.nlm.nih.gov/pubmed/23122576.

[40] Daly AF, Lysy PA, Desfilles C, Rostomyan L, Mohamed A, Caberg JH, et al. GHRH excess and blockade in X-LAG syndrome. Endocr Relat Cancer 2016;23(3):161−70. Available from: https://doi.org/10.1530/ERC-15-0478. Available from: https://www.ncbi.nlm.nih.gov/pubmed/26671997.

[41] Erdheim J. Zur normalen und pathologischen histologie der glandula thyreoidea, parathyreoidea und hypophysis. Beitr Pathol Anat 1903;33:158−236.

[42] Wermer P. Genetic aspects of adenomatosis of endocrine glands. Am J Med 1954;16(3):363−71. Available from: https://doi.org/10.1016/0002-9343(54)90353-8. Available from: https://www.ncbi.nlm.nih.gov/pubmed/13138607.

[43] Steiner AL, Goodman AD, Powers SR. Study of a kindred with pheochromocytoma, medullary thyroid carcinoma, hyperparathyroidism and Cushing's disease: multiple endocrine neoplasia, type 2. Medicine (Baltimore) 1968;47(5):371−409. Available from: https://doi.org/10.1097/00005792-196809000-00001. Available from: https://www.ncbi.nlm.nih.gov/pubmed/4386574.

[44] Carney JA, Gordon H, Carpenter PC, Shenoy BV, Go VL. The complex of myxomas, spotty pigmentation, and endocrine overactivity. Medicine (Baltimore) 1985;64(4):270−83. Available from: https://doi.org/10.1097/00005792-198507000-00007. Available from: https://www.ncbi.nlm.nih.gov/pubmed/4010501.

[45] Tsay CJ, Stratakis CA, Faucz FR, London E, Stathopoulou C, Allgauer M, et al. Harvey Cushing treated the first known patient with Carney complex. J Endocr Soc 2017;1(10):1312−21. Available from: https://doi.org/10.1210/js.2017-00283. Available from: https://www.ncbi.nlm.nih.gov/pubmed/29264456.

[46] Weed LH, Cushing H, Jacobson C. Further studies on the role of the hypophysis in the metabolism of carbohydrates: the autonomic control of the pituitary gland. Lord Baltimore Press; 1913.

[47] Nerlich A, Peschel O, Lohrs U, Parsche F, Betz P. Juvenile gigantism plus polyostotic fibrous dysplasia in the Tegernsee giant. Lancet 1991;338(8771):886−7. Available from: https://www.ncbi.nlm.nih.gov/pubmed/1681239.

[48] Vogl TJ, Nerlich A, Dresel SH, Bergman C. CT of the "Tegernsee Giant": juvenile gigantism and polyostotic fibrous dysplasia. J Comput Assist Tomogr 1994;18(2):319−22. Available from: https://www.ncbi.nlm.nih.gov/pubmed/8126293.

[49] Albright F, Butler AM, Hampton AO, Smith P. Syndrome characterized by osteitis fibrosa disseminata, areas of pigmentation and endocrine dysfunction, with precocious puberty in females. N Engl J Med 1937;216(17):727−46. Available from: https://doi.org/10.1056/nejm193704292161701. Available from: https://www.nejm.org/doi/full/10.1056/NEJM193704292161701.

[50] McCune DJ, Bruch H. Osteodystrophia fibrosa: report of a case in which the condition was combined with precocious puberty, pathologic pigmentation of the skin and hyperthyroidism, with a review of the literature. Am J Dis Child 1937;54(4):806−48.

[51] Beckers A. Familial isolated pituitary adenomas. Ninth international workshop on multiple endocrine neoplasia (MEN2004). J Intern Med 2004;255(6):696−730.

[52] Beckers A, Aaltonen LA, Daly AF, Karhu A. Familial isolated pituitary adenomas (FIPA) and the pituitary adenoma predisposition due to mutations in the aryl hydrocarbon receptor interacting protein (AIP) gene. Endocr Rev 2013;34(2):239−77. Available from: https://doi.org/10.1210/er.2012-1013. Available from: http://www.ncbi.nlm.nih.gov/pubmed/23371967.

[53] Daly AF, Jaffrain-Rea ML, Ciccarelli A, Valdes-Socin H, Rohmer V, Tamburrano G, et al. Clinical characterization of familial isolated pituitary adenomas. J Clin Endocrinol Metab 2006;91(9):3316−23. Available from: https://doi.org/10.1210/jc.2005-2671. Available from: https://www.ncbi.nlm.nih.gov/pubmed/16787992.

[54] Valdes-Socin H, Jaffrain-Réa ML, Tamburrano G, Cavagnini F, Cicarelli E, Colao A, et al. Familial isolated pituitary tumors: clinical and molecular studies in 80 patients. In: The Endocrine Society's 84th Annual Meeting, 2002, p. 647 (P3-663).

[55] Valdes Socin H, Poncin J, Stevens V, Stevenaert A, Beckers A. Adénomes hypophysaires familiaux isolés non liés avec la mutation somatique nem-1. Suivi de 27 patients. Ann Endocrinol (Paris) 2000;61:301.

[56] Chandrasekharappa SC, Guru SC, Manickam P, Olufemi SE, Collins FS, Emmert-Buck MR, et al. Positional cloning of the gene for multiple endocrine neoplasia-type 1. Science 1997;276(5311):404−7. Available from: http://www.ncbi.nlm.nih.gov/pubmed/9103196.

[57] Kirschner LS, Carney JA, Pack SD, Taymans SE, Giatzakis C, Cho YS, et al. Mutations of the gene encoding the protein kinase A type I-alpha regulatory subunit in patients with the Carney complex. Nat Genet 2000;26(1):89−92. Available from: https://doi.org/10.1038/79238. Available from: https://www.ncbi.nlm.nih.gov/pubmed/10973256.

[58] Beckers A, Lodish MB, Trivellin G, Rostomyan L, Lee M, Faucz FR, et al. X-linked acrogigantism syndrome: clinical profile and therapeutic responses. Endocr Relat Cancer 2015;22(3):353−67. Available from: https://doi.org/10.1530/ERC-15-0038. Available from: http://www.ncbi.nlm.nih.gov/pubmed/25712922.

[59] Trivellin G, Daly AF, Faucz FR, Yuan B, Rostomyan L, Larco DO, et al. Gigantism and acromegaly due to xq26 microduplications and GPR101 mutation. N Engl J Med 2014;371(25):2363−74. Available from: https://doi.org/10.1056/NEJMoa1408028. Available from: https://www.ncbi.nlm.nih.gov/pubmed/25470569.

[60] Abboud D, Daly AF, Dupuis N, Bahri MA, Inoue A, Chevigné A, et al. GPR101 drives growth hormone hypersecretion and gigantism in mice via constitutive activation of gs and gq/11. Nat Commun 2020;11(1):4752. Available from: https://doi.org/10.1038/s41467-020-18500-x.

[61] Behrens LH, Barr DP. Hyperpituitarism beginning in infancy the Alton Giant. Endocrinology 1932;16(2):120−8. Available from: http://doi.org/10.1210/endo-16-2-120.

[62] Beckers A, Fernandes D, Fina F, Novak M, Abati A, Rostomyan L, et al. Paleogenetic study of ancient DNA suggestive of x-linked acrogigantism. Endocr Relat Cancer 2017;24(2):L17−20. Available from: https://doi.org/10.1530/ERC-16-0558. Available from: https://www.ncbi.nlm.nih.gov/pubmed/28049632.

[63] Petrossians P, Daly AF, Natchev E, Maione L, Blijdorp K, Sahnoun-Fathallah M, et al. Acromegaly at diagnosis in 3173 patients from the Liege Acromegaly Survey (LAS) database. Endocr Relat Cancer 2017;24(10):505−18. Available from: https://doi.org/10.1530/ERC-17-0253. Available from: https://www.ncbi.nlm.nih.gov/pubmed/28733467.

Pathology of pituitary growth hormone excess

Sylvia L. Asa[1] and Shereen Ezzat[2]

[1]Department of Pathology, University Hospitals Cleveland Medical Center and Case Western Reserve University
[2]Department of Medicine, University Health Network and University of Toronto

2.1 Introduction

The diseases known as acromegaly and gigantism result from excess growth hormone (GH) that is synthesized and secreted by somatotrophs and mammosomatotrophs of the adenohypophysis. GH has direct actions on the liver, muscle, and fat, and indeed GH receptors are ubiquitously expressed in almost every mammalian cell, but its actions are also mediated by insulin-like growth factor 1 (IGF-1), a growth factor whose systemic levels are supported mainly through hepatic synthesis.

Acromegaly and gigantism are often considered to be a spectrum of a single disease, differing only because of the age of onset of GH excess. However, they are actually the manifestations of multiple disorders that have distinct clinical and pathological features based on different molecular genetic alterations [1]. This review will summarize the many forms of pituitary pathology and their distinct clinical manifestations, molecular alterations, and therapeutic approaches.

2.2 Classification of pituitary pathology in growth hormone excess

The disorders that can give rise to GH excess include a number of hyperplastic and neoplastic disorders of the adenohypophysis as shown in Table 2.1. In contrast to these many entities, acromegaly due to ectopic GH [2,3] or IGF-1 [4] production, while exceptionally rare, can occur and does not result in any known pituitary pathology.

Some of the variation in pathologies causing GH excess are attributable to the fact that multiple cell types in the normal adenohypophysis produce this hormone. In the mature gland, GH is produced by both somatotrophs and mammosomatotrophs (also known as somatolactotrophs). Mature somatotrophs are the classical acidophils of the adenohypophysis; these large cells have densely granulated cytoplasm that is filled with large secretory granules that contain GH and also alpha-subunit of glycoprotein hormones (αSU). They represent one of the terminally differentiated cell types of PIT1-lineage. Mammosomatotrophs are capable of producing both GH and prolactin (PRL); these cells exist in the normal pituitary and represent a fluid cell population that is capable of trans-differentiating into somatotrophs during childhood growth, and into lactotrophs during gestation and lactation. They express both PIT1 and the estrogen receptor (ER) that is not expressed in somatotrophs. Other cell

Gigantism and Acromegaly. DOI: https://doi.org/10.1016/B978-0-12-814537-1.00004-X

Table 2.1 Classification of pituitary pathology in growth hormone excess.

Pathological entity	Variants	Molecular alterations	Other associations
Mammosomatotroph hyperplasia		GNAS, PRKR1A and CNC2 at 2p16, X-LAG	Ectopic GHRH excess
Somatotroph hyperplasia			Ectopic GHRH excess Hypothalamic gangliocytoma
Mammosomatotroph tumor		GNAS	
Somatotroph tumor	Densely granulated	GNAS	
	Sparsely granulated	? AIP,?GHR	Hypothalamic gangliocytoma
Mixed tumors with somatotroph components	Mixed somatotroph–lactotroph tumors		
	Multiple synchronous pituitary tumors of different lineages		
Plurihormonal PIT1-lineage tumors	Mature		
	Poorly differentiated	MEN1	
	Acidophil stem cell		
Unusual plurihormonal tumors			

types, including mature lactotrophs and thyrotrophs that are also of PIT1-lineage, do not express GH under normal conditions. Lactotrophs express both PIT1 and ER, whereas thyrotrophs express PIT1 and GATA2/3. There is evidence that somatotrophs and/or mammosomatotrophs may transdifferentiate into thyrotrophs, as shown in an experimental model of hypothyroidism [5,6], and therefore it is not surprising that tumors can produce GH, PRL, and thyrotropin (thyroid stimulating hormone, TSH) in any combination. It is considered that a PIT1 stem cell gives rise to this lineage and some tumors may reflect neoplastic proliferation of such a poorly differentiated cell. In contrast, corticotrophs and gonadotrophs are not of PIT1-lineage and therefore do not express GH.

PIT1-lineage tumors comprise approximately 30% of surgically resected pituitary neuroendocrine tumors (PitNETs) and almost 70% of those produce GH [7]. Densely and sparsely granulated somatotroph tumors are the most common tumor types, with roughly equal incidence; the next most common lesion is the poorly differentiated PIT1-lineage tumor, followed by mammosomatotroph tumors; mixed and plurihormonal tumors are far less common.

2.3 Pituitary hyperplasias associated with growth hormone excess

Pituitary hyperplasia is not a common phenomenon but in most cases, it is secondary to a stimulus outside the pituitary. In some patients, usually those with germline alterations that predispose to pituitary proliferation, there can be multifocal proliferations in the adenohypophysis that have been classified as hyperplasia, but the clonality of these lesions has not been explored to verify that they are not neoplastic.

Secondary hyperplasia is usually due to ectopic production of growth hormone-releasing hormone (GHRH). This phenomenon has been reported in neuroendocrine tumors of the pancreas, lung, and in pheochromocytomas [8−21]. The excess GHRH results in proliferation of target cells; usually these are somatotrophs but mammosomatotroph hyperplasia has also been reported.

Patients with this disorder present with acromegaly; this disorder has not been reported in children and gigantism is not known to occur. The patients exhibit the same clinical and biochemical features noted in the typical pituitary tumor-dependent forms of the disease. In contrast, however, radiographic imaging does not reveal a distinct pituitary lesion but more diffuse enlargement of the gland. Additionally, contrast enhancement fails to define the surrounding normal pituitary tissue. The suspicion of hyperplasia raised on imaging should prompt careful evaluation of the patient for a primary tumor in the chest or abdomen, and circulating GHRH levels can be measured to help confirm the diagnosis. Unfortunately, the diagnosis is often unrecognized and the patient comes to surgery where the pathology is identified as hyperplasia characterized by the diffuse proliferation of somatotrophs or mammosomatotrophs within the intact structure of the adenohypophysis. These proliferating cells are intermixed with other adenohypophysial cell types and they retain enlarged but intact acini that are delineated by a reticulin framework; this is in contrast to the reticulin breakdown of neoplastic proliferations (Fig. 2.1).

The possibility of neoplastic transformation in this setting has long been considered, and was found in mouse models [22,23] but is not usually found in human patients; it has been documented in a single case with metastatic GHRH-producing tumor in the pituitary [21], suggesting that the metastatic tumor resulting in local GHRH excess may have played a role in transformation. The difference between systemic and local GHRH is further supported by the association of somatotroph neoplasms more commonly than hyperplasia in patients with GHRH-producing hypothalamic gangliocytomas [24−28].

Therapy involves resection of the primary tumor. Removal of the source of GHRH results in gradual reversal of the pituitary hyperplasia.

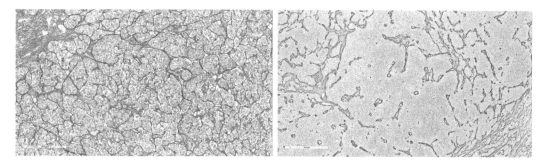

FIGURE 2.1 Adenohypophysial hyperplasia versus neoplasia.

The reticulin stain is critical to distinguish adenohypophysial hyperplasia from neoplasia. In hyperplasia (left), the acini defined by the Gordon-Sweet silver stain are intact but expanded; in this image, only the smallest acini are in the range of normal. In neoplasms (right), the reticulin network is lost, remaining only around blood vessels.

Primary somatotroph and mammosomatotroph hyperplasia is usually encountered in the setting of a germline disease such as McCune−Albright disease, X-linked acrogigantism (X-LAG), or Carney complex.

X-linked acrogigantism is a very rare disorder characterized by gigantism with onset in infants and very young children. Exaggerated growth has been documented as early as at 2−3 months of age and the median age of onset is 12 months; all children described have had onset of disease before the age of 5 years [29]. They develop very severe gigantism and subsequently acromegalic features. In addition to elevated GH and IGF-1, the majority of patients have hyperprolactinemia. Patients have germline Xq26 microduplication and/or mutation of the *GPR101* gene at that locus [30].

The morphology of this disorder has been variably reported as mammosomatotroph hyperplasia and neoplasia [31].

Surgery for this disorder requires extensive anterior pituitary resection potentially leading to significant postoperative hypopituitarism. Somatostatin analogues do not usually achieve control, and most patients have also required pegvisomant to control IGF-1 levels [29].

McCune−Albright Syndrome is a sporadic disorder caused by mosaic somatic−activating mutation of *GNAS* [32,33]. Initially it was described as the triad of polyostotic fibrous dysplasia of bone (FD), café au-lait skin pigmentation, and precocious puberty (PP). However, it subsequently became apparent that patients also exhibit proliferation and hyperactivity of many endocrine tissues, resulting in hyperthyroidism, growth hormone excess, renal phosphate wasting with or without rickets/osteomalacia, and Cushing syndrome. These manifestations are all attributed to the elevated cyclic AMP levels that result from the *GNAS* mutation that causes constitutive activation of Gsα. The endocrine manifestations are all attributable to increased signaling of the seven transmembrane domain G protein-coupled receptors that mediate synthesis and secretion of gonadotropins through GnRH, thyroid hormones through TSH, adrenal cortical glucocorticoid production through ACTH, and GH through GHRH.

Approximately 20% of affected patients develop pituitary acromegaly or gigantism [34]. They present with accelerated growth in the pediatric group and acromegaly in adults with this disorder. Hyperprolactinemia is also common. The pituitary can be enlarged with various imaging features; this is because some patients have diffuse disease while others have true neoplasms. The patients may also have fibrous dysplasia of the bone that can complicate the interpretation of scans (Fig. 2.2). Histologic examination of the pituitary has sometimes shown diffuse hyperplasia, mainly mammosomatotroph hyperplasia [35,36] but there are also cases of frank neoplasia that may be multifocal [34,36]. This is not surprising since the other endocrine glands also develop neoplasms.

Because of the multifocal nature of the disease, isolated resection of a tumor is not likely to cure patients with this disorder. In addition, the presence of highly vascular fibrous dysplasia that involves the skull base makes surgery technically difficult. For that reason, medical treatment is often preferred.

Somatostatin analogues are the first line modality and usually improve GH and IGF-1 levels, however, the addition of pegvisomant may be required. Because there is a high risk of malignant transformation of fibrous dysplasia, pituitary irradiation is typically not recommended for these patients.

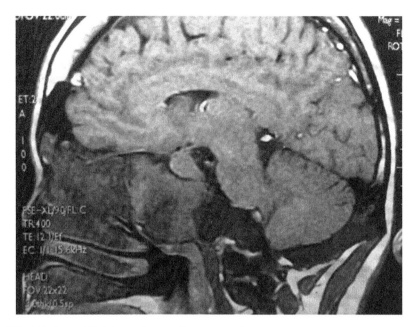

FIGURE 2.2 MRI of McCune—Albright syndrome with acromegaly.

A patient with McCune—Albright Syndrome has both a pituitary mass causing acromegaly and fibrous dysplasia of the cranial and facial bones.

Carney complex is an autosomal dominant familial tumor syndrome in which patients develop nevi that can even affect mucosal surfaces, melanotic schwannomas, myxomas (mainly cardiac), and endocrine dysfunction. In this syndrome, the adrenal manifestation is an unusual hyperfunction of the cortex associated with pigmented nodular adrenocortical disease that is micronodular, with minimal proliferative capacity, however, carcinoma may arise in this setting. Testicular tumors are another feature of the syndrome. Pituitary involvement is associated with gigantism or acromegaly [37]. In the majority of patients, this disease is due to increased cAMP signaling due to germline mutations that inactivate the *PRKAR1A* gene at the CNC1 locus (17p22—24) that encodes the PKA regulatory subunit 1Aα and is responsible for terminating cAMP activation [38,39]. A second locus, CNC2 at 2p16, is associated with this syndrome but the causative underlying genetic alteration remains unknown.

About 75% of patients with Carney complex present with abnormal GH, IGF-1, or prolactin levels, either basal or on dynamic testing, however, pituitary pathology is reported to occur in about 10% [33]. There are no unique pituitary imaging findings that characterize this syndrome.

Morphologic studies have shown both somatotroph and mammosomatotroph hyperplasia and tumor formation [33,40,41]. The data suggest that the primary process is diffuse hyperplasia that progresses to microtumors, which may be multiple [41]. It seems relatively rare to develop large tumors in this clinical setting and molecular studies show that these have progressive accumulations of alterations on comparative genomic hybridization [41].

2.4 Pituitary neuroendocrine tumors causing growth hormone excess

Somatotroph tumors occur in two variants that have completely different clinical, biochemical, morphological, and molecular features. Each variant represents about 40% of tumors responsible for adult acromegaly [7].

Densely granulated somatotroph tumors are composed of cells that resemble normal somatotrophs. These highly hormonally active cells give rise to high-circulating GH and IGF-1 and cause florid acromegaly [42−44]. GH and IGF-1 levels tend to be high and proportional to tumor size. Patients do not usually have hyperprolactinemia and if it is present, it is modest and attributable to the stalk effect, that is, tumor mass causing interruption of the tonic inhibition of hypothalamic dopamine that represses nontumorous lactotrophs.

Imaging identifies a well-defined tumor that is distinct from the surrounding adenohypophysial parenchyma that enhances with gadolinium. There is often evidence of florid acromegaly on imaging; the scalp may be thickened, the skin of the neck is often redundant, and the tongue may be enlarged. On T2-weighted MRI, these tumors are hypointense [45−47], a feature that correlates with density of granules in tumor cells (Fig. 2.3).

Densely granulated somatotroph tumors are very similar to mammosomatotroph tumors with the single exception that they do not express ER or PRL [48]. The tumor cells have abundant acidophilic cytoplasm (Fig. 2.4) and relatively bland, monotonous nuclei. They have strong nuclear positivity for PIT1 and cytoplasmic reactivity for GH and αSU. The keratin staining pattern is perinuclear with variable intensity. By electron microscopy, the tumor cells are large, round to polygonal, with well-developed rough endoplasmic reticulum, large Golgi complexes and numerous large, round generally homogenous electron-dense secretory granules. These tumor cells do not show misplaced exocytoses.

The treatment of choice for patients with these tumors is surgery. Because these patients have florid acromegaly, some may have tumors that are purely intrasellar and can be successfully resected. However, a significant percentage of patients are diagnosed with large tumors that have extrasellar extension, and when there is invasion of the cavernous sinuses, the tumor cannot be cured by surgery. These tumors are likely to harbor *GNAS* mutations and therefore have high intracellular cAMP levels [49]; this explains the frequent expression of αSU that is regulated by cAMP response element-binding protein (CREB) interaction with a binding site in its promoter [50]. Because of this, they are more likely to respond to somatostatin analogues [42−44,51−56] that are the second line of therapy. Some experts also advocate the administration of somatostatin analogues prior to surgery; while this is unlikely to result in significant tumor shrinkage, it does ameliorate many of the effects of IGF-1 excess, including soft tissue swelling in the surgical field. Thus some studies show improved outcomes of surgery in patients pretreated with somatostatin analogues [57].

Sparsely granulated somatotroph tumors are composed of cells that do not resemble any normal cell of the mature adenohypophysis. These tumors are not as hormonally active as the densely granulated and mammosomatotroph tumors and in patients with these tumors, circulating GH and IGF-1 levels are not as elevated [42−44]. In fact, the diagnosis of acromegaly may be missed if not carefully assessed, as patients do not have florid signs and symptoms. Patients may present at a younger age and with larger tumors, because these tend

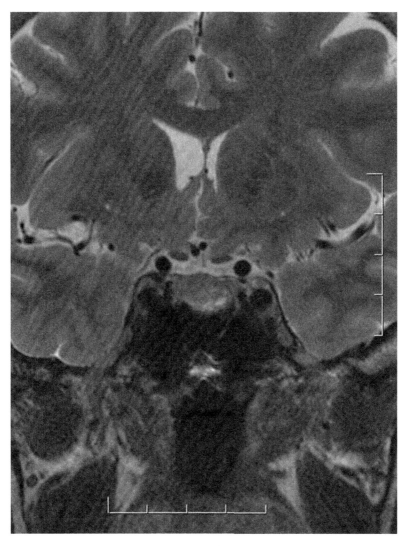

FIGURE 2.3 MRI of acromegaly due to a densely granulated somatotroph tumor.

T2-weighted imaging of a densely granulated somatotroph tumor shows characteristic hypointensity of the tumor mass.

to be more rapidly growing than densely granulated somatotroph tumors [43,44,55,58]. Hyperprolactinemia, if it occurs, is due to stalk effect, as these tumors do not synthesize or secrete PRL.

Imaging identifies a large and well-defined tumor that is delineated from the surrounding adenohypophysis that takes up gadolinium whereas the tumor does not enhance. The patients may not

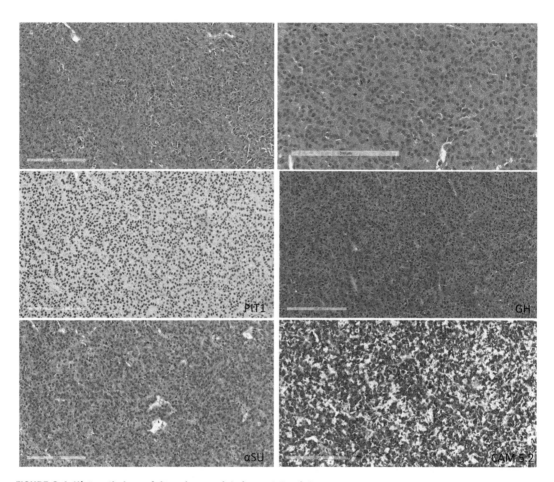

FIGURE 2.4 Histopathology of densely granulated somatotroph tumor.

The tumor has characteristic cytoplasmic acidophilia (top left); the tumor cells are relatively monotonous with bland nuclei (top right). Immunohistochemistry identifies strong nuclear PIT1-reactivity (middle left) and diffuse strong cytoplasmic staining for GH (middle right). These tumors usually express abundant alpha-subunit (bottom left). The staining pattern for keratins 8 and 18, identified with the CAM 5.2 antibody, is diffuse and perinuclear (bottom right).

have florid features of acromegaly. T2-weighted MRI identifies marked hyperintensity (Fig. 2.5) that is a more characteristic feature of sparsely granulated tumors [42,45−47,59].

Sparsely granulated tumors have a highly characteristic morphology even on routine histology. The tumor cells are discohesive and pleomorphic (Fig. 2.6), with eccentric or lobulated nuclei, chromophobic cytoplasm and a prominent hyaline globule in the cytoplasm. They stain strongly for PIT1 but GH reactivity may be scant, focal, and weak; staining for other pituitary transcription factors and hormones, including αSU, are negative. The characteristic feature of

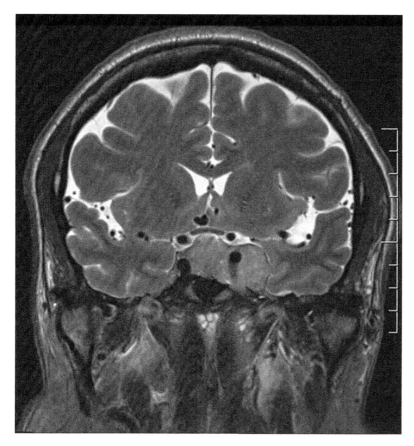

FIGURE 2.5 MRI of acromegaly due to a sparsely granulated somatotroph tumor.

T2-weighted imaging identifies a hyperintense large sellar tumor that invades the left cavernous sinus.

these tumors is the identification of fibrous bodies, globules of keratin filaments that are found with CAM 5.2 or antibodies to CK18; these correspond to the hyaline globules seen with hematoxylin and eosin. These structures are almost always found in the vast majority of tumor cells and there is no perinuclear keratin. In contrast, occasional tumors that have both perinuclear keratins and scattered fibrous bodies, classified as "intermediate" tumors, resemble densely granulated tumors in their clinical, biochemical, and therapeutic responses [58], and therefore are properly classified within the category of densely granulated somatotroph tumors. Another striking feature of sparsely granulated tumors is the loss of E-cadherin that explains their cellular dehiscence [60]. Consistent with their rapid clinical growth, these tumors have higher levels of Ki67 labeling than densely granulated tumors [7]. By electron microscopy, these tumors are composed of round, discohesive cells with scant cytoplasmic organelles and lobulated nuclei. Their cytoplasm is filled with aggregated intermediate filaments that form the fibrous body, which traps the few secretory granules and other organelles.

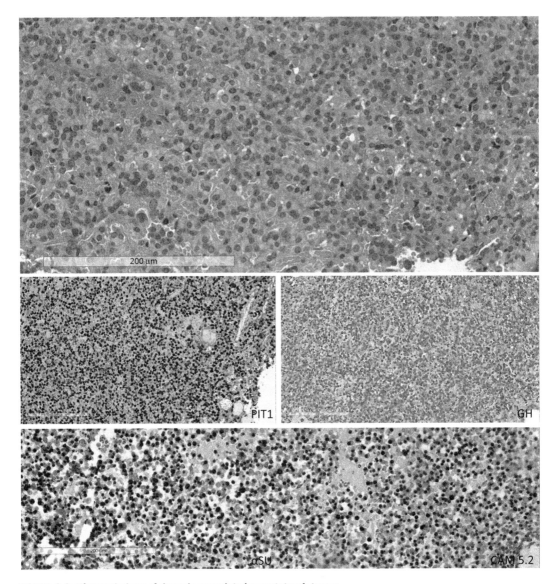

FIGURE 2.6 Histopathology of densely granulated somatotroph tumor.

The tumor is composed of large, discohesive cells with chromophobic cytoplasm and nuclear pleomorphism (top); scattered tumor cells have pale hyaline cytoplasmic inclusions that distort the nuclei.
Immunohistochemistry identifies strong nuclear PIT1-reactivity (middle left) and weak cytoplasmic staining for GH (middle right). The characteristic feature of these tumors is the juxtanuclear globular staining pattern for keratins 8 and 18, identified with the CAM 5.2 antibody, that identifies fibrous bodies in almost all tumors cells (bottom).

In contrast to densely granulated somatotroph tumors, the sparsely granulated variant is less likely to respond to somatostatin analogues [42−44,51−56]. These tumors do not harbor mutations of *GNAS* and do not have elevated cAMP levels, as evidenced by their lack of expression of αSU. Instead, they are more likely to harbor somatic or germline mutations of the aryl hydrocarbon receptor-interacting protein (AIP) that is implicated in the familial isolated pituitary tumor (FIPA) syndrome [61−64]. Mutations of AIP are not known to occur in sporadic tumors of this type [65,66] but there may be epigenetic downregulation of AIP instead [67]. Other pathogenetic mechanisms reported include mutation of the GH receptor [53], implicating the possibility of lack of GH autoregulation [68]. These tumors often require the addition of pegvisomant that blocks the GH receptor in target tissues.

Mixed tumors with somatotroph components include all composite tumors with a somatotroph element [48,69]. The commonest variants are the mixed somatotroph−lactotroph tumors that are composed of two distinct populations of cells, one that makes GH and one that makes PRL (Fig. 2.7). More rare are the mixed tumors of different lineages, including mixed somatotroph−gonadotroph or somatotroph and corticotroph tumors [69]. Each of these tumor types have several variants, since the somatotroph component may be sparsely or densely granulated. The clinical manifestations of GH excess, biochemistry, and imaging depend on the variant as described above as well as the proportion of the tumor that is somatotroph. Similarly, the therapeutic approach will depend on the same parameters with the addition of dopamine agonist therapy for tumors with PRL secretion [70,71]. In one study, mixed somatotroph−lactotroph tumors were larger and less successfully resected than mammosomatotroph or pure somatotroph tumors [72], however, most of the mixed tumors had a sparsely granulated somatotroph component that would account for this. In another study, tumors that stained for both GH and PRL were more likely to be recurrent even when mean tumor size was similar to pure GH-secreting tumors, patients more often required combination therapies with somatostatin analogues, dopamine agonists, and GH antagonist therapy, and higher doses of these agents were required [73] when compared to monohormonal tumors. However, it is not really possible to rely on much of this published literature since the reports mix together multiple pathological entities that produce both GH and prolactin, including mammosomatotroph tumors, mixed somatotroph and lactotroph tumors, as well as plurihormonal and poorly differentiated PIT1-lineage tumors that all have a distinct pathophysiology.

Mammosomatotroph tumors represent only about 2% of surgically resected pituitary neuroendocrine tumors and about 13% of tumors that are associated with GH excess [7]. They give rise to florid acromegaly or gigantism and patients usually also have hyperprolactinemia; they are common in young patients [74,75]. Growth hormone and IGF-1 levels are high and usually proportional to tumor size; because of the development of florid symptoms, these tumors are often diagnosed when the tumor is small [72].

Imaging usually identifies a well-defined tumor that can be readily distinguished from the surrounding adenohypophysial parenchyma, since the tumor does not take up gadolinium whereas the normal tissue enhances with this contrast agent. On T2-weighted MRI, these tumors are hypointense [45−47], a feature that correlates with density of granules in tumor cells.

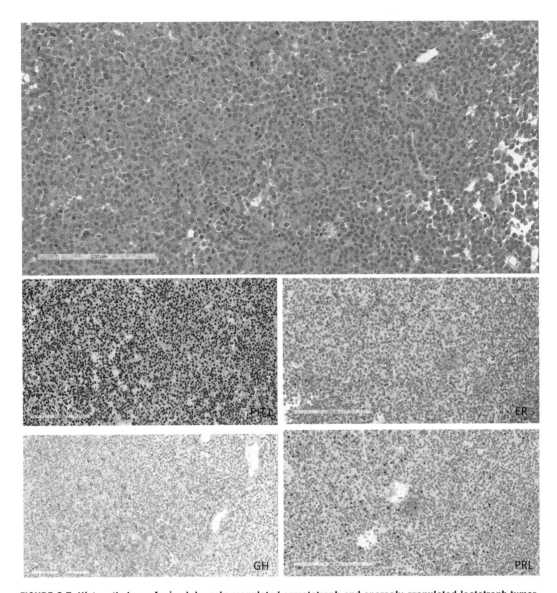

FIGURE 2.7 Histopathology of mixed densely granulated somatotroph and sparsely granulated lactotroph tumor.

This tumor has an obvious admixture of two tumor cell types, one with acidophilic cytoplasm and the other with chromophobic cytoplasm (top). Immunohistochemistry identifies PIT1 nuclear reactivity in all tumor cells (middle left) but only scattered cells corresponding to lactotrophs express nuclear ER (middle right). Staining for GH localizes the somatotrophs (bottom left) whereas prolactin is found in the lactotrophs (bottom right).

These tumors are composed of large, acidophilic cells (similar to those shown in Fig. 2.4) that have intense nuclear reactivity for PIT1 and ER, strong cytoplasmic positivity for GH and variable PRL staining [48]. They are also characteristically positive for αSU. Staining for keratins 8 and 18 using the CAM 5.2 antibody yields a pattern of positivity that is similar to that of normal cells, with a variable amount and perinuclear distribution of keratin filaments. By electron microscopy, the tumor cells are large, with relatively monotonous nuclei, prominent rough endoplasmic reticulum, well-formed Golgi complexes, and numerous secretory granules with electron-dense contents. The granules are peculiar in that there are often irregular shapes, and there may be evidence of misplaced exocytosis, that is, secretion on the lateral cell border, a unique feature of prolactin-secreting cells.

The treatment of choice for any pituitary tumor causing GH excess is surgical resection. Since these tumors are more often small and intrasellar, surgery is frequently successful [72]. However in patients with invasion of the cavernous sinuses that cannot be resected, medical therapy is required to normalize GH and IGF-1 levels and prevent the long-term sequelae of acromegaly. Mammosomatotroph tumors usually respond to somatostatin analogues and the addition of a dopamine agonist may be effective [70,71].

Plurihormonal PIT1-lineage tumors are a mixed family of tumors that express multiple hormones. The literature has been confused in this area, since there are several variants. In some classifications, these include mammosomatotroph tumors but since mammosomatotrophs are a normal cell type, this category should be reserved for tumors that also make TSH.

Mature plurihormonal PIT1-lineage tumors are the rare well-differentiated tumors that resemble mammosomatotrophs but also synthesize and even may secrete TSH in addition to GH and PRL [48]. These tumors are clinically like mammosomatotroph tumors with the addition of clinical hyperthyroidism. The clinical and biochemical features of GH excess are generally quite pronounced and as they are densely granulated acidophilic tumors, they are usually hypointense on T2-weighted MRI. These tumors are morphologically identical to mammosomatotroph tumors with the addition of GATA2/3 and TSH immunoreactivity.

Poorly differentiated PIT1-lineage tumors represent a unique clinical and morphological tumor type. These were initially classified as "Silent subtype 3" tumors by Horvath et al. when they were first described [76,77]; this is because they were thought to be clinically silent and similar to silent corticotroph tumors that were described to have two variants. Subsequent studies confirmed several unusual features including potential relationships to thyrotrophs [78,79]. However, the application of transcription factors to the armamentarium of diagnosis [80,81] led to the recognition that these were indeed PIT1-lineage tumors. They are now recognized to be composed of poorly differentiated cells that may express one or several hormones that are expressed by all cells of PIT1-lineage, and they may not be clinically silent; in fact, they may be associated with acromegaly, hyperprolactinemia, and/or hyperthyroidism [82].

The imaging features of these tumors may be distinctive since they tend to be large and infiltrative, with a peculiar tendency to invade into the cavernous sinuses and downward in addition to or even in the absence of extension upward into the suprasellar region [82]; they may therefore present as sinonasal masses [83].

The morphology of these tumors is characterized by somewhat pleomorphic cells that may be polygonal or elongated rather than round (Fig. 2.8) [82]; the tumor cell nuclei tend to have prominent nucleoli and a peculiar nuclear density known as a "spheridium." The only consistent immunohistochemical feature is intense nuclear PIT1 positivity. Staining for GH, ER, PRL, GATA2/3,

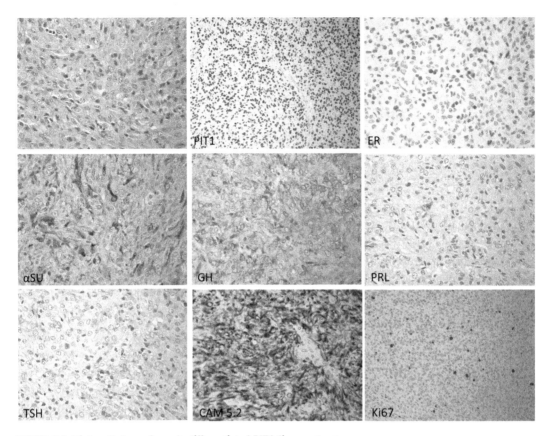

FIGURE 2.8 Histopathology of poorly differentiated PIT1-lineage tumor.

These tumors often have an unusual histological appearance; they may be composed of elongated or polygonal cells and their nuclei have prominent nucleoli (top left). They have strong and diffuse nuclear reactivity for PIT1 (top middle) that may be the only biomarker to indicate pituitary differentiation. In this case, scattered cells also express ER (top right). Immunohistochemistry for hormones identifies focal positivity for alpha subunit (middle left), some tumor cells expressing GH (middle) and only scattered single cells expressing PRL (middle right). TSH reactivity is focal and weak (bottom left). The CAM 5.2 antibody identifies cytokeratins 8 and 18 in a mixed pattern with predominant perinuclear staining but focal prominent fibrous bodies (bottom middle). These tumors often have a high Ki67 labeling index (bottom right).

TSH, and αSU is variable and these may be found in any combination; occasional tumors are negative for all hormones. Keratins are also variable in reactivity; there is usually positive staining but this may be variable in intensity throughout the lesion and the pattern also may vary, mainly as perinuclear filaments but occasionally in the form of scattered fibrous bodies.

The pathogenesis of these tumors is not known. The number of patients is relatively small but include patients with MEN1 [82].

The management of patients with these tumors is not entirely clear. Because of their invasive nature, they rarely can be completely resected at surgery [76,79,82]. They have been treated with somatostatin analogues and dopamine agonists. At least two patients have developed metastatic disease [78,84].

Acidophil stem cell tumors are another variant of poorly differentiated PIT1-lineage cell tumor that are restricted to the GH and PRL lineage without features of thyrotrophs. They are usually strongly positive for PRL with weak or focal GH staining.

These tumors usually present with hyperprolactinemia and the GH excess is often subclinical; it was initially described as "fugitive acromegaly" [85,86]. The clue to the diagnosis is based on a clear clinical discrepancy: in patients with lactotroph tumors, the degree of hyperprolactinemia is proportional to the tumor size whereas patients with acidophil stem cell tumors have large macrotumors with hyperprolactinemia that is greater than what can be attributed to stalk effect, but not as high as would be expected in relation to the tumor size on imaging. These tumors also do not respond well to dopamine agonist therapy [48,87,88], unlike lactotroph tumors. GH hypersecretion may be subtle and may only be detected on dynamic testing and features of acromegaly may be mild or absent.

The tumor cells are characterized by marked oncocytic change, that is, an accumulation of spherulated mitochondria that fill the cytoplasm, creating a granular eosinophilic appearance punctuated with chromophobic areas (Fig. 2.9). The mitochondrial abnormality can be so striking that giant swollen mitochondria can yield a bubbly appearance to the cytoplasm. The tumors express PIT1 and ER, and usually have abundant PRL that may be present in a juxtanuclear location, as in sparsely granulated lactotroph tumors, but more often shows an unusual and variable pattern with more diffuse cytoplasmic positivity. Staining for GH is weak and often focal. The keratin pattern of these tumors is irregular with areas of perinuclear positivity, scattered fibrous bodies, and some cells that are negative for keratins. Staining for mitochondrial antigens can be helpful to verify the oncocytic nature of the tumor cells, but it is usually evident even on routine microscopy.

The pathogenesis of these unusual tumors is also not known. They have been reported in patients with germline mutations of the succinate dehydrogenase enzymes (SDHx) [89].

Patients with this lesion are usually treated initially with dopamine agonists but while they may lower PRL levels, these agents do not result in the tumor shrinkage that is characteristic of lactotroph tumors [48,87,88]. Because they are usually large and invasive tumors, surgery is not usually curative. The addition of somatostatin analogues has not been formally examined in these tumors.

Unusual plurihormonal PitNETs are extremely rare tumors that may be associated with GH excess [7,48]. These are defined as monomorphous tumors that produce GH as well as hormones of cells from the other non-PIT1-lineages, that is, gonadotrophs and/or corticotrophs. There are many cases reported in the literature but most antedate the development and implementation of highly specific monoclonal antibodies to pituitary hormones. Since many GH-producing tumors express the αSU, and since many polyclonal antisera to gonadotropins and TSH cross react with αSU, many of the old reports of plurihormonal tumors producing GH and gonadotropins are likely artefactual. Very rarely, GH excess has been associated with a tumor-producing ACTH but often these are collision tumors that represent multiple synchronous lesions [90].

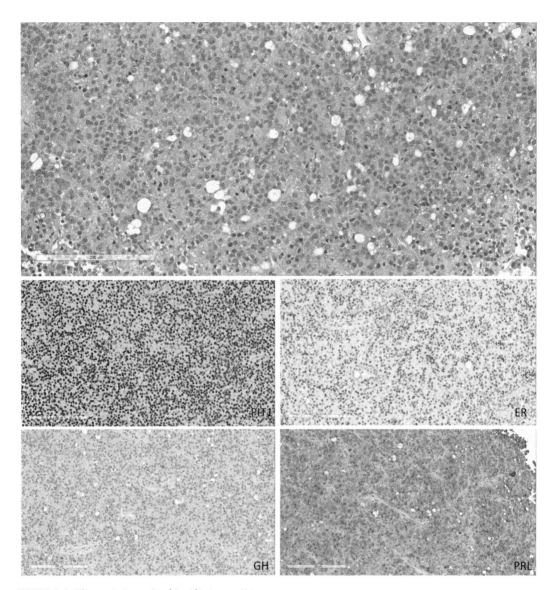

FIGURE 2.9 Histopathology of acidophil stem cell tumor.

This unusual tumor is composed of oncocytic cells that have abundant granular acidophilic cytoplasm due to the accumulation of mitochondria, some of which become swollen giant cytoplasmic globules (top). Immunohistochemistry confirms PIT1-lineage with diffuse strong nuclear reactivity (middle left), and there is usually fairly abundant ER reactivity (middle right). These tumors may be associated with subtle acromegaly, known as "fugitive acromegaly," since they produce some GH (bottom left) but their principle hormone product is PRL (bottom right).

2.5 Clinical implications of pathologic classification

As reviewed above, pituitary pathology provides critical information that should be used to drive postoperative decision-making. In the setting of somatotroph hyperplasia, an underlying source of ectopic GHRH should be sought to correct the problem. In the setting of pituitary tumors, the GH subclassification can also have significant ramifications. The densely granulated phenotype should support the use of somatostatin analogues. In contrast, sparsely granulated somatotroph tumors are significantly less likely to respond to somatostatin analogues, supporting the choice of an alternative agent. Importantly, the detection of less common forms than either of these tumor types should raise heightened vigilance in surveillance. This would include more regular pituitary imaging, reconsideration of the role of growth hormone antagonism, and possibly reassessment of radiosurgery. More precise pituitary tumor classification should also assist a much needed improvement in clinical trial design to better delineate therapeutic opportunities.

References

[1] Asa SL, Kucharczyk W, Ezzat S. Pituitary acromegaly: not one disease. Endocr Relat Cancer 2017; 24(3):C1−4.
[2] Ezzat S, Ezrin C, Yamashita S, Melmed S. Recurrent acromegaly resulting from ectopic growth hormone gene expression by a metastatic pancreatic tumor. Cancer 1993;71:66−70.
[3] Beuschlein F, Strasburger CJ, Siegerstetter V, et al. Acromegaly caused by secretion of growth hormone by a non-Hodgkin's lymphoma. N Engl J Med 2000;342(25):1871−6.
[4] Krug S, Boch M, Rexin P, et al. Acromegaly in a patient with a pulmonary neuroendocrine tumor: case report and review of current literature. BMC Res Notes 2016;9:326.
[5] Vidal S, Horvath E, Kovacs K, Cohen SM, Lloyd RV, Scheithauer BW. Transdifferentiation of somatotrophs to thyrotrophs in the pituitary of patients with protracted primary hypothyroidism. Virchows Arch 2000;436(1):43−51.
[6] Horvath E, Lloyd RV, Kovacs K. Propylthiouracyl-induced hypothyroidism results in reversible transdifferentiation of somatotrophs into thyroidectomy cells. A morphologic study of the rat pituitary including immunoelectron microscopy. Lab Invest 1990;63:511−20.
[7] Mete O, Cintosun A, Pressman I, Asa SL. Epidemiology and biomarker profile of pituitary adenohypophysial tumors. Mod Pathol 2018;31(6):900−9.
[8] Dabek JT. Bronchial carcinoid tumour with acromegaly in two patients. J Clin Endocrinol Metab 1974;38:329−33.
[9] Shalet SM, Beardwell CG, MacFarlane IA, et al. Acromegaly due to production of a growth hormone releasing factor by a bronchial carcinoid tumour. Clin Endocrinol (Oxf) 1979;10:61−7.
[10] Uz Zafar MS, Mellinger RC, Fine G, Szabo M, Frohman LA. Acromegaly associated with a bronchial carcinoid tumor: evidence for ectopic production of growth hormone-releasing activity. J Clin Endocrinol Metab 1979;48:66−71.
[11] Thorner MO, Perryman RL, Cronin MJ, et al. Somatotroph hyperplasia: successful treatment of acromegaly by removal of a pancreatic islet tumor secreting a growth hormone-releasing factor. J Clin Invest 1982;70:965−77.
[12] Rivier J, Spiess J, Thorner M, Vale W. Characterization of a growth hormone-releasing factor from a human pancreatic islet tumour. Nature 1982;300:276−8.

[13] Guillemin R, Brazeau P, Böhlen P, Esch F, Ling N, Wehrenberg WP. Growth hormone-releasing factor from a human pancreatic tumor that caused acromegaly. Science 1982;218:585−7.

[14] Webb CB, Thominet L, Frohman LA. Ectopic growth hormone releasing factor stimulates growth hormone release from human somatotroph adenomas in vitro. J Clin Endocrinol Metab 1983;56:417−19.

[15] Berger G, Trouillas J, Bloch B, et al. Multihormonal carcinoid tumor of the pancreas secreting growth hormone-releasing factor as a cause of acromegaly. Cancer 1984;54:2097−108.

[16] Ramsay JA, Kovacs K, Asa SL, Pike MJ, Thorner MO. Reversible sellar enlargement due to growth hormone-releasing hormone production by pancreatic endocrine tumors in an acromegalic patient with multiple endocrine neoplasia type I syndrome. Cancer 1988;62:445−50.

[17] Sano T, Asa SL, Kovacs K. Growth hormone-releasing hormone-producing tumors: clinical, biochemical, and morphological manifestations. Endocr Rev 1988;9:357−73.

[18] Ezzat S, Asa SL, Stefaneanu L, et al. Somatotroph hyperplasia without pituitary adenoma associated with a long standing growth hormone-releasing hormone-producing bronchial carcinoid. J Clin Endocrinol Metab 1994;78:555−60.

[19] Othman NH, Ezzat S, Kovacs K, et al. Growth hormone-releasing hormone (GHRH) and GHRH receptor (GHRH-R) isoform expression in ectopic acromegaly. Clin Endocrinol (Oxf) 2001;55(1) 135−40.

[20] Doga M, Bonadonna S, Burattin A, Giustina A. Ectopic secretion of growth hormone-releasing hormone (GHRH) in neuroendocrine tumors: relevant clinical aspects. Ann Oncol 2001;12(Suppl. 2):S89−94.

[21] Nasr C, Mason A, Mayberg M, Staugaitis SM, Asa SL. Acromegaly and somatotroph hyperplasia with adenomatous transformation due to pituitary metastasis of a growth hormone-releasing hormone-secreting pulmonary endocrine carcinoma. J Clin Endocrinol Metab 2006;91(12):4776−80.

[22] Asa SL, Kovacs K, Stefaneanu L, et al. Pituitary adenomas in mice transgenic for growth hormone-releasing hormone. Endocrinology 1992;131:2083−9.

[23] Asa SL, Kovacs K, Stefaneanu L, et al. Pituitary mammosomatotroph adenomas develop in old mice transgenic for growth hormone-releasing hormone. Proc Soc Exp Biol Med 1990;193:232−5.

[24] Puchner MJA, Lüdecke DK, Saeger W, Riedel M, Asa SL. Gangliocytomas of the sellar region - a review. Exp Clin Endocrinol 1994; in press.

[25] Asada H, Otani M, Furuhata S, Inoue H, Toya S, Ogawa Y. Mixed pituitary adenoma and gangliocytoma associated with acromegaly. Neurol Med Chir 1990;30:628−32.

[26] Bevan JS, Asa SL, Rossi ML, Esiri MM, Adams CBT, Burke CW. Intrasellar gangliocytoma containing gastrin and growth hormone- releasing hormone associated with a growth hormone-secreting pituitary adenoma. Clin Endocrinol (Oxf) 1989;30:213−24.

[27] Asa SL, Scheithauer BW, Bilbao JM, et al. A case for hypothalamic acromegaly: a clinicopathological study of six patients with hypothalamic gangliocytomas producing growth hormone-releasing factor. J Clin Endocrinol Metab 1984;58:796−803.

[28] Asa SL, Bilbao JM, Kovacs K, Linfoot JA. Hypothalamic neuronal hamartoma associated with pituitary growth hormone cell adenoma and acromegaly. Acta Neuropathol (Berl) 1980;52:231−4.

[29] Beckers A, Lodish MB, Trivellin G, et al. X-linked acrogigantism syndrome: clinical profile and therapeutic responses. Endocr Relat Cancer 2015;22(3):353−67.

[30] Trivellin G, Daly AF, Faucz FR, et al. Gigantism and acromegaly due to Xq26 microduplications and GPR101 mutation. N Engl J Med 2014;371(25):2363−74.

[31] Moran A, Asa SL, Kovacs K, et al. Gigantism due to pituitary mammosomatotroph hyperplasia. N Engl J Med 1990;323:322−7.

[32] Weinstein LS, Shenker A, Gejman PV, Merino MJ, Friedman E, Spiegel AM. Activating mutations of the stimulatory G protein in the McCune-Albright syndrome. N Engl J Med 1991;325:1688−95.

[33] Salpea P, Stratakis CA. Carney complex and McCune Albright syndrome: an overview of clinical manifestations and human molecular genetics. Mol Cell Endocrinol 2014;386(1−2):85−91.

[34] Salenave S, Boyce AM, Collins MT, Chanson P. Acromegaly and McCune-Albright syndrome. J Clin Endocrinol Metab 2014;99(6):1955−69.

[35] Kovacs K, Horvath E, Thorner MO, Rogol AD. Mammosomatotroph hyperplasia associated with acromegaly and hyperprolactinemia in a patient with the McCune-Albright syndrome. Virchows Arch [A] 1984;403:77−86.

[36] Vortmeyer AO, Glasker S, Mehta GU, et al. Somatic GNAS mutation causes widespread and diffuse pituitary disease in acromegalic patients with McCune-Albright syndrome. J Clin Endocrinol Metab 2012;97(7):2404−13.

[37] Carney JA, Gordon H, Carpenter PC, Shenoy BV, Go VL. The complex of myxomas, spotty pigmentation, and endocrine overactivity. Med (Baltim) 1985;64(4):270−83.

[38] Kirschner LS, Carney JA, Pack SD, et al. Mutations of the gene encoding the protein kinase A type I-alpha regulatory subunit in patients with the Carney complex. Nat Genet 2000;26(1):89−92.

[39] Yin Z, Williams-Simons L, Parlow AF, Asa S, Kirschner LS. Pituitary-specific knockout of the Carney complex gene prkar1a leads to pituitary tumorigenesis. Mol Endocrinol 2008;22(2):380−7.

[40] Cuny T, Mac TT, Romanet P, et al. Acromegaly in Carney complex. Pituitary 2019;22(5):456−66.

[41] Pack SD, Kirschner LS, Pak E, Zhuang Z, Carney JA, Stratakis CA. Genetic and histologic studies of somatomammotropic pituitary tumors in patients with the "complex of spotty skin pigmentation, myxomas, endocrine overactivity and schwannomas" (Carney complex). J Clin Endocrinol Metab 2000;85(10):3860−5.

[42] Fougner SL, Casar-Borota O, Heck A, Berg JP, Bollerslev J. Adenoma granulation pattern correlates with clinical variables and effect of somatostatin analogue treatment in a large series of patients with acromegaly. Clin Endocrinol (Oxf) 2012;76(1):96−102.

[43] Larkin S, Reddy R, Karavitaki N, Cudlip S, Wass J, Ansorge O. Granulation pattern, but not GSP or GHR mutation, is associated with clinical characteristics in somatostatin-naive patients with somatotroph adenomas. Eur J Endocrinol 2013;168(4):491−9.

[44] Kiseljak-Vassiliades K, Carlson NE, Borges MT, et al. Growth hormone tumor histological subtypes predict response to surgical and medical therapy. Endocrine 2015;49(1):231−41.

[45] Hagiwara A, Inoue Y, Wakasa K, Haba T, Tashiro T, Miyamoto T. Comparison of growth hormone-producing and non-growth hormone-producing pituitary adenomas: imaging characteristics and pathologic correlation. Radiology 2003;228(2):533−8.

[46] Heck A, Ringstad G, Fougner SL, et al. Intensity of pituitary adenoma on T2-weighted magnetic resonance imaging predicts the response to octreotide treatment in newly diagnosed acromegaly. Clin Endocrinol (Oxf) 2012;77(1):72−8.

[47] Potorac I, Petrossians P, Daly AF, et al. T2-weighted MRI signal predicts hormone and tumor responses to somatostatin analogs in acromegaly. Endocr Relat Cancer 2016;23(11):871−81.

[48] Asa SL. Tumors of the pituitary gland. In: Silverberg SG, editor. AFIP Atlas of Tumor Pathology, Series 4, Fascicle, 15. Silver Spring, MD: ARP Press; 2011.

[49] Spada A, Arosio M, Bochicchio D, et al. Clinical, biochemical and morphological correlates in patients bearing growth hormone-secreting pituitary tumors with or without constitutively active adenylyl cyclase. J Clin Endocrinol Metab 1990;71:1421−6.

[50] Asa SL, Ezzat S. The pathogenesis of pituitary tumors. Annu Rev Pathol 2009;4:97−126.

[51] Ezzat S, Kontogeorgos G, Redelmeier DA, Horvath E, Harris AG, Kovacs K. In vivo responsiveness of morphological variants of growth hormone-producing pituitary adenomas to octreotide. Eur J Endocrinol 1995;133:686−90.

[52] Bhayana S, Booth GL, Asa SL, Kovacs K, Ezzat S. The implication of somatotroph adenoma phenotype to somatostatin analog responsiveness in acromegaly. J Clin Endocrinol Metab 2005;90(11):6290−5.

[53] Asa SL, DiGiovanni R, Jiang J, et al. A growth hormone receptor mutation impairs growth hormone autofeedback signaling in pituitary tumors. Cancer Res 2007;67(15):7505−11.

[54] Kato M, Inoshita N, Sugiyama T, et al. Differential expression of genes related to drug responsiveness between sparsely and densely granulated somatotroph adenomas. Endocr J 2012;59(3):221−8.

[55] Brzana J, Yedinak CG, Gultekin SH, Delashaw JB, Fleseriu M. Growth hormone granulation pattern and somatostatin receptor subtype 2A correlate with postoperative somatostatin receptor ligand response in acromegaly: a large single center experience. Pituitary 2013;16(4):490−8.

[56] Ezzat S, Caspar-Bell GM, Chik CL, et al. Predictive markers for postsurgical medical management of acromegaly: a systematic review and consensus treatment guideline. Endocr Pract 2019;25(4):379−93.

[57] Yang C, Li G, Jiang S, Bao X, Wang R. Preoperative somatostatin analogues in patients with newly-diagnosed acromegaly: a systematic review and meta-analysis of comparative studies. Sci Rep 2019;9(1):14070.

[58] Obari A, Sano T, Ohyama K, et al. Clinicopathological features of growth hormone-producing pituitary adenomas: difference among various types defined by cytokeratin distribution pattern including a transitional form. Endocr Pathol 2008;19(2):82−91.

[59] Heck A, Emblem KE, Casar-Borota O, Bollerslev J, Ringstad G. Quantitative analyses of T2-weighted MRI as a potential marker for response to somatostatin analogs in newly diagnosed acromegaly. Endocrine 2016;52(2):333−43.

[60] Xu B, Sano T, Yoshimoto K, Yamada S. Downregulation of E-cadherin and its undercoat proteins in pituitary growth hormone cell adenomas with prominent fibrous bodies. Endocr Pathol 2002;13(4) 341−51.

[61] Beckers A, Aaltonen LA, Daly AF, Karhu A. Familial isolated pituitary adenomas (FIPA) and the pituitary adenoma predisposition due to mutations in the aryl hydrocarbon receptor interacting protein (AIP) gene. Endocr Rev 2013;34(2):239−77.

[62] Daly AF, Tichomirowa MA, Petrossians P, et al. Clinical characteristics and therapeutic responses in patients with germ-line AIP mutations and pituitary adenomas: an international collaborative study. J Clin Endocrinol Metab 2010;95(11):E373−83.

[63] Tahir A, Chahal HS, Korbonits M. Molecular genetics of the aip gene in familial pituitary tumorigenesis. Prog Brain Res 2010;182:229−53.

[64] Vierimaa O, Georgitsi M, Lehtonen R, et al. Pituitary adenoma predisposition caused by germline mutations in the AIP gene. Science 2006;312(5777):1228−30.

[65] Georgitsi M, De Menis E, Cannavo S, et al. Aryl hydrocarbon receptor interacting protein (AIP) gene mutation analysis in children and adolescents with sporadic pituitary adenomas. Clin Endocrinol (Oxf) 2008;69(4):621−7.

[66] DiGiovanni R, Serra S, Ezzat S, Asa SL. AIP mutations are not identified in patients with sporadic pituitary adenomas. Endocr Pathol 2007;18(2):76−8.

[67] Denes J, Kasuki L, Trivellin G, et al. Regulation of aryl hydrocarbon receptor interacting protein (AIP) protein expression by MiR-34a in sporadic somatotropinomas. PLoS One 2015;10(2):e0117107.

[68] Asa SL, Coschigano KT, Bellush L, Kopchick JJ, Ezzat S. Evidence for growth hormone (GH) autoregulation in pituitary somatotrophs in GH antagonist-transgenic mice and GH receptor-deficient mice. Am J Pathol 2000;156(3):1009−15.

[69] Mete O, Alshaikh OM, Cintosun A, Ezzat S, Asa SL. Synchronous multiple pituitary neuroendocrine tumors of different cell lineages. Endocr Pathol 2018.

[70] Sherlock M, Fernandez-Rodriguez E, Alonso AA, et al. Medical therapy in patients with acromegaly: predictors of response and comparison of efficacy of dopamine agonists and somatostatin analogues. J Clin Endocrinol Metab 2009;94(4):1255−63.

[71] Sandret L, Maison P, Chanson P. Place of cabergoline in acromegaly: a meta-analysis. J Clin Endocrinol Metab 2011;96(5):1327−35.

[72] Lv L, Jiang Y, Yin S, et al. Mammosomatotroph and mixed somatotroph-lactotroph adenoma in acromegaly: a retrospective study with long-term follow-up. Endocrine 2019;66(2):310−18.

[73] Rick J, Jahangiri A, Flanigan PM, et al. Growth hormone and prolactin-staining tumors causing acromegaly: a retrospective review of clinical presentations and surgical outcomes. J Neurosurg 2018;1−7.

[74] Horvath E, Kovacs K, Killinger DW, Smyth HS, Weiss MH, Ezrin C. Mammosomatotroph cell adenoma of the human pituitary: a morphologic entity. Virchows Arch [A] 1983;398:277−89.

[75] Felix IA, Horvath E, Kovacs K, Smyth HS, Killinger DW, Vale J. Mammosomatotroph adenoma of the pituitary associated with gigantism and hyperprolactinemia. A morphological study including immunoelectron microscopy. Acta Neuropathol (Berl) 1986;71:76−82.

[76] Horvath E, Kovacs K, Smyth HS, Cusimano M, Singer W. Silent adenoma subtype 3 of the pituitary−immunohistochemical and ultrastructural classification: a review of 29 cases. Ultrastruct Pathol 2005;29 (6):511−24.

[77] Horvath E, Kovacs K, Smyth HS, et al. A novel type of pituitary adenoma: morphological feature and clinical correlations. J Clin Endocrinol Metab 1988;66:1111−18.

[78] Roncaroli F, Scheithauer BW, Horvath E, et al. Silent subtype 3 carcinoma of the pituitary: a case report. Neuropathol Appl Neurobiol 2009.

[79] Erickson D, Scheithauer B, Atkinson J, et al. Silent subtype 3 pituitary adenoma: a clinicopathologic analysis of the Mayo Clinic experience. Clin Endocrinol (Oxf) 2009;71(1):92−9.

[80] Asa SL. Tumors of the pituitary gland Fascicle 22 In: Rosai J, editor. Atlas of Tumor Pathology. Third Series Washington, D.C: Armed Forces Institute of Pathology; 1998.

[81] Asa SL, Ezzat S. The cytogenesis and pathogenesis of pituitary adenomas. Endocr Rev 1998;19:798−827.

[82] Mete O, Gomez-Hernandez K, Kucharczyk W, et al. Silent subtype 3 pituitary adenomas are not always silent and represent poorly differentiated monomorphous plurihormonal Pit-1 lineage adenomas. Mod Pathol 2016;29(2):131−42.

[83] Hyrcza MD, Ezzat S, Mete O, Asa SL. Pituitary adenomas presenting as sinonasal or nasopharyngeal masses: a case series illustrating potential diagnostic pitfalls. Am J Surg Pathol 2017;41(4):525−34.

[84] Alshaikh OM, Asa SL, Mete O, Ezzat S. An institutional experience of tumor progression to pituitary carcinoma in a 15-year cohort of 1055 consecutive pituitary neuroendocrine tumors. Endocr Pathol 2019;30(2):118−27.

[85] Horvath E, Kovacs K, Singer W, et al. Acidophil stem cell adenoma of the human pituitary: clinicopathologic analysis of 15 cases. Cancer 1981;47:761−71.

[86] Horvath E, Kovacs K, Singer W, Ezrin C, Kerenyi NA. Acidophil stem cell adenoma of the human pituitary. Arch Pathol Lab Med 1977;101:594−9.

[87] Asa SL, Kovacs K, Horvath E, Singer W, Smyth HS. Hormone secretion in vitro by plurihormonal pituitary adenomas of the acidophil cell line. J Clin Endocrinol Metab 1992;75:68−75.

[88] Huang C, Ezzat S, Asa SL, Hamilton J. Dopaminergic resistant prolactinomas in the peripubertal population. J Pediatr Endocrinol Metab 2006;19(7):951−3.

[89] Papathomas TG, Gaal J, Corssmit EP, et al. Non-pheochromocytoma (PCC)/paraganglioma (PGL) tumors in patients with succinate dehydrogenase-related PCC-PGL syndromes: a clinicopathological and molecular analysis. Eur J Endocrinol 2014;170(1):1−12.

[90] Tordjman KM, Greenman Y, Ram Z, et al. Plurihormonal pituitary tumor of pit-1 and sf-1 lineages, with synchronous collision corticotroph tumor: a possible stem cell phenomenon. Endocr Pathol 2019;30(1) 74−80.

Gigantism: clinical diagnosis and description

3

Iulia Potorac[1], Liliya Rostomyan[2], Adrian F. Daly[1], Patrick Petrossians[2] and Albert Beckers[2]

[1]*Department of Endocrinology, Centre Hospitalier Universitaire de Liège, Liège Universite, Liège, Belgium*
[2]*Service d'Endocrinologie, Domaine Universitaire du Sart Tilman, Université de Liège, Liège, Belgium*

Gigantism has always been a subject of interest among the general public due to connotations of otherworldly strength and abilities. Until quite recently, the study of gigantism as an illness has not attracted major scientific attention [1], probably due to its very rare nature. International collaborative research has now led to significant advances in our understanding of the molecular mechanisms responsible for some forms of pituitary gigantism.

The majority of cases of gigantism can be easily recognized even without medical training. In less clear-cut cases, the diagnosis can be more difficult to establish, due to intra- and interindividual variations of growth patterns. Scientifically, tall stature is defined as a height over 2 standard deviation (SD) above the mean for the same sex, age, and population of origin or over 2 SD above the midparental height, but it can also manifest as significantly increased growth velocity or an abnormal growth pattern [2].

Excessive growth can be secondary to endocrine or to nonendocrine disturbances, although, most frequently, it is a reflection of familial tall stature or constitutionally advanced growth. This chapter is dedicated to the type of tall stature that is due to excessive growth hormone (GH) secretion, pituitary gigantism. In very rare cases, GH can be secreted ectopically [3], or pituitary GH hypersecretion can be secondary to ectopic GH-releasing hormone (GHRH) secretion [4]. Very few cases of gigantism due to ectopic GH and/or GHRH secretion have been reported [5].

Excessive GH secretion can commence at any age. When it begins before the closure of growth plates, the clinical phenotype will be gigantism and when it occurs afterwards, it leads to acromegaly. Acromegaly and gigantism are, therefore, manifestations of the same hormonal anomaly occurring at different periods of growth. Certain underlying molecular genetic mechanisms that cause GH hypersecretion—usually due to a pituitary adenoma—can be more frequent in younger versus older patients.

3.1 Clinical diagnosis—general aspects

Patients with pituitary gigantism are mainly males. In the largest series of patients with pituitary gigantism studied so far, nearly 60%−80% of patients were males [6,7]. Rapid growth was found to begin generally in late childhood or early puberty; this occurred at a significantly earlier age in

Gigantism and Acromegaly. DOI: https://doi.org/10.1016/B978-0-12-814537-1.00015-4

females than males (11 vs 13 years, respectively) [6]. Only in X-linked acrogigantism (X-LAG) (see below) does rapid growth always begin much earlier, almost always during the first year of life.

Despite their young age, patients with pituitary gigantism can develop typical clinical features of acromegaly, with enlargement of the extremities and facial dysmorphism being the most common [6]. Other clinical consequences of GH hypersecretion are also found in this population, such as excessive sweating, cutaneous changes, arthropathies, and carpal tunnel syndrome. Metabolic consequences such as glucose intolerance and diabetes mellitus also occur, as does sleep apnea syndrome. The cardiovascular impact of GH hypersecretion translates into hypertension, present in over a third of patients above the age of 20, as well as other cardiac anomalies, among which left ventricular hypertrophy, diastolic and/or systolic dysfunction, valvulopathies, and dilated cardiomyopathy are the most frequent [6,8]. Development of cardiovascular complications seems to be related to a longer disease duration before diagnosis and longer delays in treatment and disease control. Whether increased GH/insulin-like growth factor 1 (IGF-1) have an enhanced pathological effect on the growing heart during childhood and adolescence remains to be explored.

Mass effect symptoms caused by the developing pituitary mass include headaches, which can rarely be due to apoplexy of the pituitary adenoma, and visual field defects due to optic chiasm compression, as these tumors often exhibit suprasellar extension in pituitary gigantism [6,7]. This is contrary to the situation in acromegaly, in which case somatotropinomas generally exhibit infrasellar extension [9]. Hypopituitarism is present in around 25% of patients at diagnosis, but it becomes more common during follow-up as a consequence of the various treatment methods and affects over half of pituitary gigantism patients [6,7]. Prolactin (PRL) cosecretion can lead to galactorrhea and irregular menses. If, before epiphyseal closure, the pituitary mass becomes large enough to suppress gonadotropin secretion or if concomitant hyperprolactinemia occurs, the resulting hypogonadism will further delay growth plate fusion, thereby worsening increased vertical growth.

Despite rapid growth starting around the age of 13, diagnosis of a pituitary adenoma is generally established much later, around the age of 21.5 years in males. Female patients, however, are usually diagnosed earlier, around the age of 16 [6]. Therefore several years pass between the appearance of the first signs and symptoms of the disease (rapid growth) and its diagnosis, which impacts management and disease control. In male patients, this delay was found to be around 6 years, whereas in female patients, it was found to be 2.5 years [6]. This late diagnosis of excessive growth may be due to the fact that it is comparatively rare, and the referral/investigation systems may not be as efficient as in short stature; in males, there may also be some inbuilt bias that deems tall stature as being less "urgent" to investigate than short stature.

The delay in diagnosis allows for abnormally excessive growth to continue for longer and represents another factor contributing to the increased final height. In the largest specific pituitary gigantism series, final height was found to be >3 SD above the median height for the population, with a median difference from the predicted midparental height of 20 cm [6]. Greater final stature was found to be related to a younger age of disease onset, larger tumors, and higher GH hypersecretion.

The diagnosis of pituitary gigantism relies on the presence of GH hypersecretion. IGF-1 production physiologically varies depending on whether puberty has started, with levels rising as puberty progresses. It also differs according to gender with female patients having lower levels than males as a consequence of estrogen-induced hepatic GH resistance. It is important, therefore, when evaluating a patient with tall stature, to use adequate normal ranges, according to gender and pubertal status. We found that IGF-1 levels were on average around 2.5 × the upper limit of normal (ULN), while median GH levels at diagnosis were around 35 ng/mL [6,7]. PRL cosecretion was found in a third of patients with pituitary gigantism overall.

After biochemically confirming GH hypersecretion, its source should be located using pituitary magnetic resonance imaging (MRI). In pituitary gigantism, this is usually a pituitary macroadenoma. We found that the maximal pituitary adenoma diameter was around 22 mm [6], as later confirmed in other series [7], whereas nearly 15% of patients already had giant adenomas at diagnosis, measuring over 40 mm in maximal diameter [6]. Over half of the somatotropinomas responsible for pituitary gigantism were found to be invasive and over three-quarters exhibited extension outside the sella [6]. Rarely, an adenoma cannot be identified and diffuse hyperplasia of the pituitary is seen; this can sometimes precede adenoma formation, as detailed further below.

3.2 Clinical diagnosis—specific aspects

Pituitary gigantism can be an isolated occurrence, without other endocrine or nonendocrine manifestations related to a common pathophysiological mechanism, or it can develop in a syndromic context, present in the patient and/or their families. Depending on the underlying cause of GH hypersecretion, the clinical presentation varies as does the timing of the development of gigantism and its associated manifestations.

3.2.1 Nonsyndromic forms of GH hypersecretion

3.2.1.1 AIP *mutations*

About two decades ago, the disease entity familial isolated pituitary adenomas (FIPA) was originally reported in Liège, Belgium [10−13]. This familial condition consists of the occurrence of pituitary adenomas in two or more related members of the same family in the absence of other syndromic features or genetic anomalies corresponding to multiple endocrine neoplasia type 1 (MEN1) or Carney complex. International research revealed that unlike pituitary adenomas occurring sporadically, adenomas in FIPA patients were diagnosed at earlier ages and were larger at diagnosis [12]. Somatotropinomas are more frequent in the FIPA context than in sporadic cases, representing between 40% and 50% of FIPA cases [13,14]. In homogeneous somatotropinoma FIPA families, acromegaly can be diagnosed up to 10 years earlier than in heterogeneous FIPA families and in sporadic cases [13].

In 2006, Vierimaa et al. [15] discovered the role of the *aryl hydrocarbon receptor interacting protein (AIP)* gene in the predisposition towards the development of pituitary adenomas in FIPA kindreds. Further studies revealed that *AIP* germline mutations were responsible for 50% of the FIPA families with homogeneous acromegaly and 20% of FIPA kindreds in general [13,16]. Acromegaly patients bearing *AIP* mutations have a younger age of disease onset, with half manifesting symptoms during childhood and adolescence; most patients are males. Somatotropinomas in these patients are larger and associated with higher levels of GH hypersecretion at diagnosis than sporadic GH-secreting adenomas. Somatotropinomas of patients with *AIP* mutations generally require more surgical interventions and have a significantly worse response to somatostatin

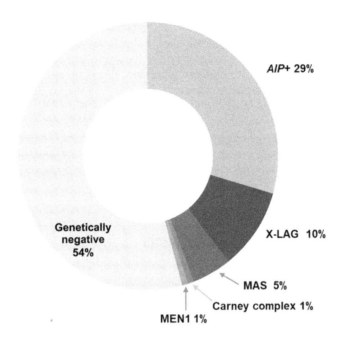

FIGURE 3.1

Prevalence of genetic forms of pituitary gigantism [6].

analog treatment (SSA) than sporadic acromegaly, both in terms of hormonal secretion and volume reduction [17].

Due to the younger age at disease manifestation, *AIP* mutation—positive somatotropinoma patients present with gigantism more frequently than *AIP*-negative somatotropinoma patients (32% vs 6.5%) [17]. Our study found that nearly a third of gigantism patients carried *AIP* mutations, thereby representing the most frequent genetic anomaly identified so far in this condition (Fig. 3.1) [6]. *AIP* mutation—positive patients with gigantism are predominantly males, with a disease onset around the age of 13. Most of the adenomas are large at diagnosis with a median maximal tumor diameter of 25 mm. Giant adenomas are found in 10% of the patients presenting with macroadenomas (Fig. 3.2). IGF-1 levels at diagnosis are around $2 \times$ ULN and nearly a third of cases exhibit PRL cosecretion.

3.2.1.2 X-linked acrogigantism

X-LAG is the most recently identified condition associated with pituitary gigantism. It manifests at a very early age, generally around the age of 1—2 years and leads to severe gigantism during adolescence if untreated. Children with this syndrome are usually born at full term, with normal height and normal weight (Fig. 3.3) [19,20]. Some X-LAG patients can show clinical features of acromegaly at diagnosis: acral enlargement, coarse facial features, prominent mandible with increased interdental spaces, and soft tissue swelling. Over a third of patients can

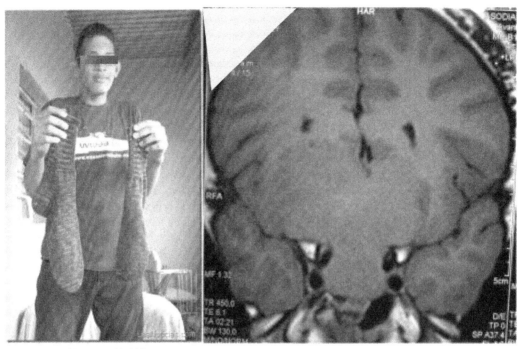

15 years old, 220 cm

FIGURE 3.2

A case of *AIP*-related gigantism secondary to a giant pituitary somatotropinoma [18].

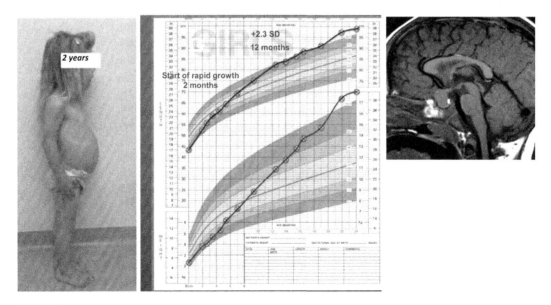

FIGURE 3.3

Overgrowth in X-LAG syndrome. Diffuse pituitary hyperplasia on MRI [19].

have an increased appetite, which seems to be uncommon among other causes of pituitary gigantism and some X-LAG patients present with acanthosis nigricans as a sign of insulin resistance [20]. X-LAG was found to be the cause of pituitary gigantism in 10% of the population studied by our group. Among X-LAG patients, unlike other forms of pituitary gigantism, the majority of patients (>70%) are females [6,19].

On MRI, X-LAG patients exhibit either pituitary macroadenomas (with a median tumor diameter of 18 mm) or diffuse pituitary enlargement corresponding to hyperplasia. However, dramatic cases with giant adenomas have been described (Fig. 3.4) [21]. Adenoma extension is mostly found to be suprasellar, with rare cases of cavernous sinus invasion [20].

Biologically, GH levels are generally greatly above the ULN, without suppression during the oral glucose tolerance tests, while IGF-1 is >3× ULN. Most cases have associated hyperprolactinemia, with PRL levels as high as 90× ULN [20].

X-LAG is due to microduplications in Xq26.3 including the *GPR101* gene, which encodes for an orphan G-protein-coupled receptor [19]. *GPR101* appears to be responsible for the phenotype [22]. X-LAG can occur sporadically or as FIPA in kindreds with acrogigantism [20]. The microduplication can be constitutional or occur as somatic mosaicism, which is the case in male sporadic patients [23].

3.2.1.3 Genetically negative cases

We found that about 50% of cases of pituitary gigantism have no detectable genetic anomaly explaining the phenotype (genetically negative patients) [6,7]. These patients are mostly male and older at diagnosis compared to *AIP*-positive and X-LAG patients (Fig. 3.5). They have levels of IGF-1 that are intermediate between the very high secretion in X-LAG patients and the values in *AIP*-positive cases. PRL cosecretion is also present in over a third of these patients at diagnosis [6,7]. They have large pituitary adenomas, just as in *AIP*-positive and X-LAG patients [6,7], but with disease control that is more difficult to achieve.

3.2.2 Syndromic forms of GH hypersecretion

Syndromic GH hypersecretion can appear in the context of MEN1 or MEN4; Carney complex; McCune–Albright syndrome (MAS); and 3PAs—pheochromocytoma, paraganglioma, and pituitary adenoma association.

3.2.2.1 Multiple endocrine neoplasia type 1

MEN1 is a tumor predisposition syndrome that mainly involves the parathyroids, the endocrine pancreas and digestive tract, and the pituitary. Primary hyperparathyroidism is present in over 90% of cases due to parathyroid hyperplasia [24]. Enteropancreatic tumors are found in 30%–70% of MEN1 patients [24]. Less frequently, adrenal cortical tumors, thymic and bronchial neuroendocrine tumors, cutaneous tumors (angiofibromas, collagenomas, lipomas), meningiomas and rarely, pheochromocytomas are also encountered.

In terms of pituitary involvement, the pituitary adenomas associated with MEN1 are more frequently prolactinomas, followed by somatotropinomas. Nonfunctioning and adrenocorticotrophic

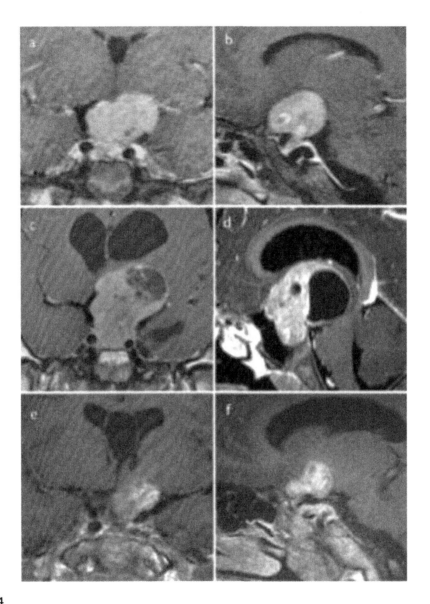

FIGURE 3.4

Natural history of somatotropinoma in X-LAG. (a, b) *First presentation* (5 years 8 months): history of excessive growth for >2 years; height 163 cm (+10 SD); noninvasive sellar mass (33 × 24 × 29 mm). *Second visit* (c, d) at the age of 10 years (no treatment): headaches, seizures, visual disturbance; height 197 cm (+7.33 SD); invasive mass (56 × 58 × 45 mm), compression of optic chiasma. Panels e, and f, show the MRI image three months after surgical debulking—no control with maximum doses of SSA + high-dose cabergoline [21].

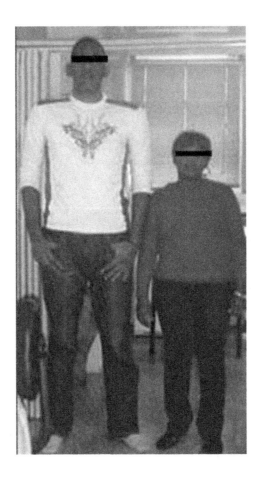

FIGURE 3.5

A case of "genetically negative" gigantism—a 24-year-old man measuring *192 cm* next to his mother [18].

hormone-secreting adenomas are rare. Cases of pituitary carcinomas have also been reported in MEN1 patients [25−27]. GH-secreting pituitary adenomas are found in approximately 10% of MEN1 patients with pituitary lesions, are diagnosed at a mean age of 43, and are generally macroadenomas at diagnosis [28]. However, as reported by Stratakis et al. [29], tall stature with growth acceleration can appear as early as 5 years in relation to a *MEN1* mutation. In that reported case, the adenoma had mixed GH/PRL secretion. Only 1% of patients in the largest series of patients with pituitary gigantism studied so far had *MEN1* gene anomalies [6].

MEN1 should be suspected in patients with gigantism in case of family or personal history of primary hyperparathyroidism, which can manifest as early as 8 years, while digestive tumors can be diagnosed as early as 5 years. In nearly 20%−30% of MEN1 cases, pituitary adenomas are the first manifestation of the syndrome [28,30,31]. According to current guidelines, in MEN1 families, screening for pituitary tumors should comprise PRL and IGF-1 measurement on an yearly basis and pituitary MRI every 3 years starting from the age of 5 [24].

3.2.2.2 Multiple endocrine neoplasia type 4

A MEN1-like phenotype was described in 2006 and found to be due to mutations in the *CDKN1B* *(cyclin-dependent kinase inhibitor 1B)* gene. It was named MEN4 when occurring in humans and MENX in rats [32]. Primary hyperparathyroidism and different types of secretory pituitary adenomas characterize MEN4 [33]. Gastrointestinal neuroendocrine, thyroid, uterine, and adrenocortical tumors have also been reported [34]. Cases of GH-secreting adenomas have been described, and cases of gigantism might, exceptionally, occur in this context [35].

Several studies have been dedicated to the identification of mutations in the *CDKN1B* gene in at-risk populations. In 124 *AIP* mutation−negative patients from FIPA families, two possibly pathogenic *CDKN1B* mutations were found. One of them occurred in a patient with a GH-secreting adenoma. However, there was no proof of segregation of the mutations with the presence of pituitary adenomas in the two FIPA families [36]. In a series of over 400 patients with acromegaly, none was found to carry *CDKN1B* or *MEN1* gene mutations, illustrating the limited role of these genetic anomalies in the pathogenesis of GH-secreting pituitary adenomas [37].

3.2.2.3 Carney complex

Carney complex is a rare tumor predisposition syndrome with both endocrine and nonendocrine manifestations, due in most cases to mutations of the *PRKAR1A* gene coding for the 1A regulatory subunit of the protein kinase A. It was first described in 1985 by J.A. Carney as the association of myxomas, spotty skin pigmentation, and endocrine overactivity [38]. The endocrine anomalies are, by order of frequency, primary pigmented nodular adrenal disease causing hypercortisolism, testicular and ovarian tumors, pituitary GH/PRL hypersecretion, and thyroid tumors [39]. Nonendocrine manifestations can be cutaneous (lentigines, blue nevi, myxomas, café-au-lait skin spots), cardiac and breast myxomas, psammommatous melanotic schwannomas, or osteochondromyxomas [40].

In terms of pituitary involvement, somatomammotroph hyperplasia appears to be the first pituitary anomaly that precedes adenoma formation [41]. Nearly 75% of patients exhibit increased IGF-1 and PRL levels and abnormal GH responses to glucose or thyrotropin-releasing hormone, but GH-secreting pituitary adenomas are only present in around 10% of cases [40]. Acromegaly appearing in a Carney complex setting can be diagnosed at a younger age compared to sporadic cases (before age 30) [42]. Microadenomas are responsible for nearly half of the cases [42], meaning that somatotropinomas in this context are generally smaller than in sporadic cases [9]. The evolution of acromegaly is slow, but the risk of recurrence after adenoma surgical removal is high due to the multifocal hyperplasia [41].

In a recent review of the literature, six cases of gigantism diagnosed in Carney complex patients have been reported. Information on these cases is incomplete, but all of the pituitary adenomas with reported dimensions at diagnosis were macroadenomas. IGF-1 levels at diagnosis were not very elevated, being between 1.4 and 1.8 of the ULN for age and sex [42].

3.2.2.4 McCune—Albright syndrome

McCune-Albright syndrome (MAS) consists of the triad *café-au-lait* skin spots, fibrous dysplasia, and hyperfunctioning endocrinopathies. The *café-au-lait* skin macules are already present at birth or appear shortly afterwards. They usually respect the body's midline and have irregular borders (similar to the coast of Maine appearance), unlike the spots in neurofibromatosis type 1 with smoother edges (coast of California aspect). Fibrous dysplasia is generally polyostotic and is the most common feature of MAS [43]. In terms of endocrinopathies, these are most often gonadal leading to precocious puberty, but thyroid (hyperthyroidism), pituitary (GH and PRL hypersecretion), adrenal (Cushing syndrome), and renal (renal phosphate wasting) lesions are well described.

Activating postzygotic mutations of the *GNAS1* gene that encodes the stimulatory α-subunit of the G protein (Gs) are the genetic basis underlying the development of MAS. These mutations lead to constitutive activation of the cAMP (cyclic adenosine monophosphate) pathway.

Around 20%−30% of MAS patients develop GH excess, with a mean age at diagnosis of acromegaly of 24 years according to a study by Salenave et al. [44]. Pituitary adenomas in patients with acromegaly and MAS were mostly macroadenomas (over two-thirds of cases). Similar to Carney complex, adenomas seem to develop from diffuse somatotroph hyperplasia [45]. As nearly all patients with acromegaly and MAS also have craniofacial fibrous dysplasia, surgery is rendered more difficult due to restricted access to the pituitary when the sphenoid bone is affected and also due to the risks of hemorrhage secondary to the hypervascular nature of fibrous dysplasia [44]. Radiotherapy is to be avoided as it is thought to increase the risk of malignant transformation [46].

Gigantism due to MAS was found in 5% of patients in our series of pituitary gigantism cases [6]. In a series of MAS acromegaly patients, Salenave et al. [44] defined gigantism as a final height over 2 m, which represented 7% of the series. Although acromegaly was diagnosed before the age of 16 in over a third of the 112 patients included in that study, 57% of these also had precocious puberty, which probably limited their final height. Gigantism in MAS can be associated with extensive craniofacial fibrous dysplasia, most likely worsened by the lack of control of GH excess, as illustrated by our case in Fig. 3.6 [47]. The thorough histological and genetic analysis of this case underlines that the *GNAS1*

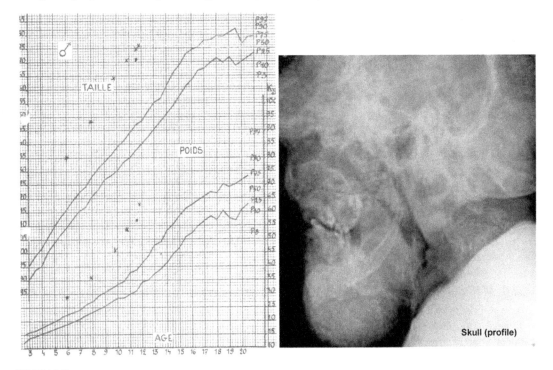

Skull (profile)

FIGURE 3.6

A case of MAS-related gigantism and severe craniofacial fibrous dysplasia [47].

mutation and tumor formation can affect both classical and less well known tissue targets in endocrine (parathyroids and pancreas), and nonendocrine nature (thymus, fat and normal skin).

3.2.2.5 Pheochromocytoma, paraganglioma, and pituitary adenoma association (3PAs)

This recently characterized multiple endocrine neoplasia association consists of pheochromocytomas and paragangliomas associated with pituitary adenomas. It is mainly due to mutations in genes encoding subunits of succinate dehydrogenase (*SDHx*), which induce a state of pseudohypoxia [48]. *MEN1* and *RET* mutations in a 3PAs context have also been described [48,49], as well as, recently, *MAX* gene defects [50]. Acromegaly in a patient with a paraganglioma was reported in 2012 due to a *SDHD* mutation [51]. SDHx-related pituitary adenomas appearing in a 3PAs context seem to more frequently be macroadenomas with an aggressive behavior as they require several forms of treatment [48].

When 3PAs occurs in a familial pheochromocytoma/paraganglioma context, *SDHx* mutations are found in 62.5%−75% of cases. For patients with *SDHx* mutations, screening for pituitary adenomas should be performed [48].

So far, cases of gigantism appearing in a 3PAs context have not been reported.

3.3 Conclusion

Pituitary gigantism is a rare cause of tall stature, developing as a consequence of early-onset GH hypersecretion. The same biological anomaly causes the development of acromegaly, but the latter occurs after the closure of growth plates, whereas gigantism occurs before epiphyseal fusion. Typical acromegaly clinical features and complications sometimes accompany the excessive growth, whereas in syndromic cases, other manifestations characteristic of an underlying genetic disorder can also be found. The condition is often diagnosed late, which complicates management and control of the excessive growth. Unlike in acromegaly, hereditary genetic anomalies frequently underlie the GH hypersecretion in pituitary gigantism, with more severe and aggressive manifestations. Better awareness of pituitary gigantism might favor an early diagnosis and rapid, effective management, which is essential for these patients.

References

[1] Herder WW. Acromegaly and gigantism in the medical literature. Case descriptions in the era before and the early years after the initial publication of Pierre Marie (1886). Pituitary 2009;12(3):236−44.

[2] Beckers A, Petrossians P, Hanson J, Daly AF. The causes and consequences of pituitary gigantism. Nat Rev Endocrinol 2018;14(12):705−20.

[3] Ozkaya M, Sayiner ZA, Kiran G, Gul K, Erkutlu I, Elboga U. Ectopic acromegaly due to a growth hormone-secreting neuroendocrine-differentiated tumor developed from ovarian mature cystic teratoma. Wien Klin Wochenschr 2015;127(11−12):491−3.

[4] Garby L, Caron P, Claustrat F, Chanson P, Tabarin A, Rohmer V, et al. Clinical characteristics and outcome of acromegaly induced by ectopic secretion of growth hormone-releasing Hormone (GHRH): a French nationwide series of 21 cases. J Clin Endocrinol Metab 2012;97(6):2093−104.

[5] Leveston SA, McKeel DW, Buckley PJ, Deschryver K, Greider MH, Jaffe BM, et al. Acromegaly and cushing's syndrome associated with a foregut carcinoid tumor. J Clin Endocrinol Metab 1981;53(4):682−9.

[6] Rostomyan L, Daly AF, Petrossians P, Nachev E, Lila AR, Lecoq AL, et al. Clinical and genetic characterization of pituitary gigantism: an international collaborative study in 208 patients. Endocr Relat Cancer 2015;22(5):745−57.

[7] Iacovazzo D, Caswell R, Bunce B, Jose S, Yuan B, Hernández-Ramírez LC, et al. Germline or somatic GPR101 duplication leads to X-linked acrogigantism: a clinico-pathological and genetic study. Acta Neuropathol Commun 2016;4(1):56.

[8] Rostomyan L, Daly AF, Shah N, Naves L, Barlier A, Jaffrain-Rea M-L, et al. Cardiovascular complications in pituitary gigantism (results of an international study). Endocr Abstr 2018;57:21.

[9] Potorac I, Petrossians P, Daly AF, Schillo F, Slama CB, Nagi S, et al. Pituitary MRI characteristics in 297 acromegaly patients based on T2-weighted sequences. Endocr Relat Cancer 2015;22(2):169−77.

[10] Valdes Socin HG, Poncin J, Vanbellinghen J-F, Jaffrain-Réa ML, Tamburrano G, Delemer B, et al. Les adénomes hypophysaires familiaux isolés non liés aux syndromes MEN1 et Carney complex: etude multicentrique. Ann Endocrinol 2001;62:322.

[11] Beckers A. Familial isolated pituitary adenomas. In: Ninth International Workshop on Multiple Endocrine Neoplasias (MEN2004), Bethesda, MD; 2004.

[12] Daly AF, Jaffrain-Rea ML, Ciccarelli A, Valdes-Socin H, Rohmer V, Tamburrano G, et al. Clinical characterization of familial isolated pituitary adenomas. J Clin Endocrinol Metab 2006;91(9):3316−23.

[13] Beckers A, Aaltonen LA, Daly AF, Karhu A. Familial isolated pituitary adenomas (FIPA) and the pituitary adenoma predisposition due to mutations in the aryl hydrocarbon receptor interacting protein (AIP) gene. Endocr Rev 2013;34(2):239−77.

[14] Hernández-Ramírez LC, Gabrovska P, Dénes J, Stals K, Trivellin G, Tilley D, et al. Landscape of familial isolated and young-onset pituitary adenomas: prospective diagnosis in AIP mutation carriers. J Clin Endocrinol Metab 2015;100(9):E1242−54.

[15] Vierimaa O, Georgitsi M, Lehtonen R, Vahteristo P, Kokko A, Raitila A, et al. Pituitary adenoma predisposition caused by germline mutations in the AIP gene. Science 2006;312(5777):1228−30.

[16] Daly AF, Vanbellinghen JF, Sok KK, Jaffrain-Rea ML, Naves LA, Guitelman MA, et al. Aryl hydrocarbon receptor-interacting protein gene mutations in familial isolated pituitary adenomas: analysis in 73 families. J Clin Endocrinol Metab 2007;92(5):1891−6.

[17] Daly AF, Tichomirowa MA, Petrossians P, Heliövaara E, Jaffrain-Rea ML, Barlier A, et al. Clinical characteristics and therapeutic responses in patients with germ-line AIP mutations and pituitary adenomas: an international collaborative study. J Clin Endocrinol Metab 2010;95(11):E373−83.

[18] Mangupli R, Rostomyan L, Castermans E, Caberg JH, Camperos P, Krivoy J, et al. Combined treatment with octreotide LAR and pegvisomant in patients with pituitary gigantism: clinical evaluation and genetic screening. Pituitary 2016;19(5):507−14.

[19] Trivellin G, Daly AF, Faucz FR, Yuan B, Rostomyan L, Larco DO, et al. Gigantism and acromegaly due to Xq26 microduplications and GPR101 mutation. N Engl J Med 2014;371(25):2363−74.

[20] Beckers A, Lodish MB, Trivellin G, Rostomyan L, Lee M, Faucz FR, et al. X-linked acrogigantism syndrome: clinical profile and therapeutic responses. Endocr Relat Cancer 2015;22(3):353−67.

[21] Naves LA, Daly AF, Dias LA, Yuan B, Zakir JCO, Barra GB, et al. Aggressive tumor growth and clinical evolution in a patient with X-linked acro-gigantism syndrome. Endocrine 2016;51(2):236−44.

[22] Abboud D, Daly AF, Dupuis N, Bahri MA, Inoue A, Chevigné A, et al. GPR101 drives growth hormone hypersecretion and gigantism in mice via constitutive activation of Gs and Gq/11. Nat Commun 2020;11(1):4752.

[23] Daly AF, Yuan B, Fina F, Caberg JH, Trivellin G, Rostomyan L, et al. Somatic mosaicism underlies X-linked acrogigantism syndrome in sporadic male subjects. Endocr Relat Cancer 2016;23(4):221−33.

[24] Thakker RV, Newey PJ, Walls GV, Bilezikian J, Dralle H, Ebeling PR, et al. Clinical practice guidelines for multiple endocrine neoplasia type 1 (MEN1). J Clin Endocrinol Metab 2012;97(9):2990−3011.

[25] Vroonen L, Jaffrain-Rea ML, Petrossians P, Tamagno G, Chanson P, Vilar L, et al. Prolactinomas resistant to standard doses of cabergoline: a multicenter study of 92 patients. Eur J Endocrinol 2012;167(5):651−62.

[26] Benito M, Asa SL, LiVolsi VA, West VA, Snyder PJ. Gonadotroph tumor associated with multiple endocrine neoplasia type 1. J Clin Endocrinol Metab 2005;90(1):570−4.

[27] Scheithauer BW, Kovacs K, Nose V, Lombardero M, Osamura YR, Lloyd RV, et al. Multiple endocrine neoplasia type 1-associated thyrotropin-producing pituitary carcinoma: report of a probable de novo example. Hum Pathol 2009;40(2):270−8.

[28] Vergès B, Boureille F, Goudet P, Murat A, Beckers A, Sassolas G, et al. Pituitary disease in MEN type 1 (MEN1): data from the France-Belgium MEN1 multicenter study. J Clin Endocrinol Metab 2002;87 (2):457−65.

[29] Stratakis CA, Schussheim DH, Freedman SM, Keil MF, Pack SD, Agarwal SK, et al. Pituitary macroadenoma in a 5-year-old: an early expression of multiple endocrine neoplasia type 1 1. J Clin Endocrinol Metab 2000;85(12):4776−80.

[30] Carty SE, Helm AK, Amico JA, Clarke MR, Foley TP, Watson CG, et al. The variable penetrance and spectrum of manifestations of multiple endocrine neoplasia type 1. Surgery 1998;124(6):1106−13.

[31] Wu Y, Gao L, Guo X, Wang Z, Lian W, Deng K, et al. Pituitary adenomas in patients with multiple endocrine neoplasia type 1: a single-center experience in China. Pituitary 2019;22(2):113−23.

[32] Pellegata NS, Quintanilla-Martinez L, Siggelkow H, Samson E, Bink K, Höfler H, et al. Germ-line mutations in p27Kip1 cause a multiple endocrine neoplasia syndrome in rats and humans. Proc Natl Acad Sci U S A 2006;103(42):15558−63.

[33] Lee M, Pellegata NS. Multiple endocrine neoplasia type 4. Front Horm Res 2013;41:63−78.

[34] Schernthaner-Reiter MH, Trivellin G, Stratakis CA. MEN1, MEN4, and Carney complex: pathology and molecular genetics. Neuroendocrinology 2016;103(1):18−31.

[35] Sambugaro S, Di Ruvo M, Ambrosio MR, Pellegata NS, Bellio M, Guerra A, et al. Early onset acromegaly associated with a novel deletion in CDKN1B 5′UTR region. Endocrine 2015;49(1):58−64.

[36] Tichomirowa MA, Lee M, Barlier A, Daly AF, Marinoni I, Jaffrain-Rea ML, et al. Cyclin-dependent kinase inhibitor 1B(CDKN1B) gene variants in AIP mutation-negative familial isolated pituitary adenoma kindreds. Endocr Relat Cancer 2012;19(3):233−41.

[37] Nachtigall LB, Guarda FJ, Lines KE, Ghajar A, Dichtel L, Mumbach G, et al. Clinical MEN-1 among a large cohort of patients with acromegaly. J Clin Endocrinol Metab 2020;105(6):e2271−81.

[38] Aidan Carney J, Gordon H, Carpenter PC, Vittal Shenoy B, Go VLW. The complex of myxomas, spotty pigmentation, and endocrine overactivity. Medicine (Baltimore) 1985;64(4):270−83.

[39] Stratakis CA, Kirschner LS, Carney JA. Clinical and molecular features of the Carney complex: diagnostic criteria and recommendations for patient evaluation. J Clin Endocrinol Metab 2001;86(9):4041−6.

[40] Correa R, Salpea P, Stratakis CA. Carney complex: an update. Eur J Endocrinol 2015;173(4):M85−97.

[41] Pack SD, Kirschner LS, Pak E, Zhuang Z, Carney JA, Stratakis CA. Genetic and histologic studies of somatomammotropic pituitary tumors in patients with the 'complex of spotty skin pigmentation, myxomas, endocrine overactivity and schwannomas' (Carney complex). J Clin Endocrinol Metab 2000;85(10):3860−5.

[42] Cuny T, Mac TT, Romanet P, Dufour H, Morange I, Albarel F, et al. Acromegaly in Carney complex. Pituitary 2019;22(5):456−66.

[43] Dumitrescu CE, Collins MT. McCune-Albright syndrome. Orphanet J Rare Dis 2008;3:12.

[44] Salenave S, Boyce AM, Collins MT, Chanson P. Acromegaly and McCune-Albright syndrome. J Clin Endocrinol Metab 2014;99(6):1955−69.

[45] Vortmeyer AO, Gläsker S, Mehta GU, Abu-Asab MS, Smith JH, Zhuang Z, et al. Somatic GNAS mutation causes widespread and diffuse pituitary disease in acromegalic patients with McCune-Albright syndrome. J Clin Endocrinol Metab 2012;97(7):2404−13.

[46] Javaid MK, Boyce A, Appelman-Dijkstra N, Ong J, Defabianis P, Offiah A, et al. Best practice management guidelines for fibrous dysplasia/McCune-Albright syndrome: a consensus statement from the FD/MAS international consortium. Orphanet J Rare Dis 2019;14(1):139.

[47] Vasilev V, Daly AF, Thiry A, Petrossians P, Fina F, Rostomyan L, et al. McCune-Albright syndrome: a detailed pathological and genetic analysis of disease effects in an adult patient. J Clin Endocrinol Metab 2014;99(10):E2029−38.

[48] Xekouki P, Brennand A, Whitelaw B, Pacak K, Stratakis CA. The 3PAs: an update on the association of pheochromocytomas, paragangliomas, and pituitary tumors. Horm Metab Res 2019;51(7):419−36.

[49] Dénes J, Swords F, Rattenberry E, Stals K, Owens M, Cranston T, et al. Heterogeneous genetic background of the association of pheochromocytoma/paraganglioma and pituitary adenoma: Results from a large patient cohort. J Clin Endocrinol Metab 2015;100(3):E531−41.

[50] Daly AF, Castermans E, Oudijk L, Guitelman MA, Beckers P, Potorac I, et al. Pheochromocytomas and pituitary adenomas in three patients with MAX exon deletions. Endocr Relat Cancer 2018;25(5): L37−42.

[51] Xekouki P, Pacak K, Almeida M, Wassif CA, Rustin P, Nesterova M, et al. Succinate dehydrogenase (SDH) D subunit (SDHD) inactivation in a growth-hormone-producing pituitary tumor: a new association for SDH? J Clin Endocrinol Metab 2012;97(3):E357−66.

Acromegaly: clinical description and diagnosis

4

Kevin C.J. Yuen[1] and Albert Beckers[2]

[1]*Barrow Pituitary Center, Barrow Neurological Institute, Departments of Neuroendocrinology and Neurosurgery, University of Arizona College of Medicine and Creighton School of Medicine, Phoenix, AZ, United States*
[2]*Service d'Endocrinologie, Domaine Universitaire du Sart Tilman, Université de Liège, Liège, Belgium*

4.1 Introduction

Acromegaly and gigantism are rare disorders caused by growth hormone (GH) hypersecretion with secondary elevation of serum insulin-like growth factor-I (IGF-I) levels, and characterized by progressive somatic disfigurement, and local and systemic clinical manifestations that are associated with increased morbidity and mortality [1,2]. The prevalence of acromegaly is approximately 40−70 cases per million [3], but recent studies have suggested that this figure may be higher in certain populations [4−6]. The average age of patients at diagnosis ranges between 40 and 45 years, while a minority of cases (up to 5%) are diagnosed before the age of 20 [3,7]. Early diagnosis and optimal treatment to achieve disease remission is crucial in reducing the risk of long-term morbidity and premature death. However, due to its insidious clinical onset, slow progression, and variable clinical spectrum, the diagnosis is often delayed by at least 6−10 years [8]. Nevertheless, recent data have shown that the delay in diagnosis has shortened to between 3 and 6 years [9−11], which may be attributable to improved GH and IGF-I assays, increased frequency of magnetic resonance imaging (MRI) for head- and neck-related complaints, and increased awareness among endocrinologists and general public with information on the internet.

More than 95% of patients with acromegaly present with an underlying GH-secreting pituitary adenoma, of which >70% are macroadenomas at diagnosis and 25% are co-secretors of prolactin, while <5% of cases are caused by excess GH-releasing hormone (GHRH) secretion from a hypothalamic tumor or neuroendocrine pancreatic or bronchial carcinoid tumors, ectopic pituitary GH-secreting tumors (e.g., sphenoidal sinus, petrous temporal bone, and nasopharyngeal cavity), and peripheral GH-secreting tumors (e.g., pancreatic islet tumor or lymphoma) [12]. The majority (>90%) of GH-secreting pituitary adenomas occur as sporadic and isolated cases [13]. However, there are increasing reports of familial cases, either associated with other genetic endocrine syndromes (e.g., multiple endocrine neoplasia types 1 and 4, Carney complex, and McCune−Albright syndrome) or as an isolated disorder (e.g., familial isolated pituitary adenoma) of which 20% of these cases harbor germline mutations of the aryl-hydrocarbon receptor−interacting protein (*AIP*) gene [14]. The X-linked acrogigantism (X-LAG) syndrome is another recently described rare condition of pituitary gigantism caused by microduplication on chromosome Xq26.3 encompassing the *GPR101* gene [15].

Gigantism and Acromegaly. DOI: https://doi.org/10.1016/B978-0-12-814537-1.00010-5

4.2 General clinical manifestations

Acromegaly affects both genders equally, and the average age at diagnosis ranges between 40 and 50 years, whereas patients under the age of 20 account for only 5% of cases [3,7]. Because GH and IGF-I can affect many organs, the clinical features in each patient are dependent on the histopathology of the tumor, duration and prevailing levels of GH and IGF-I, age of onset, and mass effects of the tumor [14,16]. The clinical manifestations of severe acromegaly are virtually pathognomonic, whereas early acromegaly may be more challenging to detect as the clinical signs and symptoms can be subtle and nonspecific. The clinical signs and symptoms of acromegaly have been reported in a series of 472 patients (Fig. 4.1) [17], and include enlarged hands and feet, facial modifications, jaw prognathism, macroglossia, excessive snoring, asthenia, hyperhidrosis, skin thickening, arthropathy, headache, visual disturbances, carpal tunnel syndrome, menstrual abnormalities, loss of libido, erectile dysfunction, sleep apnea, hypertension, hyperglycemia, and altered voice (Table 4.1). It has been reported that approximately 40% of patients are detected by nonendocrinologists, for example, internists during annual physical examinations, ophthalmologists during routine eye examinations, neurologists for chronic headaches, dentists due to mandibular prognathism and maxillary teeth separation, gynecologists due to menstrual irregularities and infertility, rheumatologists due to joint swelling and pains, or pulmonologists due to underlying obstructive sleep apnea [10,18]. Changes in cosmetic appearance derive from skeletal growth, soft tissue enlargement such as macroglossia, skin thickness, and skin tags are typically more subtle and develop insidiously in the early stages of the disease, and may be detected by the patient's internist or dermatologist. Conversely, the appearance of a GH-secreting pituitary tumor in a young patient before closure of epiphyseal bone results in accelerated linear growth and gigantism that may present to a pediatrician. Once acromegaly or gigantism is suspected, a full endocrinological workup to confirm the diagnosis and establish its etiology is warranted (Fig. 4.2).

4.3 Mass effects of the tumor

Headache and visual disturbances (e.g., bitemporal hemianopsia, ophthalmoplegia, and blindness) are generally the most frequently reported symptoms and may not correlate to the size and degree of invasion of the tumor. The precise mechanism of headache and visual disturbance is unknown but likely due to biochemical properties of the tumor itself, tumor growth stretching the dura and abutting the optic chiasm, and cavernous sinus invasion leading to trigeminal nerve irritation [19].

Other local neurological manifestations may be caused by the size and direction of tumor invasion (e.g., unilateral exophthalmos and rarely seizures) [20,21]. Tumor damage or compression to the pituitary stalk may raise prolactin levels due to loss of hypothalamic inhibitory regulation of prolactin secretion, while prolactin co-secretion by the tumor may also cause hyperprolactinemia but in these cases, the degree of hyperprolactinemia is often greater [22,23]. In females with hyperprolactinemia, galactorrhea and menstrual abnormalities may be present [24], while in males, erectile dysfunction and loss of libido have been reported [25]. Tumor damage directly to normal pituitary tissue can result in deficiencies of other anterior pituitary hormones (e.g., adrenocorticotropic hormone, thyroid-stimulating hormone, gonadotropins, and rarely vasopressin) resulting in the presentation of symptoms of adrenal insufficiency, hypothyroidism, hypogonadism, and diabetes insipidus.

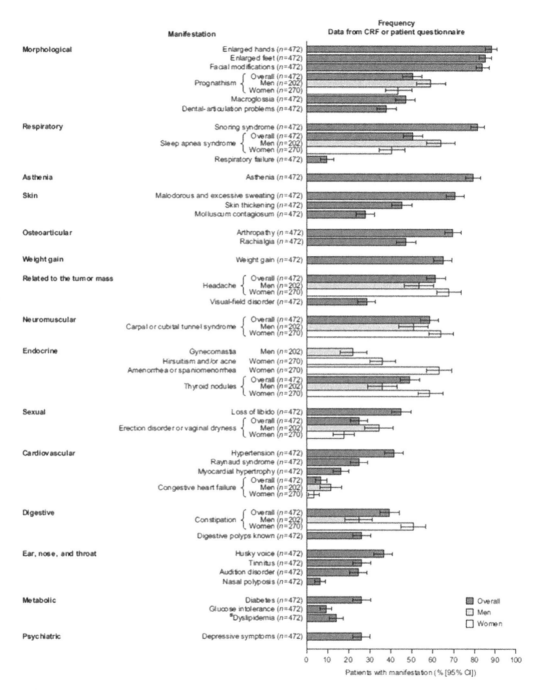

FIGURE 4.1

Frequency of signs and symptoms at diagnosis of acromegaly with gender differences.

Reproduced with permission from Caron P, Brue T, Raverot G, Tabarin A, Cailleux A, Delemer B, et al. Signs and symptoms of acromegaly at diagnosis: the physician's and the patient's perspectives in the ACRO-POLIS study. Endocrine. 2019; 63:120–9 [17].

Table 4.1 Summary of clinical features of acromegaly and gigantism.

Tumor mass effects
Headache, visual disturbances, hyperprolactinemia, pituitary stalk compression, hypopituitarism (including menstrual abnormalities)

Systemic effects of chronic GHRH/GH/IGF-I excess
  Acral growth
    Enlarged hands and feet, thickening of acral parts, skin thickening, and tissue hypertrophy
  Orofacial changes
    Forehead brow, rhinophyma, prominent check bones, thick lips, macroglossia, prognathism, maxillary widening, teeth separation, jaw malocclusion, soft tissue, and skin changes
  Skin and appendices changes
    Facial skin wrinkles, nasolabial folds, skin thickening, hyperhidrosis, oily skin, skin tags, acanthosis nigricans, Raynaud's phenomenon
  Musculoskeletal comorbidities
    Arthropathy, increased articular cartilage thickness, vertebral fractures
  Pulmonary comorbidities
    Altered voice, sleep apnea (central and obstructive), chest wall deformity, ventilator dysfunction
  Endocrinological and metabolic comorbidities
    Thyroid diseases (goiter, nodules, hyperthyroidism, thyroid cancer), hyperparathyroidism, glucose intolerance, insulin resistance, dyslipidemia
  Cardiovascular comorbidities
    Hypertension, cardiomyopathy, hypertrophy (biventricular or asymmetric septal), congestive heart failure (systolic and/or diastolic), arrhythmias
  Gastrointestinal comorbidities
    Colonic polyps, dolichocolon, diverticular disease, gallbladder polyps, colonic cancer
  Neurological and neuropsychological comorbidities
    Carpal tunnel syndrome, intracranial aneurysm, cerebellar tonsil herniation, hearing impairment, pituitary apoplexy, impaired quality of life
  Gigantism

4.4 Systemic effects of excess GHRH/GH/IGF-I

4.4.1 Acral growth

In the majority of patients, the hands and feet are enlarged and fingers widened due to soft tissue swelling [17] (Fig. 4.3). Osteoarthritic changes and joint swelling can contribute to the deformities in the hands and feet. Patients may report that their ring and shoes have become too tight and have had to increase their ring sizes or stop wearing their ring, and change to extra-wide fitting shoes due to their widened and deformed fingers and feet, respectively.

4.4.2 Orofacial changes

The patient presenting with late de novo acromegaly facies generally does not represent a diagnostic challenge, but these classic orofacial changes may be more subtle in younger patients. Typical components of coarsened facial features include pronounced forehead brow, enlarged nose, prominent cheekbones, thick lips, marked facial lines, mandibular overgrowth with prognathism, maxillary widening,

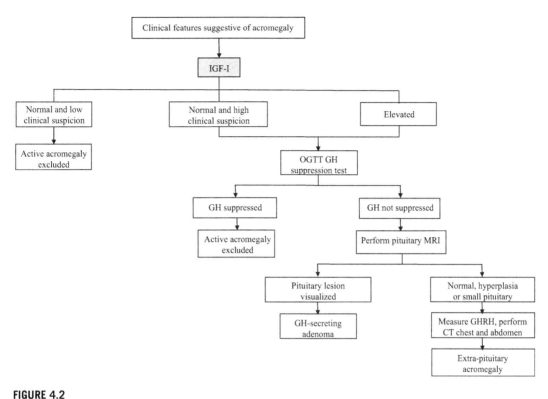

FIGURE 4.2

Suggested algorithm for the diagnosis of acromegaly.

Modified from Giustina A et al. [107].

teeth separation, and jaw malocclusion [9,26] (Fig. 4.4). The combination of macroglossia, jaw progna-thism, and swelling of pharyngeal tissues contribute to the development of upper airway obstruction that may result in sleep apnea [27,28] (Fig. 4.5). Orofacial changes are among the most burdensome and debilitating features of acromegaly that negatively impact self-esteem and body image perception [29] and may persist even after long-term remission of acromegaly is achieved [30].

4.4.3 **Skin and appendices changes**

The facial and acral deformities that characterize acromegaly are mainly related to skin changes such as facial skin wrinkles, nasolabial folds, and skin thickening in the scalp, face, palms of hands, and soles of feet [31−33]. Enlarged sweat glands cause hyperhidrosis and malodorous oily skin that are common early signs occurring in up to 70% of patients, whereas the presence of truncal skin tags and acanthosis nigricans may suggest the presence of colonic polyps and insulin resistance [34], respectively. Altered cutaneous microcirculation produces increased vasoconstriction, with Raynaud's phenomenon reported in 30% of cases [35].

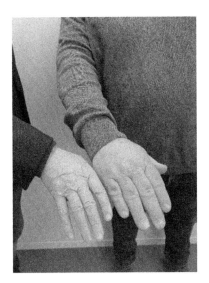 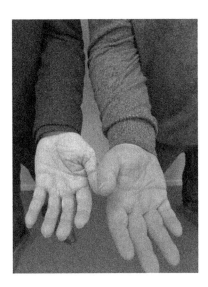

FIGURE 4.3

This patient had enlarged hands, widened fingers due to soft tissue swelling, and osteoarthritic changes with joint swelling. He was not able to wear his ring due to being too tight, and his ring had to be specifically cut.

4.4.4 Musculoskeletal comorbidities

Arthropathy is a common occurring in 72% of cases most frequently in the spine and hands [36], especially in older female patients, frequently impairs quality of life (QoL) [37], and in some cases, remains impaired despite biochemical control of the disease [38]. All joints can be affected with resultant joint swelling and deformity, hypomobility, osteophytosis, and axial arthropathy that are often irreversible [39]. Neck and back pains are common and may be due to mechanical in nature, inflammation, and nerve compression [39]. Vertebral fractures may occur despite normal bone mineral density, while fracture risk may be increased by concurrent hypogonadism that negates the anabolic effect of GH on trabecular bone [40]. Peripheral neuropathies such as symptomatic carpal tunnel syndrome [41] and ulnar nerve neuropathy at the cubital tunnel [42] are also frequently reported.

4.4.5 Pulmonary comorbidities

Altered voice is probably one of the earlier clinical manifestations of acromegaly as a result of increased vocal cord mass that frequently normalizes following induction of remission of the disease [17]. Upper airway obstruction due to soft tissue swelling causes excessive snoring and sleep apnea in more than 50% of cases [43] that may persist in up to 40% of cases even after medical or surgical treatment, whereas a minority of cases have reported either pure central sleep apnea or mixed apnea [44]. Additionally, chest wall deformity and kyphoscoliosis can predispose patients to hypoventilation and hypoxemia [44].

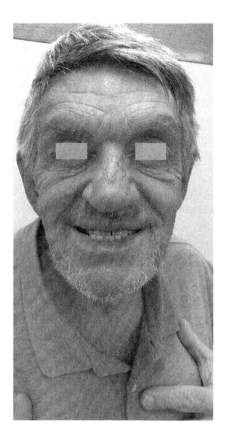

FIGURE 4.4

This patient presented with coarsened facial features including pronounced forehead brow, enlarged nose, prominent cheekbones, thick lips, marked facial lines, mandibular overgrowth with prognathism, maxillary widening, teeth separation, and jaw malocclusion.

4.4.6 Endocrinological and metabolic comorbidities

Thyroid disease is not uncommon in acromegaly, with one study reporting up to 78% of patients having concomitant thyroid disease of which 58% of patients having a diffuse or multinodular nontoxic goiter [45]. The risk of developing thyroid nodules increases with time of onset of acromegaly [46]. Hyperthyroidism and thyroid cancer have been reported in 15% [45] and 4% [47] of patients, respectively. Hypercalcemia is reported in some patients due to concurrent hyperparathyroidism that tends to persists even after successful treatment of acromegaly [48,49]. Growth hormone excess is also associated with hepatic and peripheral insulin resistance resulting in glucose intolerance and diabetes mellitus in more than 50% of cases [50] and dyslipidemia in 24%−65% of cases [51].

FIGURE 4.5

This patient had macroglossia, and swelling of pharyngeal tissues that contributed to the development of obstructive sleep apnea.

4.4.7 Cardiovascular comorbidities

Hypertension has been observed in 18%−60% of patients [52]. Its prevalence increases with duration of onset of acromegaly, age, and serum GH levels, and is considered an important prognostic factor for mortality [53]. Additionally, chronic GH/IGF-I excess may induce acromegalic cardiomyopathy that is characterized by concentric hypertrophy, diastolic dysfunction, progressive systolic impairment, and congestive heart failure [54,55]. Other cardiovascular comorbidities and risk factors include valvulopathies (e.g., aortic and mitral regurgitation) that are related to the degree of hypertrophy, arrhythmias (e.g., atrial fibrillation, supraventricular tachycardia, and ventricular arrhythmias), conduction disorders, increased intima-media thickness, hypertriglyceridemia and increased lipoprotein-a, and fibrinogen and plasminogen activator inhibitor [56]. Despite the negative cardiovascular risk profile, the prevalence of coronary artery disease in patients with acromegaly is surprisingly not increased [56], but given the excess cardiovascular comorbidities and risk factors, it is prudent that all cardiovascular risk factors are optimally managed.

4.4.8 Gastrointestinal comorbidities

Patients with acromegaly are predisposed to the development of colonic adenomatous polyps, dolichocolon, diverticular disease, and gallbladder polyps [57−59], especially in patients >50 years of age, male, have three or more skin tags, or have a family history of colon cancer [58]. Excess GH and IGF-I have been linked to colon epithelial transformation and polyposis, respectively [60,61]. However, previous epidemiological and meta-analysis studies have reported inconsistent findings on the risk of colonic cancer in acromegaly [62,63]. Some studies have reported the relative risk to be 10- to 20-fold compared with the general population, whereas in a meta-analysis of nine studies that compared colon adenoma and cancer rates in 701 acromegalic patients and 1573 control subjects, the frequency of adenomas and cancer was 2.5-fold and 4-fold higher in acromegaly patients than controls, respectively [64]. These discrepant data may be related to increased colonoscopy surveillance being performed for colonic polyps, where premalignant polyps are more frequent especially if the disease were still active. Hence, all patients with acromegaly should be offered colonoscopy at diagnosis [65]. How often to perform repeat colonoscopy will depend on several factors at screening. The presence of adenomatous polyps, normal or elevated serum IGF-I, and the age of the patient are determinants for future repeat colonoscopies, but the timing of when to perform (3 vs. 5 yearly) remains undefined. If acromegaly is in remission or controlled and no polyps were found at screening, it is unclear whether repeat colonoscopy is necessary. It is also not known whether patients found with gallbladder polyps screening are at increased risk of developing future gallbladder cancer, and whether there is a role in performing biliary ultrasound surveillance [66]. Nevertheless, it seems prudent to perform regular colonoscopies given the link between GH/IGF-I excess and cancer, and the increased prevalence of premalignant colonic polyps during active disease.

4.4.9 Neurological and neuropsychological comorbidities

Carpal tunnel syndrome is the most common peripheral neuropathy in acromegaly [17], and is thought to be related to GH-induced edema compressing the median nerve [67]. Other reported neurological comorbidities in patients with acromegaly include intracranial aneurysm [68], herniation of cerebellar tonsils [69], hearing loss [70], and pituitary apoplexy [71]. Patients with acromegaly also frequently experience impaired QoL due to multiple comorbidities and somatic changes [72]. These patients report emotional and psychological changes including body image distortion, social withdrawal, impaired self-confidence, poor interpersonal relationships, impaired cognition, depression and anxiety, with headache, chronic pain, and fatigue being major factors responsible for their QoL impairment [72].

4.4.10 Gigantism

Gigantism is rare occurring in only 5% of patients, where GH hypersecretion occurs before closure of epiphyseal cartilages. Due to recent research advances, the genetic pathophysiology of ~50% of patients with gigantism is now described [73,74]. The causative genes, when disrupted, alter the function of pituitary somatotropes, leading to hyperplasia, aggressive tumor formation, and GH hypersecretion that often responds poorly to existing treatments. Males are more affected than females, presumably due to exacerbation of the sexual dimorphic pubertal growth patterns by excess GH and IGF-I [74]. Conversely, when females are affected, they tend to be present at a younger age than males [74], presumably due to the preponderance of females with X-LAG syndrome [15]. Patients with gigantism almost invariably present with macroadenomas and a strong family history of pituitary adenomas [74]. Patients who present at an older age have a higher disease burden, likely due to a longer exposure to GH and IGF-I hypersecretion. Given the early-onset of presentation, genetic abnormalities can take the form of germline genetic mutations or pathological copy number variations, presenting either constitutionally or as somatic mosaicism. The genes encoding proteins that are involved in gigantism can also drive tumorigenesis in some patients, leading to inherited or familial forms of pituitary adenomas [75,76], including multiorgan tumor syndromes (e.g., McCune–Albright syndrome, multiple endocrine neoplasia type 1, or Carney complex) or pituitary-specific conditions, such as familial isolated pituitary adenomas (FIPAs) (e.g., *AIP* mutations or X-LAG syndrome) (Table 4.2). *AIP* mutations cause a low-penetrance disease manifesting clinically in the second decade of life, either sporadically or in the

Table 4.2 Summary of genetic causes of gigantism.

Syndrome or genetic cause	Genetic abnormalities	Clinical description	Percentage of cases
AIP gene	*AIP* mutations; exon and whole-gene deletions	Isolated pituitary adenomas; sporadic, familial gigantism and FIPA	29
X-linked acrogigantism syndrome	Duplication on chromosome Xq26.3 including *GPR101*; somatic mosaicism	Isolated pituitary adenomas; sporadic, familial gigantism and FIPA	10
McCune–Albright syndrome	*GNAS* mosaicism	Syndromic multiorgan disease; sporadic	5
Multiple endocrine neoplasia type 1	*MEN1* gene mutations, deletions; possibility of pituitary hyperplasia due to GHRH hypersecretion from neuroendocrine tumor	Syndromic multiorgan disease; sporadic and familial	1
Carney complex	*PRKAR1A* mutations and deletions	Syndromic multiorgan disease; sporadic and familial	1
Unknown genetic causes	Unidentified as yet	Isolated or syndromic; sporadic and familial	54

AIP, *Aryl-hydrocarbon receptor–interacting protein;* FIPA, *familial isolated pituitary adenomas;* GHRH, *growth hormone-releasing hormone;* GNAS, *GNAS complex locus;* GPR101, *G protein-coupled receptor 101;* MEN1; *multiple endocrine neoplasia type 1;* PRKAR1A, *protein kinase cAMP-dependent type 1 regulatory subunit-α.*
Reproduced with permission from Beckers A et al. [81].

setting of FIPA [77]. As incidental pituitary microadenomas are common in the general population [78], it may be difficult to determine whether small asymptomatic tumors presenting in *AIP* mutation carriers will eventually develop into larger adenomas [79]. *AIP* mutation tumors are poor responders to somatostatin receptor ligand (SRL) therapy compared to those with non-*AIP* mutated acromegaly [77]. X-LAG syndrome is a recently described disorder of development of pituitary macroadenomas or infrequently with pituitary hyperplasia alone [15] in young children that if left untreated, may become aggressive [80]. The syndrome is very rare (~30 individual cases described to date) [81] and is associated with duplications on chromosome Xq26.3 that involve *GPR101*, an orphan G protein-coupled receptor [15]. Patients have normal growth patterns in the initial postnatal period [82], but during early childhood, demonstrate rapid increases in height and weight, features of acromegaly (e.g., large hands and feet, facial coarsening, broad nasal bridge, prognathism, and increased interdental spaces), and in some, episodic hyperphagia. Serum GH levels are markedly elevated and not suppressible during an oral glucose tolerance test (OGTT), and hyperprolactinemia is also usually present.

4.5 Differences in clinical manifestations based on clinicopathological findings

Pituitary GH-secreting tumors exhibit several histopathological variants, including pure somatotroph (densely- or sparsely granulated) or plurihormonal (mammosomatotroph, mixed somatotroph-lactotroph, mature plurihormonal Pit1-lineage, acidophil stem cell, and poorly differentiated Pit1-lineage) tumors, which can express multiple hormones (e.g., GH, prolactin, β-thyroid−stimulating hormone, and/or α-subunit of glycoprotein hormones) [14]. Each tumor type has a distinct pathophysiology, resulting in variations in clinical features, imaging and therapeutic responses that are summarized in Table 4.3.

Therefore it is important to note that the manifestations of acromegaly are clinical spectra that vary from frank signs and symptoms to subtle features that may be quite easily missed in some patients. The spectrum of histopathological tumors that may give rise to acromegaly helps to better understand the different clinical, biochemical, radiological characteristics and variable treatment responses seen in these patients.

Table 4.3 Differences in clinical acromegaly according to the varying clinicopathological features.

Pathogenic entity	Frequency	Clinicopathologic manifestations	Biochemical features	MRI features	Treatment
Densely granulated tumors	30%−50% (usually in patients >50 years)	Slow-growing high GH-secretory lesions → more florid symptoms of acromegaly	↑↑GH and ↑↑IGF-I	T2-hypointensity	SRL achieves 65-90% response rates
Sparsely granulated tumors	15%−35% (usually in patients <50 years)	Aggressive modest GH-secretory lesions → more subtle symptoms of acromegaly	↑GH and ↑IGF-I	T2-hyperintensity	Resistant to first generation SRLs, may respond to pasireotide, pegvisomant, and SRS
Mammosomatotroph tumors	Younger patients and in cases of childhood onset-gigantism	Slow-growing high GH-secretory lesions with PRL secretion → more florid symptoms of acromegaly and some symptoms of hyperprolactinemia and rarely hyperthyroidism	↑↑GH, ↑↑IGF-I, and ↑PRL (rarely ↑TSH)	T2-hypointensity	SRL achieves 65-90% response rates, Cabergoline may be used
Mixed somatotroph-lactotroph tumors (densely or sparsely granulated)	Unknown	Increased risk of invasion into surrounding structures, difficult to treat, have a low surgical cure rate	↑↑GH, ↑↑IGF-I, and ↑PRL	T2-hypointensity	High doses of SRL, pegvisomant, and cabergoline

(Continued)

Table 4.3 Differences in clinical acromegaly according to the varying clinicopathological features. *Continued*

Pathogenic entity	Frequency	Clinicopathologic manifestations	Biochemical features	MRI features	Treatment
Acidophil stem cell tumors	Rare	More aggressive and invasive than prolactinomas, fast-growing, modest GH- and PRL-secretory lesions → symptoms of hyperprolactinemia and more subtle symptoms of acromegaly	↑↑GH, ↑IGF-I, and ↑PRL	T1-isointensity	Surgery and SRS, resistant to Cabergoline
Poorly differentiated Pit1-lineage tumor	Rare	Aggressive and invasive, with increased risk for recurrence following surgery → can be silent or manifest with subtle symptoms of acromegaly, hyperprolactinemia and hyperthyroidism	↑/- GH, ↑/-IGF-I, ↑/-PRL and ↑/-TSH	Cavernous sinus invasion, involvement of the clivus, and frequently both suprasellar and downward growth	Surgery and SRS, Cabergoline may normalize PRL but not shrink the tumor
Pituitary carcinoma	Very rare	Aggressive and only 5% are GH-secretors	↑/- GH, ↑/-IGF-I, and ↑/- PRL	Resembles pituitary adenomas and defined only by the presence of central nervous system or systemic metastases.	Surgery, medical therapy (e.g., DA agonists, SRL, chemotherapy, and temozolomide) and SRS
Pituitary hyperplasia	Very rare	Associated with GHRH-secreting tumor or an underlying germline genetic predisposition syndrome	↑GH and ↑IGF-I	Diffusely enlarged sella with no pituitary tumor	Surgery, SRS, DA agonists and pegvisomant (resistant to first generation SRLs)

Table 4.3 Differences in clinical acromegaly according to the varying clinicopathological features. *Continued*

Pathogenic entity	Frequency	Clinicopathologic manifestations	Biochemical features	MRI features	Treatment
Ectopic GH secretion	Very rare	Associated with bronchial or pancreatic neuroendocrine tumors and lymphomas and patients present with typical signs and symptoms of acromegaly and may also manifest additional symptoms from the primary tumor	↑GH and ↑IGF-I	MRI may be normal or may even show an empty sella	Surgery, DA agonists, and SRLs.
Ectopic GHRH secretion	Very rare	Caused by gangliocytomas in children and young adults, and neuroendocrine tumors of pancreas or lung, and pheochromocytomas, requires CT, octreotide scintigraphy or 68Gallium DOTATATE-PET/CT scans to localize the extra-pituitary neuroendocrine tumor	↑GHRH, ↑GH, and ↑IGF-I	Pituitary hyperplasia	Surgery, SRLs, and DA agonists.

CT, *Computerized tomography;* DA, *dopamine agonist;* GH, *growth hormone;* GHRH, *growth hormone-releasing hormone;* IGF-I, *insulin-like growth factor-I;* MRI, *magnetic resonance imaging;* PET, *positron emission tomography;* PRL, *prolactin;* SRL, *somatostatin receptor ligand;* SRS, *stereotactic radiosurgery;* TSH, *thyroid-stimulating hormone;* ↑↑, *markedly increased levels;* ↑, *increased levels; -, normal levels.*

4.6 Pseudoacromegaly

Pseudoacromegaly refers to a heterogeneous group of disorders in which patients present with clinical features resembling those of acromegaly or gigantism but without GH/IGF-I axis abnormalities [83–86]. The features include acromegaloid facies, acral enlargement, prognathism, visceromegaly, hypertension, fatigue, headaches, arthralgias, paresthesias, hyperhidrosis, oily odorous skin, hypertrichosis, hyperpigmentation, and dysphonia. It is likely that the earlier reports of patients with pseudoacromegaly [87–89] were part of a syndromic disorder that includes Cantú syndrome [90], Sotos syndrome [84], overgrowth syndrome [91], nonislet cell tumor–induced hypoglycemia syndrome [92], double male syndrome [93], multiple hamartoma syndrome [94], and severe insulin resistance syndrome [95]. Other nonsyndromic conditions that present with acromegaloid features include pachydermoperiostosis [96], X-tetrasomy [97], primary hypothyroidism [85], Fabry disease [98], or drug-related (e.g., minoxidil or phenytoin) [99,100]. Some conditions mimic acromegaly more closely, while others may demonstrate some acromegaloid features but are less likely to have underlying acromegaly (Table 4.4).

Pseudoacromegaly cases are often referred for evaluation of possible GH excess, and the diagnosis of many of pseudoacromegaly conditions, especially the syndromic conditions, is established by pediatric endocrinologists and experienced geneticists. However, the differential diagnosis of pseudoacromegaly can be challenging due to the long list of overlapping and rare conditions. Key facial features characterize Sotos syndrome [91], distinctive forehead skin folds and joint abnormalities help the diagnosis of pachydermoperiostosis [96], severe hypoglycemia would be typical in IGF-2-secreting tumors [101], and generalized infant-onset hirsutism indicates Cantú syndrome [90].

Pathological accelerated growth and/or tall stature can be unrelated to the GH axis, and may occur in isolation or as part of a syndrome, such as the overgrowth syndromes or Marfan syndrome [102]. Overgrowth syndromes are typically characterized by accelerated growth, macrocephaly, distinctive facial appearance, and impaired intellectual ability [103]. Due to the overlapping features and because the underlying genetic basis of many overgrowth conditions has now been identified, genetic testing should now be offered to all these patients [102].

Table 4.4 The likelihood of pseudoacromegaly conditions to have underlying acromegaly.

High likelihood	Low likelihood
Antiretroviral-induced lipodystrophy	Borjeson–Forssman–Lehmann syndrome
Baraquer–Simons syndrome	Coffin–Lowry syndrome
Beckwith–Wiedemann syndrome	Cornella de Lange syndrome
Cantú syndrome	Fabry syndrome
Chromosome 11 pericentric inversion	Fragile *X* syndrome
Coffin–Siris syndrome	*HMGA2*-related overgrowth
Insulin-mediated pseudoacromegaly	Klippel–Trenaunay syndrome
Long-standing primary hypothyroidism	Malan syndrome
Minoxidil-induced changes	Marfan syndrome
Nonislet cell tumor–induced hypoglycemia	Marshall–Smith syndrome
Pachydermoperiostosis	Nicolaides–Baraitser syndrome
Pallister–Killian syndrome	Osteopetrosis
Phenytoin-induced changes	Phelan–McDermid syndrome
Simpson–Golabi–Behmel syndrome	Sclerosteosis
Sotos syndrome	*SETD5*-related syndrome
Sotos-like *SETD2*-related syndrome	X-tetrasomy
Tatton–Brown–Rahman syndrome	15q26 microduplications
Weaver syndrome	
Weaver-like *EED*-related syndrome	

Reproduced with permission from Marques P, Korbonits M. Pseudoacromegaly. Front Neuroendocrinol. 2019; 52:113–43 [86].

4.7 **Clinical description and diagnosis**

Because the clinical spectrum of acromegaly varies from large stature and disfigurement to subtle nonspecific features, the diagnosis may be missed in some patients who are thought to have clinically nonfunctioning pituitary incidentalomas or may not even be suspected to have a pituitary disorder at all. Patients with chronic dental problems, osteoarthritis, sleep apnea, and other seemingly unrelated medical issues may be undiagnosed until an astute clinician raises the possibility of acromegaly. The clinicopathological spectrum of tumors may provide an explanation for the varying clinical, biochemical, and radiological characteristics of patients with acromegaly. However, once acromegaly is suspected, making the diagnosis is usually seamless involving a series of basal and dynamic biochemical testing of the GH/IGF-I axis. Proper interpretation of the test results often relies on the appreciation of the strengths and limitations of the GH and IGF-I assays used, and allows the diagnosis to be made or excluded relatively simply in a timely manner.

The first and best screening test for acromegaly is the measurement of serum IGF-I levels (Fig. 4.2) because these levels are elevated in virtually all patients [16], although it should be borne in mind that elevated serum IGF-I levels are also seen in pregnancy [104] and adolescence [105]. A normal serum IGF-I level excludes the diagnosis of acromegaly. There are several conditions where low-serum IGF-I levels have been reported, including hypothyroidism, malnutrition, poorly controlled diabetes mellitus, liver failure, renal failure, and oral estrogen use [106]. In these situations, it is possible that the diagnosis of acromegaly could be missed. Hence, if acromegaly is suspected, the next step is to perform a 2-hour OGTT with serum GH measurements. Inadequate suppression of serum GH levels to <1 µg/L during an OGTT confirms the diagnosis [16]. The sensitivity of the test is reported to be improved at a GH cutoff of 0.4 ng/mL [107]. Because the GH assay method can impact the absolute GH level reported by a laboratory [108], this may also impact the cutoff for GH suppression following an OGTT [109]. Currently used cutoffs for GH after an OGTT are 1.0 and 0.4 ng/dL, but may not be accurate for all commercial assays; hence method-specific values for GH cutoffs must be reported when available [16,107].

After biochemical confirmation of acromegaly, pituitary MRI is recommended to ascertain tumor size, location, and invasiveness [16]. Macroadenomas are detected in up to 77% of patients [110]. A CT scan is reserved for subjects with a contraindication to MRI. Rarely encountered, a patient with biochemically confirmed acromegaly with a normal or small pituitary gland on MRI might pose a diagnostic challenge [111,112]. Although the tumor may be microscopic and not visualized on a routine MRI scan, further testing, including measurement of serum GHRH as well as imaging (e.g., somatostatin receptor scintigraphy and thoracic and abdominal imaging) may be considered to evaluate for ectopic disease [113,114]. If there is evidence of tumor abutment on the optic chiasm, formal visual field testing should then be performed. Less frequently, tumor involvement in the cavernous sinus can cause other cranial nerve abnormalities leading to diplopia and blurred vision [115]. Hence, the presence of visual impairment usually dictates the choice and timeliness of surgical intervention.

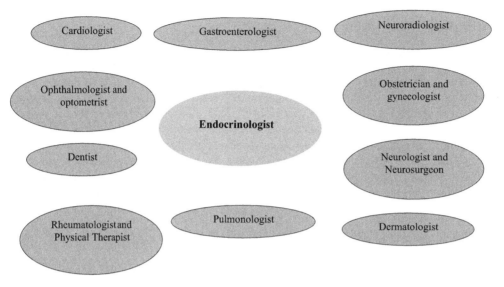

FIGURE 4.6

Various specialties that potential patients with acromegaly may present to.

4.7.1 Case illustrations

The following two cases demonstrate the common presenting features that are not typically unique to acromegaly but may prompt the patient to seek medical assistance from various medical specialties (Fig. 4.6).

4.7.1.1 Case 1

A 56-year-old man presented to his rheumatologist with long-standing lower back pain and generalized joint pains, which prompted an initial diagnosis of osteoarthritis. The patient also complained of muscle weakness and worsening headaches and, over the course of the preceding year, experienced increasing voice hoarseness. His wife endorses him having excessive snoring, episodic apnea, and daytime somnolence. His ring and shoe size had gradually increased during the previous 3 years. The patient was diagnosed with type 2 diabetes mellitus 15 months prior and despite no dietary changes, his blood glucose levels continued to worsen. His rheumatologist suspected that he had acromegaly and ordered a serum IGF-I level which returned elevated at 661 ng/mL (reference range: 81−225 ng/mL), prompting a referral to endocrinology.

4.7.2 Examination and diagnostic workup

On examination, the patient had a blood pressure of 154/89 mmHg, and his body mass index (BMI) was 35 kg/m^2. He had mild jaw protrusion and macroglossia, large spade-like hands, axillary skin tags (Fig. 4.7), and acanthosis nigricans on the nape of his neck and axillae. His visual fields were grossly normal on confrontation.

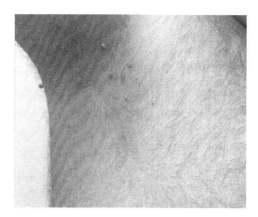

FIGURE 4.7

Case 1: axillary skin tags.

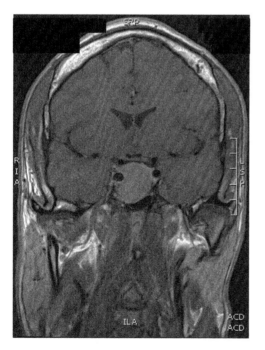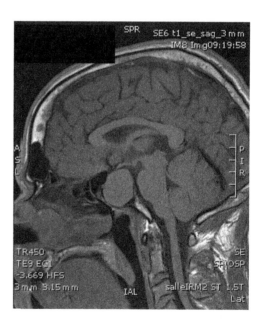

FIGURE 4.8

Case 1: MRI findings.

Additional hormonal evaluations included confirmation of elevation of serum IGF-I level of 583 ng/mL (reference range: 81−225 ng/mL), and elevated GH nadir level during an OGTT of 1.9 ng/dL. Fasting morning serum testosterone level was low at 204 ng/dL (reference range: 300−890 ng/dL); so were his luteinizing hormone (LH) and follicle-stimulating hormone (FSH) levels at 3 mIU/dL (reference range: 1−10 mIU/dL) and 5 mIU/dL (reference range: 1−18 mIU/dL), respectively. Thyroid-stimulating hormone (TSH), free T4, and prolactin were within normal ranges. A normal peak cortisol level of 25.7 μg/dL was observed following a 250-μg adrenocortico-trophic hormone (ACTH) stimulation test. Fasting glucose level was elevated at 127 mg/d, with a hemoglobin A1C of 7.4%. Magnetic resonance imaging revealed a 1.1 cm pituitary macroadenoma with no impingement to the optic chiasm (Fig. 4.8).

The diagnosis of acromegaly was confirmed in this patient, who also had concurrent hypogona-dotropic hypogonadism and secondary diabetes mellitus, but normal thyroid function and adrenal reserve. The patient was referred to a neurosurgeon for transsphenoidal resection, and a diabetes educator and dietitian for dietary advice, focusing on carbohydrate counting and healthy eating habits to help improve his overall glycemic control.

4.7.2.1 Case 2

A 59-year-old male was referred to a neuro-ophthalmologist by his optometrist, whom he had con-sulted for his deteriorating vision and persistent headaches, which he had attributed to his gradual deterioration in vision. Visual field testing confirmed bilateral homonymous hemianopia using a Goldmann visual examination. The patient also endorses a reduction in morning erections, decreased libido and bilateral gynecomastia (Fig. 4.9). With the combination of these clinical mani-festations, his neuro-ophthalmologist referred him to endocrinology for further diagnostic workup.

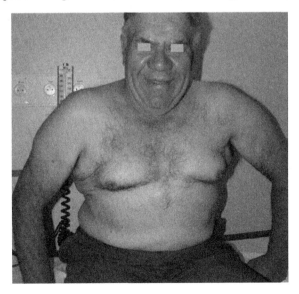

FIGURE 4.9

Case 2: bilateral gynecomastia.

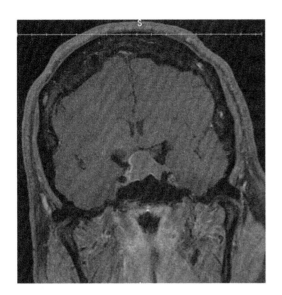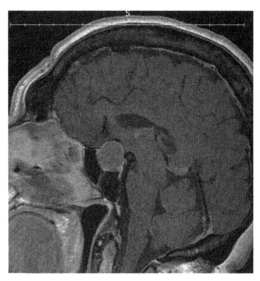

FIGURE 4.10

Case 2: MRI findings.

4.7.3 Examination and diagnostic workup

The patient had mildly elevated blood pressure of 143/91 mmHg and a BMI of 33.6 kg/m^2. This patient had spaces between his teeth, a large forehead brow, macroglossia, and spade-like hands. A visual field defect on confrontation was confirmed.

Hormonal measurements ordered by his endocrinologist revealed a serum IGF-I level of 470 ng/ mL (reference range: 178−295 ng/mL), and a basal morning serum GH level of 18.9 ng/mL. The nadir serum GH levels did not decrease from baseline at any time point during an OGTT, remaining significantly above the 0.4 ng/mL cutoff. Additional laboratory investigations revealed a low testosterone level of 115 ng/dL (reference range: 300−890 ng/dL), LH of 2 mIU/dL (reference range: 1−10 mIU/dL), FSH of 7 mIU/dL (reference range: 1−18 mIU/dL), TSH of 0.62 mIU/dL (reference range: 0.34−5.6 mIU/L), and free T4 of 0.6 ng/L (reference range: 0.6−1.2 ng/L), whereas prolactin level was raised at 85 ng/mL (reference range: 3−13 ng/mL). The 250 µg ACTH stimulation test produced a peak cortisol level of 15.2 µg/dL. Magnetic resonance imaging revealed a 1.6 cm pituitary macroadenoma compressing the optic chiasm (Fig. 4.10).

Acromegaly was confirmed and the patient also had concurrent hypogonadotropic hypogonadism, impaired fasting glucose, central hypothyroidism, central hypoadrenalism, and hyperprolactinemia. His visual field deficits warranted an urgent referral to neurosurgery for surgical resection of his pituitary tumor.

4.8 Summary of the two cases

These two patients sought medical care primarily for joint pains and visual abnormalities, both common medical issues in the general population. Their specialists referred these patients to an

Table 4.5 Diagnostic referral sources for patients with acromegaly.

Diagnostic source	Percentage of diagnoses
Internist/family medicine	39−44
Endocrinologist	13−25
Emergency room staff	10
Gynecologist	<1
Ophthalmologist/optometrist	3−6
Relative/self	6−7
Neurologist	5−6
Dentist	3−5
Other	14

Reproduced with permission from Drange MR, Fram NR, Herman-Bonert V, Melmed S. Pituitary tumor registry: a novel clinical resource. J Clin Endocrinol Metab. 2000; 85:168−74 [18].

endocrinologist when they had noted that the patients had signs and symptoms suggestive of acromegaly. Both patients had undergone physical changes characteristic of acromegaly; however, these were not striking features to either the patients or to their family members that would prompt them to seek medical attention sooner. This commonly occurs in the setting of very gradual changes, which may not be noticed by the patients themselves nor by those in contact with the patients on a daily basis. In a study by Drange et al. [18] utilizing a pituitary tumor registry found that patients with acromegaly were initially referred to an endocrinologist mainly by internists, gynecologists, ophthalmologists, optometrists, neurologists, and dentists (Table 4.5). Referrals can also originate from anesthesiologists in pain clinics, physical therapists, dermatologists, diabetes educators and dietitians, and gastroenterologists.

4.9 Conclusion

Acromegaly is a rare disorder of chronic GH hypersecretion usually caused by a pituitary adenoma and can present with a wide variety of clinical manifestations ranging from subtle to severe. Therefore measurement of serum IGF-I is important in cases of incidentally detected pituitary adenomas to rule out acromegaly. Rarely, some patients may have pseudoacromegaly, where their clinical features resemble those of acromegaly yet their GH/IGF-I axis is normal, and ascertaining the differential diagnosis in these patients can be challenging due to its variability and significant overlap between some of these conditions that can only be confirmed following genetic testing. Once acromegaly is diagnosed, usually with elevated serum IGF-I levels and unsuppressed serum GH levels during an OGTT, detailed workup of the various organs potentially involved is important to improve the prognosis, QoL, and life expectancy of these patients. Furthermore, obtaining detailed clinicopathological information will be useful for the clinician to personalize treatment regimen in order to achieve optimal therapeutic efficacy.

References

[1] Petrossians P, Daly AF, Natchev E, Maione L, Blijdorp K, Sahnoun-Fathallah M, et al. Acromegaly at diagnosis in 3173 patients from the Liege Acromegaly Survey (LAS) Database. Endocr Relat Cancer 2017;24:505–18.

[2] Ritvonen E, Loyttyniemi E, Jaatinen P, Ebeling T, Moilanen L, Nuutila P, et al. Mortality in acromegaly: a 20-year follow-up study. Endocr Relat Cancer 2016;23:469–80.

[3] Holdaway IM, Rajasoorya C. Epidemiology of acromegaly. Pituitary 1999;2:29–41.

[4] Cannavo S, Ferrau F, Ragonese M, Curto L, Torre ML, Magistri M, et al. Increased prevalence of acromegaly in a highly polluted area. Eur J Endocrinol 2010;163:509–13.

[5] Fernandez A, Karavitaki N, Wass JA. Prevalence of pituitary adenomas: a community-based, cross-sectional study in Banbury (Oxfordshire, UK). Clin Endocrinol (Oxf) 2010;72:377–82.

[6] Lavrentaki A, Paluzzi A, Wass JA, Karavitaki N. Epidemiology of acromegaly: review of population studies. Pituitary 2017;20:4–9.

[7] Etxabe J, Gaztambide S, Latorre P, Vazquez JA. Acromegaly: an epidemiological study. J Endocrinol Invest 1993;16:181–7.

[8] Reid TJ, Post KD, Bruce JN, Nabi Kanibir M, Reyes-Vidal CM, Freda PU. Features at diagnosis of 324 patients with acromegaly did not change from 1981 to 2006: acromegaly remains under-recognized and under-diagnosed. Clin Endocrinol (Oxf) 2010;72:203–8.

[9] Kreitschmann-Andermahr I, Siegel S, Kleist B, Kohlmann J, Starz D, Buslei R, et al. Diagnosis and management of acromegaly: the patient's perspective. Pituitary 2016;19:268–76.

[10] Nachtigall L, Delgado A, Swearingen B, Lee H, Zerikly R, Klibanski A. Changing patterns in diagnosis and therapy of acromegaly over two decades. J Clin Endocrinol Metab 2008;93:2035–41.

[11] Ohno H, Yoneoka Y, Jinguji S, Watanabe N, Okada M, Fujii Y. Has acromegaly been diagnosed earlier? J Clin Neurosci 2018;48:138–42.

[12] Colao A, Grasso LFS, Giustina A, Melmed S, Chanson P, Pereira AM, et al. Acromegaly. Nat Rev Dis Prim 2019;5:20.

[13] Dworakowska D, Korbonits M, Aylwin S, McGregor A, Grossman AB. The pathology of pituitary adenomas from a clinical perspective. Front Biosci (Sch Ed) 2011;3:105–16.

[14] Akirov A, Asa SL, Amer L, Shimon I, Ezzat S. The clinicopathological spectrum of acromegaly. J Clin Med 2019;8.

[15] Trivellin G, Daly AF, Faucz FR, Yuan B, Rostomyan L, Larco DO, et al. Gigantism and acromegaly due to Xq26 microduplications and GPR101 mutation. N Engl J Med 2014;371:2363–74.

[16] Katznelson L, Laws Jr. ER, Melmed S, Molitch ME, Murad MH, Utz A, et al. Acromegaly: an Endocrine Society Clinical Practice Guideline. J Clin Endocrinol Metab 2014;99:3933–51.

[17] Caron P, Brue T, Raverot G, Tabarin A, Cailleux A, Delemer B, et al. Signs and symptoms of acromegaly at diagnosis: the physician's and the patient's perspectives in the ACRO-POLIS study. Endocrine 2019;63:120–9.

[18] Drange MR, Fram NR, Herman-Bonert V, Melmed S. Pituitary tumor registry: a novel clinical resource. J Clin Endocrinol Metab 2000;85:168–74.

[19] Levy MJ. The association of pituitary tumors and headache. Curr Neurol Neurosci Rep 2011;11:164–70.

[20] Castro Cabezas M, Zelissen PM, Jansen GH, Van Gils AP, Koppeschaar HP. Acromegaly: report of two patients with an unusual presentation. Neth J Med 1999;54:163–6.

[21] Daita G, Yonemasu Y, Hashizume A. Unilateral exophthalmos caused by an invasive pituitary adenoma. Neurosurgery 1987;21:716–18.

[22] Rick J, Jahangiri A, Flanigan PM, Chandra A, Kunwar S, Blevins L, et al. Growth hormone and prolactin-staining tumors causing acromegaly: a retrospective review of clinical presentations and surgical outcomes. J Neurosurg 2018;1−7.

[23] Vilar L, Fleseriu M, Bronstein MD. Challenges and pitfalls in the diagnosis of hyperprolactinemia. Arq Bras Endocrinol Metab 2014;58:9−22.

[24] Kaltsas GA, Mukherjee JJ, Jenkins PJ, Satta MA, Islam N, Monson JP, et al. Menstrual irregularity in women with acromegaly. J Clin Endocrinol Metab 1999;84:2731−5.

[25] Raju JA, Shipman KE, Inglis JA, Gama R. Acromegaly presenting as erectile dysfunction: case reports and review of the literature. Rev Urol 2015;17:246−9.

[26] Kreitschmann-Andermahr I, Kohlmann J, Kleist B, Hirschfelder U, Buslei R, Buchfelder M, et al. Orodental pathologies in acromegaly. Endocrine 2018;60:323−8.

[27] Davi MV, Dalle Carbonare L, Giustina A, Ferrari M, Frigo A, Lo Cascio V, et al. Sleep apnoea syndrome is highly prevalent in acromegaly and only partially reversible after biochemical control of the disease. Eur J Endocrinol 2008;159:533−40.

[28] Dostalova S, Sonka K, Smahel Z, Weiss V, Marek J, Horinek D. Craniofacial abnormalities and their relevance for sleep apnoea syndrome aetiopathogenesis in acromegaly. Eur J Endocrinol 2001;144:491−7.

[29] Pantanetti P, Sonino N, Arnaldi G, Boscaro M. Self image and quality of life in acromegaly. Pituitary 2002;5:17−19.

[30] Roerink SH, Wagenmakers MA, Wessels JF, Sterenborg RB, Smit JW, Hermus AR, et al. Persistent self-consciousness about facial appearance, measured with the Derriford appearance scale 59, in patients after long-term biochemical remission of acromegaly. Pituitary 2015;18:366−75.

[31] Braham C, Betea D, Pierard-Franchimont C, Beckers A, Pierard GE. Skin tensile properties in patients treated for acromegaly. Dermatology 2002;204:325−9.

[32] Centurion SA, Schwartz RA. Cutaneous signs of acromegaly. Int J Dermatol 2002;41:631−4.

[33] Quatresooz P, Hermanns-Le T, Ciccarelli A, Beckers A, Pierard GE. Tensegrity and type 1 dermal dendrocytes in acromegaly. Eur J Clin Invest 2005;35:133−9.

[34] Ben-Shlomo A, Melmed S. Skin manifestations in acromegaly. Clin Dermatol 2006;24:256−9.

[35] Chanson P, Salenave S. Acromegaly. Orphanet J Rare Dis 2008;3:17.

[36] Wassenaar MJ, Biermasz NR, van Duinen N, van der Klaauw AA, Pereira AM, Roelfsema F, et al. High prevalence of arthropathy, according to the definitions of radiological and clinical osteoarthritis, in patients with long-term cure of acromegaly: a case-control study. Eur J Endocrinol 2009;160:357−65.

[37] Kropf LL, Madeira M, Vieira Neto L, Gadelha MR, de Farias ML. Functional evaluation of the joints in acromegalic patients and associated factors. Clin Rheumatol 2013;32:991−8.

[38] Claessen KM, Ramautar SR, Pereira AM, Smit JW, Roelfsema F, Romijn JA, et al. Progression of acromegalic arthropathy despite long-term biochemical control: a prospective, radiological study. Eur J Endocrinol 2012;167:235−44.

[39] Killinger Z, Kuzma M, Sterancakova L, Payer J. Osteoarticular changes in acromegaly. Int J Endocrinol 2012;2012:839282.

[40] de Azevedo Oliveira B, Araujo B, Dos Santos TM, Ongaratti BR, Rech C, Ferreira NP, et al. The acromegalic spine: fractures, deformities and spinopelvic balance. Pituitary 2019;22:601−6.

[41] Zoicas F, Kleindienst A, Mayr B, Buchfelder M, Megele R, Schofl C. Screening for acromegaly in patients with carpal tunnel syndrome: a prospective study (ACROCARP). Horm Metab Res 2016;48:452−6.

[42] Tagliafico A, Resmini E, Nizzo R, Derchi LE, Minuto F, Giusti M, et al. The pathology of the ulnar nerve in acromegaly. Eur J Endocrinol 2008;159:369−73.

[43] van Haute FR, Taboada GF, Correa LL, Lima GA, Fontes R, Riello AP, et al. Prevalence of sleep apnea and metabolic abnormalities in patients with acromegaly and analysis of cephalometric parameters by magnetic resonance imaging. Eur J Endocrinol 2008;158:459−65.

[44] Rodrigues MP, Naves LA, Casulari LA, Silva CA, Araujo RR, Viegas CA. Using clinical data to predict sleep hypoxemia in patients with acromegaly. Arq Neuropsiquiatr 2007;65:234−9.

[45] Gasperi M, Martino E, Manetti L, Arosio M, Porretti S, Faglia G, et al. Acromegaly Study Group of the Italian Society of E. Prevalence of thyroid diseases in patients with acromegaly: results of an Italian multi-center study. J Endocrinol Invest 2002;25:240−5.

[46] Reverter JL, Fajardo C, Resmini E, Salinas I, Mora M, Llatjos M, et al. Benign and malignant nodular thyroid disease in acromegaly. Is a routine thyroid ultrasound evaluation advisable? PLoS One 2014;9: e104174.

[47] Wolinski K, Czarnywojtek A, Ruchala M. Risk of thyroid nodular disease and thyroid cancer in patients with acromegaly−meta-analysis and systematic review. PLoS One 2014;9:e88787.

[48] Shah R, Licata A, Oyesiku NM, Ioachimescu AG. Acromegaly as a cause of 1,25-dihydroxyvitamin D-dependent hypercalcemia: case reports and review of the literature. Pituitary 2012;15(Suppl. 1):S17−22.

[49] Ueda M, Inaba M, Tahara H, Imanishi Y, Goto H, Nishizawa Y. Hypercalcemia in a patient with primary hyperparathyroidism and acromegaly: distinct roles of growth hormone and parathyroid hormone in the development of hypercalcemia. Intern Med 2005;44:307−10.

[50] Alexopoulou O, Bex M, Kamenicky P, Mvoula AB, Chanson P, Maiter D. Prevalence and risk factors of impaired glucose tolerance and diabetes mellitus at diagnosis of acromegaly: a study in 148 patients. Pituitary 2014;17:81−9.

[51] Portocarrero-Ortiz LA, Vergara-Lopez A, Vidrio-Velazquez M, Uribe-Diaz AM, Garcia-Dominguez A, Reza-Albarran AA, et al.Mexican Acromegaly Registry G The Mexican Acromegaly Registry: clinical and biochemical characteristics at diagnosis and therapeutic outcomes. J Clin Endocrinol Metab 2016;101:3997−4004.

[52] Bondanelli M, Ambrosio MR, degli Uberti EC. Pathogenesis and prevalence of hypertension in acromegaly. Pituitary 2001;4:239−49.

[53] Holdaway IM, Rajasoorya RC, Gamble GD. Factors influencing mortality in acromegaly. J Clin Endocrinol Metab 2004;89:667−74.

[54] Mizera L, Elbaum M, Daroszewski J, Bolanowski M. Cardiovascular complications of acromegaly. Acta Endocrinol (Buchar) 2018;14:365−74.

[55] Mosca S, Paolillo S, Colao A, Bossone E, Cittadini A, Iudice FL, et al. Cardiovascular involvement in patients affected by acromegaly: an appraisal. Int J Cardiol 2013;167:1712−18.

[56] Gadelha MR, Kasuki L, Lim DST, Fleseriu M. Systemic complications of acromegaly and the impact of the current treatment landscape: an update. Endocr Rev 2019;40:268−332.

[57] Bogazzi F, Cosci C, Sardella C, Costa A, Manetti L, Gasperi M, et al. Identification of acromegalic patients at risk of developing colonic adenomas. J Clin Endocrinol Metab 2006;91:1351−6.

[58] Renehan AG, Shalet SM. Acromegaly and colorectal cancer: risk assessment should be based on population-based studies. J Clin Endocrinol Metab 2002;87:1909 author reply 1909.

[59] Terzolo M, Reimondo G, Gasperi M, Cozzi R, Pivonello R, Vitale G, et al. Colonoscopic screening and follow-up in patients with acromegaly: a multicenter study in Italy. J Clin Endocrinol Metab 2005;90:84−90.

[60] Chesnokova V, Zonis S, Barrett R, Kameda H, Wawrowsky K, Ben-Shlomo A, et al. Excess growth hormone suppresses DNA damage repair in epithelial cells. JCI Insight 2019;4.

[61] Chesnokova V, Zonis S, Zhou C, Recouvreux MV, Ben-Shlomo A, Araki T, et al. Growth hormone is permissive for neoplastic colon growth. Proc Natl Acad Sci USA 2016;113:E3250−9.

[62] Boguszewski CL, Boguszewski M. Growth hormone's links to cancer. Endocr Rev 2019;40:558−74.

[63] Dal J, Leisner MZ, Hermansen K, Farkas DK, Bengtsen M, Kistorp C, et al. Cancer incidence in patients with acromegaly: a cohort study and meta-analysis of the literature. J Clin Endocrinol Metab 2018;103:2182−8.

[64] Rokkas T, Pistiolas D, Sechopoulos P, Margantinis G, Koukoulis G. Risk of colorectal neoplasm in patients with acromegaly: a meta-analysis. World J Gastroenterol 2008;14:3484−9.

[65] Giustina A, Barkan A, Beckers A, Biermasz N, Biller BMK, Boguszewski C, et al. A consensus on the diagnosis and treatment of acromegaly comorbidities: an update. J Clin Endocrinol Metab 2019;.

[66] Annamalai AK, Gayton EL, Webb A, Halsall DJ, Rice C, Ibram F, et al. Increased prevalence of gallbladder polyps in acromegaly. J Clin Endocrinol Metab 2011;96:E1120−5.

[67] Jenkins PJ, Sohaib SA, Akker S, Phillips RR, Spillane K, Wass JA, et al. The pathology of median neuropathy in acromegaly. Ann Intern Med 2000;133:197−201.

[68] Manara R, Maffei P, Citton V, Rizzati S, Bommarito G, Ermani M, et al. Increased rate of intracranial saccular aneurysms in acromegaly: an MR angiography study and review of the literature. J Clin Endocrinol Metab 2011;96:1292−300.

[69] Manara R, Bommarito G, Rizzati S, Briani C, Della Puppa A, Citton V, et al. Herniation of cerebellar tonsils in acromegaly: prevalence, pathogenesis and clinical impact. Pituitary 2013;16:122−30.

[70] Carvalho MA, Montenegro Junior RM, Freitas MR, Vilar L, Mendonca AT, Montenegro RM. Sensorineural hearing loss in acromegalic patients under treatment. Braz J Otorhinolaryngol 2012;78:98−102.

[71] Glezer A, Bronstein MD. Pituitary apoplexy: pathophysiology, diagnosis and management. Arch Endocrinol Metab 2015;59:259−64.

[72] Kyriakakis N, Lynch J, Gilbey SG, Webb SM, Murray RD. Impaired quality of life in patients with treated acromegaly despite long-term biochemically stable disease: results from a 5-years prospective study. Clin Endocrinol (Oxf) 2017;86:806−15.

[73] Rostomyan L, Daly AF, Beckers A. Pituitary gigantism: causes and clinical characteristics. Ann Endocrinol (Paris) 2015;76:643−9.

[74] Rostomyan L, Daly AF, Petrossians P, Nachev E, Lila AR, Lecoq AL, et al. Clinical and genetic characterization of pituitary gigantism: an international collaborative study in 208 patients. Endocr Relat Cancer 2015;22:745−57.

[75] Daly AF, Beckers A. Familial isolated pituitary adenomas (FIPA) and mutations in the aryl hydrocarbon receptor interacting protein (AIP) gene. Endocrinol Metab Clin North Am 2015;44:19−25.

[76] Hannah-Shmouni F, Trivellin G, Stratakis CA. Genetics of gigantism and acromegaly. Growth Horm IGF Res 2016;30-31:37−41.

[77] Daly AF, Tichomirowa MA, Petrossians P, Heliovaara E, Jaffrain-Rea ML, Barlier A, et al. Clinical characteristics and therapeutic responses in patients with germ-line AIP mutations and pituitary adenomas: an international collaborative study. J Clin Endocrinol Metab 2010;95:E373−83.

[78] Ezzat S, Asa SL, Couldwell WT, Barr CE, Dodge WE, Vance ML, et al. The prevalence of pituitary adenomas: a systematic review. Cancer 2004;101:613−19.

[79] Vasilev V, Rostomyan L, Daly AF, Potorac I, Zacharieva S, Bonneville JF, et al. Management of endocrine disease: pituitary 'incidentaloma': neuroradiological assessment and differential diagnosis. Eur J Endocrinol 2016;175:R171−84.

[80] Naves LA, Daly AF, Dias LA, Yuan B, Zakir JC, Barra GB, et al. Aggressive tumor growth and clinical evolution in a patient with X-linked acro-gigantism syndrome. Endocrine 2016;51:236−44.

[81] Beckers A, Petrossians P, Hanson J, Daly AF. The causes and consequences of pituitary gigantism. Nat Rev Endocrinol 2018;14:705−20.

[82] Iacovazzo D, Caswell R, Bunce B, Jose S, Yuan B, Hernandez-Ramirez LC, et al. Germline or somatic GPR101 duplication leads to X-linked acrogigantism: a clinico-pathological and genetic study. Acta Neuropathol Commun 2016;4:56.

[83] Chakraborty PP, Bhattacharjee R, Roy A, Chowdhury S. Pseudoacromegaly: an unusual presenting manifestation of long-standing undiagnosed primary hypothyroidism. Postgrad Med J 2017;93:639−40.

[84] Dahlqvist P, Spencer R, Marques P, Dang MN, Glad CAM, Johannsson G, et al. Pseudoacromegaly: a differential diagnostic problem for acromegaly with a genetic solution. J Endocr Soc 2017;1:1104−9.

[85] Kumar KV, Shaikh A, Anwar I, Prusty P. Primary hypothyroidism presenting as pseudoacromegaly. Pituitary 2012;15(Suppl. 1):S49−52.

[86] Marques P, Korbonits M. Pseudoacromegaly. Front Neuroendocrinol 2019;52:113−43.

[87] Girard PF, Guinet P, Devic M, Mornex. Aneurysm of the internal carotid and acromegaloid syndrome. Rev Neurol (Paris). 1953; 89:279−280.

[88] Marabini B, Matteini M, Bigozzi U. A case of multiple congenital malformations associated with partial acromegaloid manifestations. Rass Neurol Veg 1956;11:358−69.

[89] Robecchi A, Einaudi G. Case of polyarthritis with hypertrophic osteopathy caused by pulmonary tumor with acromegaloid changes of the face & of extremities. Minerva Med 1957;48:3389−99.

[90] Marques P, Spencer R, Morrison PJ, Carr IM, Dang MN, Bonthron DT, et al. Cantu syndrome with coexisting familial pituitary adenoma. Endocrine 2018;59:677−84.

[91] Baujat G, Cormier-Daire V. Sotos syndrome. Orphanet J Rare Dis 2007;2:36.

[92] Trivedi N, Mithal A, Sharma AK, Mishra SK, Pandey R, Trivedi B, et al. Non-islet cell tumour induced hypoglycaemia with acromegaloid facial and acral swelling. Clin Endocrinol (Oxf) 1995;42:433−5.

[93] Yamane Y, Okamoto S, Fukui H, Matsumura Y, Yoshikawa M, Tsujita S, et al. 48, XXYY syndrome associated with acromegaloidism. Intern Med 1993;32:160−5.

[94] Korkij W, Plengvidhya CS. Multiple hamartoma syndrome with acromegaloidism. Int J Dermatol 1991;30:48−50.

[95] Dib K, Whitehead JP, Humphreys PJ, Soos MA, Baynes KC, Kumar S, et al. Impaired activation of phosphoinositide 3-kinase by insulin in fibroblasts from patients with severe insulin resistance and pseudoacromegaly. A disorder characterized by selective postreceptor insulin resistance. J Clin Invest 1998;101:1111−20.

[96] Karimova MM, Halimova ZY, Urmanova YM, Korbonits M, Cranston T, Grossman AB. Pachydermoperiostosis masquerading as acromegaly. J Endocr Soc 2017;1:109−12.

[97] Alvarez-Vazquez P, Rivera A, Figueroa I, Paramo C, Garcia-Mayor RV. Acromegaloidism with normal growth hormone secretion associated with X-tetrasomy. Pituitary 2006;9:145−9.

[98] Hogarth V, Hughes D, Orteu CH. Pseudoacromegalic facial features in Fabry disease. Clin Exp Dermatol 2013;38:137−9.

[99] Mishriki YY. When the face tells the tale. Course facial features caused phenytoin use. Postgrad Med 1998;103:49−50.

[100] Nguyen KH, Marks Jr. JG. Pseudoacromegaly induced by the long-term use of minoxidil. J Am Acad Dermatol 2003;48:962−5.

[101] Ries M, Moore DF, Robinson CJ, Tifft CJ, Rosenbaum KN, Brady RO, et al. Quantitative dysmorphology assessment in Fabry disease. Genet Med 2006;8:96−101.

[102] Albuquerque EVA, Scalco RC, Jorge AAL. Management of endocrine disease: diagnostic and therapeutic approach of tall stature. Eur J Endocrinol 2017;176:R339−53.

[103] Tlemsani C, Luscan A, Leulliot N, Bieth E, Afenjar A, Baujat G, et al. SETD2 and DNMT3A screen in the Sotos-like syndrome French cohort. J Med Genet 2016;53:743−51.

[104] Asvold BO, Eskild A, Jenum PA, Vatten LJ. Maternal concentrations of insulin-like growth factor I and insulin-like growth factor binding protein 1 during pregnancy and birth weight of offspring. Am J Epidemiol 2011;174:129−35.

[105] Lofqvist C, Andersson E, Gelander L, Rosberg S, Blum WF, Albertsson WK, et al. Reference values for IGF-I throughout childhood and adolescence: a model that accounts simultaneously for the effect of gender, age, and puberty. J Clin Endocrinol Metab 2001;86:5870−6.

[106] Puche JE, Castilla-Cortazar I. Human conditions of insulin-like growth factor-I (IGF-I) deficiency. J Transl Med 2012;10:224.

[107] Giustina A, Chanson P, Bronstein MD, Klibanski A, Lamberts S, Casanueva FF, et al.Acromegaly Consensus G A consensus on criteria for cure of acromegaly. J Clin Endocrinol Metab 2010;95:3141−8.

[108] Clemmons DR. Clinical laboratory indices in the treatment of acromegaly. Clin Chim Acta 2011;412:403−9.

[109] Arafat AM, Mohlig M, Weickert MO, Perschel FH, Purschwitz J, Spranger J, et al. Growth hormone response during oral glucose tolerance test: the impact of assay method on the estimation of reference values in patients with acromegaly and in healthy controls, and the role of gender, age, and body mass index. J Clin Endocrinol Metab 2008;93:1254−62.

[110] Mestron A, Webb SM, Astorga R, Benito P, Catala M, Gaztambide S, et al. Epidemiology, clinical characteristics, outcome, morbidity and mortality in acromegaly based on the Spanish Acromegaly Registry (Registro Espanol de Acromegalia, REA). Eur J Endocrinol 2004;151:439−46.

[111] Daud S, Hamrahian AH, Weil RJ, Hamaty M, Prayson RA, Olansky L. Acromegaly with negative pituitary MRI and no evidence of ectopic source: the role of transphenoidal pituitary exploration? Pituitary 2011;14:414−17.

[112] Kurowska M, Tarach JS, Zgliczynski W, Malicka J, Zielinski G, Janczarek M. Acromegaly in a patient with normal pituitary gland and somatotropic adenoma located in the sphenoid sinus. Endokrynol Pol 2008;59:348−51.

[113] Borson-Chazot F, Garby L, Raverot G, Claustrat F, Raverot V, Sassolas G, et al. Acromegaly induced by ectopic secretion of GHRH: a review 30 years after GHRH discovery. Ann Endocrinol (Paris) 2012;73:497−502.

[114] Garby L, Caron P, Claustrat F, Chanson P, Tabarin A, Rohmer V, et al. Clinical characteristics and outcome of acromegaly induced by ectopic secretion of growth hormone-releasing hormone (GHRH): a French nationwide series of 21 cases. J Clin Endocrinol Metab 2012;97:2093−104.

[115] Kan E, Kan EK, Atmaca A, Atmaca H, Colak R. Visual field defects in 23 acromegalic patients. Int Ophthalmol 2013;33:521−5.

GPR101, an orphan G-protein coupled receptor, with roles in growth, puberty, and possibly appetite regulation

Fady Hannah-Shmouni and Constantine A. Stratakis

Section on Endocrinology and Genetics, Eunice Kennedy Shriver National Institute of Child Health and Human Development (NICHD), Bethesda, MD, United States

5.1 Introduction

In 2014 [1] we described X-linked acrogigantism (X-LAG, MIM #300942) [2], a disorder that presents with infantile overgrowth and leads to growth hormone (GH) over secretion and gigantism in children and acromegaly in late adolescents and young adults [3,4]. The clinical characteristics of patients with X-LAG are described in the accompanying review by Vasilev et al. [5].

In this review, we will focus on the identification of what appears to be the causative gene in X-LAG, GPR101, a class A, rhodopsin-like orphan guanine nucleotide-binding protein (G protein) coupled receptor (GPCR) (https://www.guidetopharmacology.org/GRAC/GPCRListForward?class = A) with a yet unknown endogenous ligand [6], the delineation of its expression, and ongoing work in animal models, as well as the identification of potential ligands.

5.2 Identification of GPR101 in X-LAG: is it the causative gene?

To date, more than 35 patients have been confirmed to have X-LAG due to Xq26.3 microduplications, recently reviewed by Trivellin et al. [3]. In more than 30 of these cases, array comparative genomic hybridization (aCGH) has been performed and confirmed that the smallest region of overlap (SRO) by these nonrecurrent genomic rearrangements contains a single fully coded gene within a 73 kb sequence that is known to code for GPR101 (SRO2) [3]. Patients also share an 8.3 kb sequence that includes the last two exons of the *VGLL1* gene and the micro-RNA *miR-934*(SRO1).

The originally reported SRO contained at least four coding genes and a number of other sequences but subsequent work narrowed the X-LAG-linked region to SRO1 and SRO2. Perhaps the most important among them were two cases: one subject with X-LAG in whom the SRO contained only *GPR101* [6] and another boy with various developmental defects, who carried a Xq26.3

Gigantism and Acromegaly. DOI: https://doi.org/10.1016/B978-0-12-814537-1.00011-7

duplication that stopped short of including GPR101 and did not have overgrowth or any other abnormality that would indicate X-LAG [7].

The duplications are generated by replication errors at regions of microhomology, which, in most of the cases, can be explained by the mitotic DNA replication-based mechanism of fork stalling and template switching/microhomology-mediated break-induced replication (FoSTeS/MMBIR) [1,3,8]. Only in one case, an Alu−Alu mediated rearrangement was reported [6,8]. In the cases so far analyzed, the duplications were germline in females and somatic in sporadic males, with variable levels of mosaicism ranging from about 15% to 60% [6,9]. However, the clinical characteristics of X-LAG patients are similar in both sexes [10]. This suggests that even a small fraction of cells carrying the duplication in specific tissues is sufficient to cause the phenotype, as we have suggested elsewhere [3]. Moreover, the impact of X chromosome inactivation (XCI) in females should be taken into consideration, as this could alter the expression of the duplicated genes, similarly to what happens in mosaic males.

Indeed, we observed skewed XCI with the duplicated allele being preferentially methylated in blood-derived DNA from a familial case [1]: this patient also carried duplication of the proximal to *GPR101* gene, *CD40LG*. Most recently, it was reported that patients with autoimmunity and *CD40LG* duplications preferentially inactivate their duplicated X chromosome [11]. Although *CD40LG* is commonly duplicated in patients with X-LAG, no autoimmune phenotypes have been observed to date [10]. This likely suggests that Xq26.3 microduplications associated with X-LAG have a different impact on local chromatin domains and consequently on gene regulation, therefore causing completely unrelated phenotypes. We have now extended our analysis to 5 additional patients with X-LAG and showed that only 2 of 12 patients (17%) exhibited skewed XCI [12]; no significant differences in clinical phenotypes were observed in these 2 patients compared with the rest, consistent with our observation in male patients with X-LAG and mosaicism, in whom only 16% of blood cells with the duplicated X chromosome led to full-blown acrogigantism [9]. Although we cannot fully exclude the well-known effect of age on the rate of skewing in our patients (the two patients showing skewed XCI were diagnosed, on average, 26 years later than those with random XCI), our data suggested caution in assuming skewed XCI as a general phenomenon in patients with *CD40LG* duplication-associated autoimmune disease [12].

Several lines of evidence support a crucial role for GPR101 as being the causative gene in X-LAG. Besides being the only entire gene that is invariably duplicated in all reported X-LAG patients, as stated above, it is also expressed at very high levels in their pituitary lesions (up to 1000 times the expression levels seen in somatotroph tumors not harboring Xq26.3 duplications or in normal pituitary tissues) [1,13], and it activates the cAMP signaling pathway [1,3], which is well-known to stimulate proliferation of pituitary cells as well as GH and prolactin secretion [14,15].

Although GPR101 was identified as a GPCR linked to $Gs\alpha$ [16−18], one study reported that GPR101 did not constitutively activate the cAMP pathway, as it did not meet the authors' criteria for constitutive activity [200% elevation over baseline cAMP-dependent response element (CRE) reporter expression] in their experimental setting. In the same study, GPR101 was also found to inhibit forskolin-stimulated CRE reporter activity, raising the possibility that it may be coupled to both G stimulatory and inhibitory (Gi) proteins [19], as reported for other G protein-coupled receptors (GPCRs) [20]. Further studies are thus necessary to validate these findings. In the setting of X-LAG, GPR101 seems to behave as a putative oncogene, but it remains to be elucidated how GPR101 is involved in the molecular pathogenesis of X-LAG.

5.3 **The GPR101 gene and protein: structure and expression**

5.3.1 **Structure**

The human *GPR101* gene has been discovered and mapped to the X chromosome in 2001 [21], while the mouse *Gpr101* gene was characterized only in 2006 [16]. Until recently, the complete structure of human *GPR101* was unknown, as only the coding sequence (CDS) had been characterized (NM_054021.1/ENST00000298110.1). GPR101 encodes a 508 amino acid-long orphan GPCR; the human GPR101 CDS is specified by a single 1527 bp-long protein-coding exon that is 80%, 79%, and 55% identical to the mouse, rat, and zebrafish orthologues, respectively (http://www.ncbi.nlm.nih.gov/homologene/). Interestingly, some parts of the gene seem to have diverged more than others during evolution, particularly the intracellular loop 3 (ICL3).

We recently further characterized the structure of the GPR101 gene by combining in silico and in vitro approaches [13]. We identified four transcripts generated by alternative splicing events at the 5'-untranslated region (UTR); all isoforms share a common 6.1 kb-long 3'UTR, which could potentially harbor binding sites for regulatory molecules (miRNAs and lncRNAs). One of the isoforms is expressed at much higher levels than the others in the pituitary tumor cells of X-LAG patients, suggesting that this is the main isoform, at least in the pituitary gland [3,13]. We also predicted using bioinformatics that the GPR101 promoter is large and TATA-less; it may also overlap with a CpG island located approximately 2 kb upstream of the START codon [1,3,13]. Estradiol was recently reported to stimulate *Gpr101* expression in rats, although no classical estrogen response elements (EREs) were detected in the promoter region [22].

GPCRs are integral membrane proteins that share a common architecture consisting of an extracellular N-terminal domain, seven hydrophobic transmembrane domains (TM1-TM7) linked by three extracellular loops (ECL1-ECL3) and three intracellular loops (ICL1-ICL3), as well as an intracellular C-terminal domain. While the TM domains of different GPCRs show great similarity in overall architecture and secondary structure, there is considerable variation among loop regions, in particular regarding the length of the N- and C-terminal domains and ICL3 [23]. According to most recent estimates, the human genome contains 825−826 GPCRs [24,25], phylogenetically categorized into five major classes: class A (rhodopsin-like), class B1 (secretin-like), class B2 (adhesion-like), class C (glutamate-like), and the Frizzled/Taste2 family. About half GPCRs have sensory functions, while the remaining are nonsensory GPCRs that mediate signaling upon ligand binding. Class A comprises the majority of GPCRs, 719, of which 284−297 are nonsensory [5,24]; about 87 class A GPCRs are still orphan receptors (meaning that they have a largely unknown ligand) [5]. GPCRs can vary greatly in length, with some consisting of more than 1000 amino acids; most, however, are 200−400 residues long [5,25,26].

Phylogenetic analysis showed that GPR101 belongs to class A [5,27], which includes the prototypical and well-studied rhodopsin and β2-adrenergic receptor. Within class A, GPR101 is distantly related to amine-binding GPCRs [adrenergic, serotonin (5-HT), dopamine, and histamine receptors], as well as to the muscarinic receptors, with similarity values just above 30% across the full protein sequence [analysis performed with the GPCRd (http://gpcrdb.org/similaritysearch/referenceselection)]. When the analysis is limited to the TM domains, GPR101 shares 50% similarity with alpha1A-adrenergic receptor and 49% with GPR161. GPR161 is an orphan GPCR associated with primary cilia [28,29]. Interestingly, GPR161 defects were linked to a pituitary developmental disorder: a homozygous LOF missense mutation was observed in a consanguineous family presenting with pituitary stalk interruption [30].

A bioinformatic analysis of GPR101 amino acid sequence [3] revealed that out of the three highly conserved structural motifs that are involved in G protein coupling/recognition and activation, GPR101 has only an intact E/DRY motif located at the interface of ECL2-TM3 (DRY residues 128–130), but lacks a complete BBXXB motif (where B represents a basic residue and X a nonbasic residue) at the interface of ICL3-TM6 (CKAAK residues 395–399) as well as a complete D/NPXXY motif in TM7 (HPYVY residues 450–454). Mutations of the BBXXB motif can lead to constitutive activation in GPCRs [31,32]. It can thus be speculated that the absence of an intact BBXXB motif could possibly contribute to the high cAMP activity characteristic of GPR101. Most GPCRs also contain two conserved cysteine residues located in TM3 and ECL2 that form a disulfide bond; in GPR101, these cysteines are located in position 104 (TM3) and 182 (ECL2). GPCR kinases (GRKs) and second messenger-associated kinases usually phosphorylate GPCRs at serine or threonine residues at the ICL3 and the C-terminus of the active receptor. This event promotes the binding of β-arrestin, which then leads to desensitization and internalization of the receptor [33]. Four phosphorylation sites at serine residues 15 (N-terminus), 27 (TM1), and 309–310 (ICL3) have been functionally annotated for GPR101 [34]. Interestingly, the p.E308D variant of GPR101 is located next to the ICL3 phosphorylation site and might interfere with this posttranslational modification.

Out of the more than 800 human GPCRs, only 40 have been crystalized so far; 33 of them belong to class A [35]. The limited number of solved three-dimensional (3D) structures is mainly due to challenges in GPCRs expression, purification, stability, and crystallization [36]. We recently generated a bioinformatic structural model of GPR101 in complex with a G stimulatory (Gs) protein heterotrimer [3] using the crystallized structure of the β2 adrenergic receptor (β2 AR) in its active state. Another structure is now available at GPCRdb (http://gpcrdb.org/structure/homology_models/gp101_human_intermediate). That model shows GPR101 in its intermediate state (a conformation assumed by a GPCR in-between the inactive and active states) [37] and has been constructed using 6 backbone and 18 rotamer templates, with the 5-HT2B receptor as main template. On both 3D models the remarkably long IL3 connecting TM5 and TM6 (157/162 amino acids based on the model, representing about 30% of the total length) could not be modeled because of the absence of a template. ICL3 is indeed one of the least conserved regions in GPCRs, being very heterogeneous in sequence and length. In particular, the difference in size (which can range from as short as five residues in CXCR4 to up to hundreds in other GPCRs) has been proposed to account for selectivity on G proteins coupling [38]. Even structurally related receptors, like GPR101 and GPR161, that activate the same type of G proteins (Gs for both), exhibit very divergent ICL3s, making thus impossible to in silico predict function and G protein coupling based solely on primary structure.

5.3.2 Structure and tissue distribution

Different groups have extensively studied the distribution of GPR101 expression at the mRNA and protein level in mammalian and fish tissues. Several studies have shown that GPR101 is predominantly expressed in the brain across different species (zebrafish, rodents, and humans), suggesting a conservation of function through evolution for this GPCR. Within the brain, GPR101 mRNA is found at high levels in the hypothalamus, particularly the arcuate nucleus, in rodents [3,13,39–43], and in the nucleus accumbens, in humans [3,44]. The detection of GPR101 in the nucleus accumbens is an interesting finding, especially in relation to X-LAG. This region is the main component of ventral striatum and is involved in the regulation of motivated behavior by playing an important role as the reward center. About one-third of X-LAG patients show increased food seeking, a behavior that we speculate might be controlled by GPR101. Functional studies are necessary to test this hypothesis. Other human tissues showing moderate expression of *GPR101* mRNA are fat, optic nerve, and lymphocytes [3,13], while contrasting results concerning GPR101 expression in the human hypothalamus have been reported. In mice, food deprivation increases *Gpr101* mRNA expression in the posterior hypothalamus [39], while in rats, *Gpr101* mRNA expression in the arcuate nucleus increases in an age-dependent manner [3,22]. Within the arcuate nucleus, GPR101 is expressed in about half the neuronal population that expresses the anorexigenic peptide proopiomelanocortin (POMC), a finding that suggests a possible role for GPR101 in the control of energy homeostasis [18,39].

Similarly, GPR101 expression also shows an age-dependent pattern in the normal human anterior pituitary gland. GPR101 is highly expressed during fetal development (starting at around the gestational age of 19 weeks) and moderately during adolescence during the so-called growth spurt, while very low expression is observed during adulthood. This expression pattern suggests that GPR101 might play an important physiological role as the pituitary proliferates and matures. Moreover, the strong expression localized in the lateral wings of the fetal anterior lobe, where most GH- and prolactin-secreting cells reside, suggests that GPR101 might regulate, directly or indirectly, the differentiation of mammosomatotroph cells. In other species, GPR101 expression in the adult pituitary gland showed differences in cell type expression, with gonadotroph cells being immunostained in rhesus macaque and GH-secreting cells in rat. These results suggest that GPR101 might exert different functions in the pituitary of different species [3,13].

5.4 GPR1O1 possible ligands: GnRH-(1−5) and Rvd5n-3 DPA

GPCRs may be activated by an extremely diverse array of ligands, such as photons, nucleotides, lipids, neurotransmitters, small peptides, and hormones [5]. GnRH-(1−5), a processed pentapeptide derived from the decapeptide gonadotrophin-releasing hormone (GnRH), was recently reported as a putative GPR101 ligand [45]. Upon GPR101 binding, GnRH-(1−5) induces epidermal growth factor (EGF) release, followed by EGF receptor (EGFR) phosphorylation with consequent activation of downstream signaling pathways; this ultimately leads to increased proliferation, migration, and invasion of endometrial cancer cells [46,47]. However, the real efficacy of this ligand in pituitary cells has not been shown. Indeed, very little, if any, effects on cAMP pathway activation and on hormone secretion were observed in in vitro studies using pituitary tumor cell lines as well as primary X-LAG tumor cells [3]. Since the treatment of endometrial cells with GnRH-(1−5) resulted in no significant changes in cAMP levels [48], it is possible that this molecule activates another intracellular signal transduction pathway in a cell type-dependent manner.

Most recently, N-3 docosapentaenoic acid-derived resolvin D5 (RvD5n-3 DPA), an autacoid, pro-resolving mediator, regulating inflammatory responses like arachidonic acid was found to be a strong candidate molecule binding specifically to GPR101 in macrophages and other monocytes

[49]. RvD5n-3 DPA was found to bind to GPR101 after an orphan GPCR screening; RvD5n-3 DPA bound to GPR101 with "high selectivity and stereospecificity" [49]. Interestingly, knockdown of GPR101 reversed the protective actions of RvD5n-3 DPA in limiting inflammatory arthritis, suggesting a possible role for GPR101 in the regulation of inflammation in leukocytes [49]. It remains unclear how these findings translate to the pituitary gland and hypothalamic GPR101 functions but the discovery has already sparked interest.

GPCRs are the most successful class of druggable targets, with estimates of around 30%–50% of available pharmaceutical drugs targeting these proteins [50]. Our group is currently performing a high-throughput screening of small molecule libraries to identify additional GPR101-binding molecules. Clearly, the discovery of inverse agonists might be very beneficial to study the physiological function of GPR101's constitutive activity and hopefully for the treatment of X-LAG patients.

5.5 Is GPR101 involved in other diseases?

Like we have done elsewhere [3], in this section, we will discuss several rare single nucleotide variants [minor allele frequency (MAF) <1%] that have been reported in GPR101 and studied in diseases other than X-LAG. Single nucleotide polymorphisms (SNPs) commonly observed in healthy controls, including p.V124L (MAF = 36.7%), p.T293I (MAF = 2.1%), and p.L376P (MAF = 16.7%), will not be discussed.

A missense variant affecting residue 308 (p.E308D), located within ICL3, was identified in about 4% of patients with sporadic acromegaly, in one case occurring de novo at the pituitary tumor level [1,3]. This variant is found in healthy controls with a MAF of 0.36% according to the Genome Aggregation Database (gnomAD). A synonymous variant affecting the same codon is also observed with a MAF of 0.18% [51]. p.E308D moderately increased GH secretion and cell proliferation when over-expressed in a cell line model for GH- and prolactin-secreting tumors (rat GH3 cells) [1]. However, further studies in separate cohorts of patients with acromegaly did not find an increased prevalence of this variant, suggesting that it probably does not play a major role in the pathogenesis of this disease [52–54]. Another missense variant located within ICL3, p.D366E, was also observed in a patient with sporadic acromegaly [54]. This variant, which has not been functionally tested in vitro yet, is not found in controls. A structural hypothesis on the effect of these and other variants located on ICL3 cannot be proposed currently, because this domain has not been properly modeled. Solving that region (as well as the entire receptor) would be a great step forward to better understand the impact of GPR101 mutations, especially because ICL3 appears to be a hot spot for gain of function missense mutations [55].

Other missense variants have also been observed in different pituitary tumor types, including prolactinomas and corticotropinomas. Two missense variants affecting highly conserved amino acids, p.T325I and p.P432S, have been reported in one prolactinoma patient each (in a cohort of 256 patients). The former is a novel variant located in ICL3, while the latter is located in ECL3 and is rarely seen in controls (MAF = 0.04%). The impact of these variants on GPR101 structure/function remains to be determined. Two heterozygous missense variants were also described in female patients with Cushing disease. A novel variant located in TM3, p.I122T, was reported in a cohort of 68 patients [54], while a rare variant located in TM1, p.G31S (MAF = 0.11%), was detected in a separate cohort of 36 patients [56]. p.I122T affects a highly conserved residue and is in silico predicted to be deleterious, while p.G31S was evaluated in vitro in the mouse pituitary corticotroph AtT-20 cell line but was not found to increase ACTH secretion or cell proliferation [56].

GPR101 has been screened for LOF mutations and deletions in 41 patients with unexplained congenital isolated GH deficiency (GHD) [57]. While no deletions were detected, a novel missense variant in TM5, p.V197L, was present in the heterozygote state in one female patient. Although predicted to be deleterious by three in silico algorithms, when expressed in GH3 cells this variant did not lead to a statistically significant decrease in GH secretion or cAMP activity [57].

Heterotaxy is a condition resulting from abnormalities of the proper specification of left−right patterning during embryonic development. Whole exome sequencing of four families with X-linked heterotaxy revealed a missense GPR101 variant, p.V409M, in one pedigree with four affected males. This very rare variant (MAF = 0.07%) is located in TM6 and is predicted to be damaging. Functional studies in mice and zebrafish supported a causative role in this disease [58]. Interestingly, on Xq26.3 GPR101 is located in close proximity to ZIC3, a gene known to be involved in X-linked heterotaxy [59], and GPR101's closely related GPCR, GPR161, has also been reported to be required for left−right patterning in zebrafish [60].

Finally, GPR101 promoter was identified as being hypermethylated in about 40% of patients with colorectal cancer (in a cohort of 77 patients) [61]; this event was shown to have prognostic value in stage IV male patients by being correlated with a longer time to disease progression. Most recently, a variant within the *GPR101* gene was among the 10 most closely associated with smoking behavior genetic loci [62].

Funding

This study was funded by the NICHD Intramural Research Program, Bethesda, MD, United States.

References

[1] Trivellin G, Daly AF, Faucz FR, Yuan B, Rostomyan L, Larco DO, et al. Gigantism and acromegaly due to Xq26 microduplications and GPR101 mutation. N Engl J Med 2014;371(25):2363−74.

[2] Online Mendelian Inheritance in Man, OMIM®. McKusick-Nathans Institute of Genetic Medicine, Johns Hopkins University (Baltimore, MD), January 7, 2021. World Wide Web URL: https://omim.org/.

[3] Trivellin G, Hernández-Ramírez LC, Swan J, Stratakis CA. An orphan G-protein-coupled receptor causes human gigantism and/or acromegaly: molecular biology and clinical correlations. Best Pract Res Clin Endocrinol Metab 2018;32(2):125−40.

[4] Hannah-Shmouni F, Stratakis CA. An update on the genetics of benign pituitary adenomas in children and adolescents. Curr Opin Endocr Metab Res 2018;1:19−24.

[5] Vasilev V, Daly AF, Trivellin G, Stratakis CA, Zacharieva S, Beckers A. HEREDITARY ENDOCRINE TUMOURS: CURRENT STATE-OF-THE-ART AND RESEARCH OPPORTUNITIES: The roles of AIP and GPR101 in familial isolated pituitary adenomas (FIPA). Endocr-Relat Cancer 2020;27:77−86. Available from: https://doi.org/10.1530/ERC-20-0015.

 [6] Iacovazzo D, Caswell R, Bunce B, Jose S, Yuan B, Hernández-Ramírez LC, et al. Germline or somatic GPR101 duplication leads to X-linked acrogigantism: a clinico-pathological and genetic study. Acta Neuropathol Commun 2016;4(1):56.

 [7] Trivellin G, Sharwood E, Hijazi H, Carvalho CMB, Yuan B, Tatton-Brown K, et al. Xq26.3 Duplication in a boy with motor delay and low muscle tone refines the X-linked acrogigantism genetic locus. J Endocr Soc 2018;2(10):1100–8.

 [8] Rodd C, Millette M, Iacovazzo D, Stiles CE, Barry S, Evanson J, et al. Somatic GPR101 duplication causing X-linked acrogigantism (XLAG)-diagnosis and management. J Clin Endocrinol Metab 2016;101 (5):1927–30.

 [9] Daly AF, Yuan B, Fina F, Caberg JH, Trivellin G, Rostomyan L, et al. Somatic mosaicism underlies X-linked acrogigantism syndrome in sporadic male subjects. Endocr Relat Cancer 2016;23(4):221–33.

[10] Beckers A, Lodish MB, Trivellin G, Rostomyan L, Lee M, Faucz FR, et al. X-linked acrogigantism syndrome: clinical profile and therapeutic responses. Endocr Relat Cancer 2015;22(3):353–67.

[11] Le Coz C, Trofa M, Syrett CM, Martin A, Jyonouchi H, Jyonouchi S, et al. CD40LG duplication-associated autoimmune disease is silenced by nonrandom X-chromosome inactivation. J Allergy Clin Immunol 2018;141(6):2308–2311.e7.

[12] Trivellin G, Stratakis CA. CD40LG duplications in patients with X-LAG syndrome commonly undergo random X-chromosome inactivation. J Allergy Clin Immunol 2019;143(4):1659.

[13] Trivellin G, Bjelobaba I, Daly AF, Larco DO, Palmeira L, Faucz FR, et al. Characterization of GPR101 transcript structure and expression patterns. J Mol Endocrinol 2016;57(2):97–111.

[14] Hernández-Ramírez LC, Trivellin G, Stratakis CA. Cyclic 3',5'-adenosine monophosphate (cAMP) signaling in the anterior pituitary gland in health and disease. Mol Cell Endocrinol 2018;463:72–86.

[15] Peverelli E, Mantovani G, Lania AG, Spada A. cAMP in the pituitary: an old messenger for multiple signals. J Mol Endocrinol 2014;52:R67–77.

[16] Bates B, Zhang L, Nawoschik S, Kodangattil S, Tseng E, Kopsco D, et al. Characterization of Gpr101 expression and G-protein coupling selectivity. Brain Res 2006;1087:1–14.

[17] Murphy AJ, Shanker YG, Croll-Kalish S, Torres R. Assay methods for identifying RE2-like antagonists, methods of use, and non-human transgenic animals, <http://www.google.com.pg/patents/US20060123502>. United States; 2006.

[18] Bagnol D. Use of gpr101 receptor in methods to identify modulators of hypothalamic proopiomelanocortin (POMC)-derived biologically active peptide secretion useful in the treatment of pomc-derived biologically active peptide-related disorders, <http://www.google.com/patents/US8142762?cl = en>. United States; 2010.

[19] Martin AL, Steurer MA, Aronstam RS. Constitutive activity among orphan class-A G protein coupled receptors. PLoS One 2015;10:e0138463.

[20] Michal P, Lysikova M, Tucek S. Dual effects of muscarinic M(2) acetylcholine receptors on the synthesis of cyclic AMP in CHO cells: dependence on time, receptor density and receptor agonists. Br J Pharmacol 2001;132:1217–28.

[21] Lee DK, Nguyen T, Lynch KR, Cheng R, Vanti WB, Arkhitko O, et al. Discovery and mapping of ten novel G protein-coupled receptor genes. Gene 2001;275:83–91.

[22] Bauman BM, Yin W, Gore AC, Wu TJ. Regulation of gonadotropin-releasing hormone-(1-5) signaling genes by estradiol is age dependent. Front Endocrinol 2017;8:282.

[23] Moreira IS. Structural features of the G-protein/GPCR interactions. Biochim Biophys Acta 2014;1840:16–33.

[24] The UniProt C. UniProt: the universal protein knowledgebase. Nucleic Acids Res. 2017;45:D158–D69.

[25] Lv X, Liu J, Shi Q, Tan Q, Wu D, Skinner JJ, et al. In vitro expression and analysis of the 826 human G protein-coupled receptors. Protein Cell 2016;7:325–37.

[26] Tao YX, Conn PM. Chaperoning G protein-coupled receptors: from cell biology to therapeutics. Endocr Rev 2014;35:602−47.

[27] Kakarala KK, Jamil K. Sequence-structure based phylogeny of GPCR class A rhodopsin receptors. Mol Phylog Evol 2014;74:66−96.

[28] Mukhopadhyay S, Wen X, Ratti N, Loktev A, Rangell L, Scales SJ, et al. The ciliary G-protein-coupled receptor Gpr161 negatively regulates the Sonic hedgehog pathway via cAMP signaling. Cell 2013;152:210−23.

[29] Bachmann VA, Mayrhofer JE, Ilouz R, Tschaikner P, Raffeiner P, Rock R, et al. Gpr161 anchoring of PKA consolidates GPCR and cAMP signaling. Proc Natl Acad Sci USA 2016;113:7786−91.

[30] Karaca E, Buyukkaya R, Pehlivan D, Charng WL, Yaykasli KO, Bayram Y, et al. Whole-exome sequencing identifies homozygous GPR161 mutation in a family with pituitary stalk interruption syndrome. J Clin Endocrinol Metab 2015;100:E140−7.

[31] Pauwels PJ, Wurch T. Review: amino acid domains involved in constitutive activation of G-protein-coupled receptors. Mol Neurobiol 1998;17:109−35.

[32] Muroi T, Matsushima Y, Kanamori R, Inoue H, Fujii W, Yogo K. GPR62 constitutively activates cAMP signaling but is dispensable for male fertility in mice. Reproduction 2017;154:755−64.

[33] Yang Z, Yang F, Zhang D, Liu Z, Lin A, Liu C, et al. Phosphorylation of G protein-coupled receptors: from the barcode hypothesis to the flute model. Mol Pharmacol 2017;92:201−10.

[34] Hornbeck PV, Zhang B, Murray B, Kornhauser JM, Latham V, Skrzypek E. PhosphoSitePlus, 2014: mutations, PTMs and recalibrations. Nucleic Acids Res 2015;43:D512−20.

[35] Isberg V, Mordalski S, Munk C, Rataj K, Harpsoe K, Hauser AS, et al. GPCRdb: an information system for G protein-coupled receptors. Nucleic Acids Res 2016;44:D356−64.

[36] Salon JA, Lodowski DT, Palczewski K. The significance of G protein-coupled receptor crystallography for drug discovery. Pharmacol Rev 2011;63:901−37.

[37] Bouvier M. Unraveling the structural basis of GPCR activation and inactivation. Nat Struct Mol Biol 2013;20:539−41.

[38] Katritch V, Cherezov V, Stevens RC. Diversity and modularity of G protein-coupled receptor structures. Trends Pharmacol Sci 2012;33:17−27.

[39] Nilaweera KN, Ozanne D, Wilson D, Mercer JG, Morgan PJ, Barrett P. G protein-coupled receptor 101 mRNA expression in the mouse brain: altered expression in the posterior hypothalamus and amygdala by energetic challenges. J Neuroendocr 2007;19:34−45.

[40] Nilaweera KN, Wilson D, Bell L, Mercer JG, Morgan PJ, Barrett P. G protein-coupled receptor 101 mRNA expression in supraoptic and paraventricular nuclei in rat hypothalamus is altered by pregnancy and lactation. Brain Res 2008;1193:76−83.

[41] Regard JB, Sato IT, Coughlin SR. Anatomical profiling of G protein-coupled receptor expression. Cell 2008;135:561−71.

[42] Ronnekleiv OK, Fang Y, Zhang C, Nestor CC, Mao P, Kelly MJ. Research resource: gene profiling of G protein-coupled receptors in the arcuate nucleus of the female. Mol Endocrinol 2014;28:1362−80.

[43] Vassilatis DK, Hohmann JG, Zeng H, Li F, Ranchalis JE, Mortrud MT, et al. The G protein-coupled receptor repertoires of human and mouse. Proc Natl Acad Sci USA 2003;100:4903−8.

[44] Consortium GT. The genotype-tissue expression (GTEx) project. Nat Genet 2013;45:580−5.

[45] Larco DO, Cho-Clark M, Mani SK, Wu TJ. The metabolite GnRH-(1-5) inhibits the migration of immortalized GnRH neurons. Endocrinology 2013;154:783−95.

[46] Cho-Clark M, Larco DO, Semsarzadeh NN, Vasta F, Mani SK, Wu TJ. GnRH-(1-5) transactivates EGFR in Ishikawa human endometrial cells via an orphan G protein-coupled receptor. Mol Endocrinol 2014;28:80−98.

[47] Cho-Clark M, Larco DO, Zahn BR, Mani SK, Wu TJ. GnRH-(1-5) activates matrix metallopeptidase-9 to release epidermal growth factor and promote cellular invasion. Mol Cell Endocrinol 2015;415:114−25.

[48] Baldwin EL, Wegorzewska IN, Flora M, Wu TJ. Regulation of type II luteinizing hormone-releasing hormone (LHRH-II) gene expression by the processed peptide of LHRH-I, LHRH-(1-5) in endometrial cells. Exp Biol Med 2007;232:146−55.

[49] Flak MB, Koenis DS, Sobrino A, Smith J, Pistorius K, Palmas F, et al. GPR101 mediates the pro-resolving actions of RvD5n-3 DPA in arthritis and infections. J Clin Invest 2020;130(1):359−73.

[50] Fang Y, Kenakin T, Liu C. Editorial: Orphan GPCRs as emerging drug targets. Front Pharmacol 2015;6:295.

[51] Lek M, Karczewski KJ, Minikel EV, Samocha KE, Banks E, Fennell T, et al. Analysis of protein-coding genetic variation in 60,706 humans. Nature 2016;536:285−91.

[52] Ferrau F, Romeo PD, Puglisi S, Ragonese M, Torre ML, Scaroni C, et al. Analysis of GPR101 and AIP genes mutations in acromegaly: a multicentric study. Endocrine 2016;.

[53] Matsumoto R, Izawa M, Fukuoka H, Iguchi G, Odake Y, Yoshida K, et al. Genetic and clinical characteristics of Japanese patients with sporadic somatotropinoma. Endocr J 2016;63:953−63.

[54] Lecoq AL, Bouligand J, Hage M, Cazabat L, Salenave S, Linglart A, et al. Very low frequency of germline GPR101 genetic variation and no biallelic defects with AIP in a large cohort of patients with sporadic pituitary adenomas. Eur J Endocrinol 2016;174:523−30.

[55] Fukami M, Suzuki E, Igarashi M, Miyado M, Ogata T. Gain-of-function mutations in G-protein-coupled receptor genes associated with human endocrine disorders. Clin Endocrinol (Oxf) 2017;.

[56] Trivellin G, Correa RR, Batsis M, Faucz FR, Chittiboina P, Bjelobaba I, et al. Screening for GPR101 defects in pediatric pituitary corticotropinomas. Endocr Relat Cancer 2016;.

[57] Castinetti F, Daly AF, Stratakis CA, Caberg JH, Castermans E, Trivellin G, et al. GPR101 mutations are not a frequent cause of congenital isolated growth hormone deficiency. Horm Metab Res 2016;.

[58] Tariq M, Cast A, Belmont J, Ware S. Identification of a novel cause of x-linked heterotaxy. Boston, MA: American Society of Human Genetics; 2013.

[59] Paulussen AD, Steyls A, Vanoevelen J, van Tienen FH, Krapels IP, Claes GR, et al. Rare novel variants in the ZIC3 gene cause X-linked heterotaxy. Eur J Hum Genet 2016;24:1783−91.

[60] Leung T, Humbert JE, Stauffer AM, Giger KE, Chen H, Tsai HJ, et al. The orphan G protein-coupled receptor 161 is required for left-right patterning. Dev Biol 2008;323:31−40.

[61] Kober P, Bujko M, Olędzki J, Tysarowski A, Siedlecki JA. Methyl-CpG binding column-based identification of nine genes hypermethylated in colorectal cancer. Mol Carcinog 2011;50:846−56.

[62] Erzurumluoglu AM, Liu M, Jackson VE, Barnes DR, Datta G, Melbourne CA, et al. Meta-analysis of up to 622,409 individuals identifies 40 novel smoking behaviour associated genetic loci. Mol Psychiatry 2020;25(10):2392−409. Available from: https://doi.org/10.1038/s41380-018-0313-0.

The role of the aryl hydrocarbon receptor interacting protein in pituitary tumorigenesis

6

Laura C. Hernández-Ramírez

Section on Endocrinology and Genetics, Eunice Kennedy Shriver National Institute of Child Health and Human Development (NICHD), National Institutes of Health (NIH), Bethesda, MD, United States

6.1 Introduction

The term *familial isolated pituitary adenoma* (FIPA) defines a clinical entity characterized by the development of pituitary neuroendocrine tumors (PitNETs) in two or more members of the same family in the absence of other clinical features, such as those encountered in the syndromes of multiple endocrine neoplasia [1]. This diagnosis includes also cases previously reported in the literature under the terms *pituitary adenoma predisposition* and *isolated familial somatotropinoma* [2−5]. FIPA is an autosomal dominant disease with incomplete penetrance and variable clinical presentation that accounts for an estimated 2%−4% of all PitNETS [6,7].

Most FIPA families include two to five affected members who are related in first degree in three-quarters of the cases [8]. Although any type of PitNET can be found in this setting, somatotropinomas and prolactinomas are the most common tumor types (75% of cases) [9−11]. FIPA families are classified as *homogeneous* (60% of the reported families), when all the affected family members carry the same pituitary tumor type, or *heterogeneous* (40%), when two or more different pituitary tumor types are found within a family [6]. Families with only somatotropinomas or with somatotropinomas and prolactinomas account for the vast majority (90%) of the homogeneous FIPA kindreds [8]. PitNETs occurring in the setting of FIPA are diagnosed on average 4 years earlier than their nonfamilial counterparts, and in families homogeneous for somatotropinomas, 70% of the PitNETs are diagnosed before the age of 30 years [3,4,8,9].

6.2 *AIP* variants in FIPA

In 2006 linkage analysis identified truncating germline variants of the aryl hydrocarbon receptor interacting protein gene (*AIP*, located in 11q13.2) as the cause of FIPA in Finnish and Italian families [5]. Thereafter, studies in other populations identified many other novel *AIP* likely pathogenic or pathogenic variants (hereafter collectively referred to as *AIP mutations*) leading to loss-of-function (LOF) of the AIP protein in individuals from FIPA families with gigantism, acromegaly,

Gigantism and Acromegaly. DOI: https://doi.org/10.1016/B978-0-12-814537-1.00003-8

and prolactinomas, and also in patients with apparently sporadic PitNETs [10−13]. Disease penetrance is incomplete and highly variable among *AIP* mutation carriers, ranging from 15% to 54% among the large kindreds studied to date; to simplify, it is usually considered that only every fifth mutation carrier manifests the disease [5,14−17]. Germline *AIP* mutations have been identified in around 15%−20% of the FIPA cases, with a higher prevalence among families with only cases of somatotropinomas, where mutations are found in one-quarter to half of the cases [7,8,12,13,16]. Half of the PitNET patients carrying a germline *AIP* mutation have a positive family history of FIPA [16].

6.3 *AIP* variants in sporadic PitNETs

In addition to FIPA, *AIP* mutations cause 3%−4% of the sporadic PitNETs in the general population, and 8% of the cases of acromegaly resistant to the treatment with somatostatin analogues (SSAs) [18−20]. *AIP* mutations are strongly associated with PitNETs occurring in young patients, being found in 6%−8% of all sporadic cases diagnosed at ≤ 30 years, in 12% of patients with sporadic macroadenomas <30 years at diagnosis, as well as in one-fifth of pediatric patients with functional PitNETs and in one-third of cases of gigantism [16,21−26]. Since the occurrence of de novo *AIP* mutations has so far been proven in only two patients, the probable reason why *AIP* mutations are not infrequent among young PitNET patients with sporadic presentation is the low penetrance of the disease among variant carriers [22,27,28].

6.4 *AIP* variants associated with PitNETs

Besides PitNETs, no other tumor types have been consistently associated with *AIP* mutations, and such genetic defects have not been detected at the somatic level in any tumor type [13,18,29]. Interestingly, translocations resulting in concomitant loss of *AIP* and the multiple endocrine neoplasia type 1 gene (*MEN1*) at the somatic level cause benign brown fat tumors known as hibernomas [30]. Germline *AIP* mutations in patients with PitNETs are always found in heterozygosis and are almost always inherited from one of the parents. Large genomic deletions (in only 10% of the cases), segmental duplications as well as nonsense, missense, splice site and promoter mutations of the *AIP* gene have been found in PitNET patients [12,16,31]. Three-quarters of the *AIP* gene defects leading to PitNETs result in loss of the carboxy (C)-terminal end of the protein (truncating mutations) [16].

There are three mutational hotspots in *AIP*, corresponding to C-phosphate-G (CpG) islands: c.910C > T (p.R304*), c.811C > T (p.R271W), and c.241C > T (p.R81*) [32]. A founder effect has been described for a few *AIP* mutations in certain populations: p.R304* (which is the most common *AIP* mutation so far reported among PitNET patients) in Northern Irish and Italian populations, c.40C > T (p.Q14*) in the Finnish population, c.805−825dup (p.F269_H275dup) among English individuals, and c.350del, p.E117Afs*39 in Comoros islanders [5] [24,33−36].

6.5 Physiological roles of the AIP protein

The product of the *AIP* gene is a 37 kDa and 330 amino acids protein, composed of two domains: an amino (N)-terminal peptidylprolyl cis-trans isomerase (PPIase)-like domain and a C-terminal tetratricopeptide repeat (TPR) domain composed of three TPR motifs (residues 179−212, 231−264, and 265−298 for TPR1, 2, and 3, respectively) and a final α-helix; the latter is crucial for the interaction of AIP with multiple molecular partners [37−40]. AIP has a predominantly cytoplasmic subcellular localization and is an abundant and stable protein; its intracellular levels are regulated, at least in part, by the ubiquitin-proteasome system via the SKP1-CUL1-FBXO3 E3 ubiquitin ligase [37,41]. AIP is widely expressed in human tissues and is highly conserved among species [42−44].

The main function of AIP is to serve as a co-chaperone for the heat shock proteins HSPA5, HSPA8, HSPA9, and HSP90, as well as the mitochondrial membrane translocase TOMM20, thereby regulating the interaction between these molecular chaperones and their client proteins [40,45−47]. Direct and indirect interactions with many other proteins, with still unclear biological roles, have been described for AIP [47−50]. The most studied role of AIP in a signaling pathway consists of regulating the nuclear translocation of the dioxin-induced transcription factor aryl hydrocarbon receptor, as part of a complex including HSP90 and translationally controlled tumor protein (TCTP, also known as p23) [37,51−57]. Another function of AIP is to regulate the turnover of the B cell lymphoma-6 (BCL6) protein, a master regulator of the germinal center B cell phenotype. AIP accomplishes this function by directly binding to the deubiquitinase UCHL1 [58].

AIP also directly interacts with the phosphodiesterase PDE4A5, a negative regulator of the intracellular levels of 3′,5′-cyclic adenosine monophosphate (cAMP), inhibiting its enzymatic activity, increasing its sensitivity to inhibition by rolipram, and attenuating its phosphorylation by the cAMP-dependent protein kinase (PKA) [59]. Also pertaining to the cAMP signaling pathway, a recently described interaction of AIP with PKA, probably mediated by HSP90, might play a role in regulating the subcellular localization of this enzyme [47,50]. Other interacting partners of AIP are the cytoskeletal proteins TUBB and TUBB2A, suggesting that AIP might act as a regulator of the microtubule network, and NME1, a negative regulator of cell cycle progression and cell motility [47]. The biological roles for these interactions have not yet been characterized.

In the normal human pituitary, AIP is expressed in the somatotroph and lactotroph cells, localized on the surface of the hormone secreting vesicles, but not in gonadotroph and corticotroph cells [13,60]. Nevertheless, AIP is particularly abundant in clinically nonfunctioning PitNETs (NF-PitNETs) and it has also been detected in corticotropinomas [13]. Interestingly, a mutation affecting the AIP-related 1 protein (AIPR-1), encoded by the *Caenorhabditis elegans* ortholog of *AIP*, causes an increase in neurotransmitter release due to the unrestricted function of a presynaptic ryanodine receptor. It has been proposed that, via recruitment of HSP90 to the ryanodine receptor, AIP might stabilize the closed state of such receptor, thus regulating neurotransmitter release [44]. AIP is detectable in human serum, but its levels are independent of GH secretion, glycemia, body mass index, age, and sex. The physiological role of circulating AIP remains obscure, and it is not yet clear whether its main source is the pituitary gland [61].

6.6 Pathophysiology of *AIP* LOF-associated pituitary tumorigenesis

In the somatotroph cells, AIP has a complex function as a mediator of various signals that negatively regulate the activity of the cAMP pathway, possibly acting in combination with other cAMP regulators (reviewed in [62]). The cAMP pathway is the main intracellular signaling cascade activated upon binding of the growth hormone releasing hormone (GHRH) to its receptor, and it stimulates both cell growth and GH expression and secretion [63]. In rat somatotropinoma GH3 cells, *Aip* overexpression attenuates the cAMP rise when the cells are exposed to the cAMP agonist drug forskolin, while *Aip* knockdown (KD) enhances the effects of the drug [64]. *Aip* deficiency results in increased intracellular cAMP in the mouse anterior pituitary; in contrast with these findings, there is reduced phosphorylation of the mitogen-activated protein kinases 3 and 1 and the cAMP response element binding protein (CREB) in *AIP* mutation positive human somatotropinoma tissues, and *Aip* knockout (KO) reduces CREB phosphorylation in mouse embryonic fibroblasts [65]. From these results, it is not clear which cAMP effectors are regulated by AIP. Interestingly, *GNAS* activating mutations (*gsp* oncogene), the most common somatic genetic defects in acromegaly, are absent in somatotropinomas from *AIP* mutation positive patients, suggesting that insufficient cAMP negative regulation predominates over increased cAMP production as a pathogenic mechanism in these tumors [16,66].

Impaired inhibitory G protein (Gi) signaling (due to $G\alpha_{i2}/G\alpha_{i3}$ downregulation) has been observed in pituitary cells from *Aip* KO mice and, likewise, somatotropinomas from *AIP* mutation positive patients display reduced $G\alpha_{i2}$ immunostaining [65,66]. Upregulation of the microRNA miR-34a, which reduces the expression of *AIP*, has been observed in aggressive somatotropinomas and in the setting of *AIP* mutations [67,68]. Experimental overexpression of miR-34a results in downregulation of $G\alpha_{i2}$, activation of the cAMP pathway, and impaired response to octreotide [68]. These results suggest that AIP might regulate the downstream effects of a Gi protein-coupled receptor in the somatotroph cells.

Via activation of multiple antiproliferative and antisecretory responses, including inhibition of GH exocytosis, Gi-mediated inactivation of adenylyl cyclase, and dephosphorylation of cAMP effectors by phosphotyrosine phosphatases, somatostatin receptors (SSTs) are the main negative regulators of the cAMP effects in somatotroph cells [69]. Activation of the SSTs by SSAs results in *AIP* upregulation, both in GH3 cells and in human somatotropinomas [70,71]. Along these lines, somatotropinomas expressing high AIP levels respond better to the treatment with SSAs than those with low AIP expression, and aggressive, SSA-unresponsive somatotropinomas, display reduced AIP expression [13,60,71−75]. By immunohistochemistry, membranous SST_{2A} expression is low in some, but not all, AIP-deficient tumors and, indeed, AIP and SST_{2A} are independent markers of response to first-generation SSAs [70,73,75]. The effect of SST activation on *AIP* expression is independent of SST_{2A} expression and of the presence or absence of the *gsp* oncogene in the tumor, indicating that an additional mechanism, not related with Gi-protein signaling, might take place [71,73].

The glycogen synthase kinase 3b (GSK3β), which activates the expression of genes involved in cell cycle control, as well as β-catenin dephosphorylation, is a Gi-independent effector of SST activation [69]. Among the GSK3β targets, the transcription factor ZAC1 is a known mediator of the antiproliferative effect of SSAs in the somatotroph cells, and this transcription factor, often downregulated in PitNETs, is upregulated by *AIP* overexpression [70,76]. Presurgical treatment of somatotropinomas with SSAs increases ZAC1 expression, and silencing of this tumor suppressor blocks the antiproliferative effect of octreotide in pituitary cells in vitro [76,77]. AIP overexpression upregulates ZAC1 expression in vitro, but the exact mechanism underlying this effect is unclear [70]. ZAC1 immunoreactivity in somatotropinomas positively correlates with the biochemical and tumoral responses to the treatment with SSAs [77]. Therefore linking SSTs to their effector ZAC1 might represent a Gi-independent tumor suppressor effect of AIP.

Although AIP is considered to act as tumor suppressor in the pituitary gland, loss of heterozygosity (LOH) is not a universal finding among *AIP* mutation positive patients [13]. In cases with LOH, the extent of the affected chromosomal region varies, and it frequently includes both *AIP* and *MEN1* [78]. Alternatively, the loss of *AIP* function could occur through various other mechanisms. The foremost importance of the C-terminal α-helix of AIP is supported by the finding that loss-of-function of this structure is sufficient to cause PitNET predisposition [40]. In addition to the loss of crucial molecular interactions, missense mutations affecting the TPR domain of the protein, as well as the truncating mutation p.R304*, result in unstable AIP mutants that are rapidly degraded [41]. Along these lines, cells from *AIP* mutation carriers express a significantly reduced AIP protein content compared with noncarriers [36,41,79].

A particular characteristic of somatotropinomas in *AIP* mutation positive patients is their infiltration with a large number of macrophages. Upregulation of the cytokine CCL5, which exerts chemotaxis on macrophages via interaction with CCR5, has been observed in such tumors. Treatment of *AIP* KD GH3 cells with bone marrow-derived macrophage-conditioned media induced a partial epithelial-to-mesenchymal transition-like phenotype, which is characteristic of invasive tumors. The same phenotype was also observed in *AIP* mutation positive patients and in a mouse model, supporting that these tumor features are due to *AIP* deficiency [80].

6.7 Animal models of *AIP* loss-of-function

The *AIP* gene ortholog in the mouse, *Aip*, is necessary for early development: its homozygous inactivation results in embryonic lethality due to cardiovascular malformations [81,82]. A hypomorphic mouse model with a heterozygous partial *Aip* deletion developed tumors of the *pars distalis* (most frequently mammosomatotroph adenomas) in adulthood, with full penetrance by the age of 15 months [83]. In a different study, no PitNETs were observed in these animals by the age of 12 months, and only a few animals developed tumors at a later age. It was observed, however, that AIP deficiency caused a leaner phenotype and an increase in total GH secretion, with preserved GH pulsatility and normal IGF-1 levels but no gigantism. By the age of 12 months, areas of hyperplasia were noted in one-third of the animals, and pituitary explants displayed increased GH secretion after stimulation with GHRH [84]. Interestingly, a recent study showed partial reversal of the $Aip^{+/-}$-associated acromegalic phenotype when the animals were crossed with $Prkar1a^{+/-}$ mice [85].

The study of a somatotroph-specific conditional KO mouse model showed that, by the age of 40 weeks, >80% of these animals developed somatotropinomas and a phenotype that recapitulated the features of acromegaly in humans. These findings suggest that loss of both *Aip* alleles is necessary for such phenotype to occur. Somatotroph hyperplasia with features of cell cycle dysregulation (cytosolic localization of Cdkn1b and perinuclear localization of Cdk4) were observed in pretumorigenic somatotrophs, while somatotropinomas displayed reduced SST_5 and impaired GH suppression with octreotide [86]. A different recently reported mouse model, with pituitary-specific *Aip* KO, developed somatotropinomas with significant macrophage infiltration observed at the age of 15 months, which recapitulates the phenotype recently described in *AIP* mutation positive PitNETs [80].

The *AIP* ortholog gene in *Drosophila melanogaster, CG1847*, located on the X chromosome, is predicted to encode a protein that shares 38% identity with the human AIP protein, with an N-terminal PPIase domain and a C-terminal TPR domain. Full *CG1847* KO is lethal at an early larval stage, and this phenotype can be rescued by the expression of the human *AIP* full-length cDNA, but not by a construct missing the sequence that encodes the C-terminal α-helix, supporting functional conservation. This model has been employed as a functional assay to assess the pathogenicity of *AIP* variants found in PitNET patients [87].

6.8 Clinical features

The typical phenotype associated with *AIP* mutations consists of familial gigantism arising during adolescence, due to a pituitary macroadenoma, as it has been described for the historical case of the Northern Irish giant Charles Byrne, who carried the p.R304* variant [34]. Aside from this presentation, a number of clinical and histopathological features distinguish *AIP* mutation positive individuals among other PitNET patients. Predominance of male patients has been shown in the largest cohorts so far reported, which is likely due to the inclusion of a large number of patients with gigantism, a condition that is more common in men [10,88]. There are usually more affected individuals per family among *AIP* mutation positive FIPA kindreds, compared with families with no *AIP* mutations; however, 25%−30% of the *AIP* mutation positive patients in a recent study presented as apparently sporadic (*simplex*) cases, due to the incomplete disease penetrance [11,16,28]. The variability in disease penetrance might be explained by disease modifying genetic factors not yet unveiled.

PitNETs arise during the second and third decades of life in three-quarters of *AIP* mutation positive patients; the youngest age at disease onset so far reported is 4 years and only 4% of patients develop the disease after the fifth decade of life [10,16,89]. Genotype−phenotype correlations have been described: patients with truncating mutations are younger at disease onset and at diagnosis, and there is a direct correlation between the age at diagnosis and the experimental protein half-life for patients with GH excess carrying missense mutations [16,41]. Ninety-three percent of patients carrying *AIP* pathogenic variants have macroadenomas, which often display extrasellar extension [10,16].

For reasons still unknown, patients with this genetic defect present an increased risk of developing clinically evident apoplexy of the PitNET. This life-threatening emergency can be the first manifestation of disease and is estimated to occur in 8% of the patients with *AIP* mutations, with familial presentation described in five kindreds in the literature [13,16,90,91]. Although extrapituitary tumors have been reported in some *AIP* mutation positive patients, their rate does not seem to be increased in this setting, and a causal association has not been firmly proven [16,92,93]. A recent study, however, reported loss of heterozygosity for an *AIP* variant in follicular thyroid carcinoma tissue in a patient who was part of a FIPA family, raising the possibility of pathogenic association [94]. Currently, screening for additional tumors, as it is done for patients with syndromes of multiple endocrine neoplasia, is not indicated in *AIP* mutation carriers.

The clinical phenotype is growth hormone excess in 80% of cases and, remarkably, 30%−40% of all patients develop gigantism [10,16]. Among kindreds with FIPA phenotype, *AIP* mutations are the most common cause of familial gigantism (i.e., two or more cases of gigantism in the same family); this phenotype is usually associated with truncating mutations. Two-thirds of *AIP* mutation positive patients with gigantism are males, and the features of gigantism in these patients typically arise during the second decade of life [16]. Patients with GH excess and *AIP* mutations often display impaired response to the treatment with SSAs and dopamine agonists (DA) regarding tumor size and hormone secretion, usually requiring multimodal treatment [10,13,14,88,95]. A good response to the combined treatment with an SSA and pegvisomant has been described in a small number of patients with gigantism and *AIP* pathogenic variants reported in the literature [96]. In one study, AIP-deficient somatotropinomas were equally as responsive to pasireotide and displayed similar SST_5 expression by immunohistochemistry, compared with AIP proficient tumors, and good response to pasireotide has been reported in *AIP* mutation positive patients as well [75,97]. Prolactinomas are the second most common type of PitNET among *AIP* mutation positive patients, although only one *AIP* mutation positive FIPA kindred homogeneous for prolactinoma has been reported in the literature; other tumors are much less frequent [16,88,98]. Larger tumor size and poorer response to treatment with cabergoline have been described in *AIP* mutation positive patients with prolactinomas, compared with their *AIP* mutation negative counterparts, although this has not been confirmed in other cohorts [88,99,100].

In general, PitNETs in *AIP* mutation positive patients may have a highly variable clinical presentation: while the best-known clinical picture is that of aggressive and difficult to treat tumors, some patients present with apparently indolent PitNETs detected only by targeted screening tests in mutation carriers, which might respond well to standard treatments [16,88,101]. There are no specific clinical guidelines established for the treatment of PitNETs in the setting of *AIP* mutations, and although these patients should be treated according to the established guidelines for the PitNET type diagnosed, the possibility of suboptimal response should be kept in mind.

6.9 Histopathological and immunohistochemical features

At the histopathological analysis, the majority of PitNETs in *AIP* mutation positive patients are somatotropinomas or mammosomatotroph adenomas, and more than 90% of them are of the sparsely granulated type [16,90]. Most of the NF-PitNETs associated with *AIP* pathogenic variants stain positive for GH by immunohistochemistry, therefore accounting for silent somatotropinomas, while 17% of patients develop plurihormonal tumors [16,90]. One study described areas or somatotroph hyperplasia surrounding the somatotropinomas in two patients from a FIPA family carrying a truncating *AIP* mutation. LOH was demonstrated in the PitNET, but not in the foci of hyperplasia, suggesting a potential progression from hyperplasia to adenoma, in which the loss of the normal *AIP* allele might be a late event [90].

Macrophage infiltration has only recently been described as a characteristic feature of *AIP* mutation positive tumors, but it has not yet been correlated with clinical parameters [80]. Along these lines, macrophages were shown to be significantly more abundant in densely granulated versus sparsely granulated somatotropinomas in a previous study [102]. AIP immunostaining is not a good predictor of *AIP* mutations, but is an excellent marker of invasiveness and response to first-generation SSAs in somatotropinomas, independently of SST_{2A} expression and perhaps with greater sensitivity than Ki-67 and TP53, although it does not predict the response to treatment with pasireotide [72−75,93].

6.10 Genetic testing and *AIP* variant characterization

Although there are no well-established guidelines available as yet, expert recommendations are available in the literature to assist the clinicians in the screening of individuals for *AIP* variants, their genetic counseling, and their required baseline and follow-up clinical screening [8,16,103−105]. Genetic testing for *AIP* variants is strongly suggested in probands from FIPA families, as well as in individuals affected with familial or sporadic gigantism [8,103,105]. Cases of young-onset (age ≤ 30 years) PitNETs should also be tested, especially patients with disease presentation during the second decade of life and those with macroadenomas, in particular those secreting GH and/or prolactin [8,28,103−106]. In addition, the occurrence of pituitary apoplexy in a young patient with clinical manifestations of GH excess should raise suspicion for the possibility of *AIP* pathogenic variants. A risk-categorization system has recently been proposed to determine the likelihood of a PitNET patient to carry an *AIP* pathogenic variant, based on a multivariate analysis that included a variety of clinical features from a large cohort of patients. Four independent predictors of *AIP* pathogenic variants were identified: age at disease onset 0−18 years, positive family history, GH excess, and large tumor size [107]. The application of this tool might improve the cost-efficiency of genetic testing, although it requires further validation in other cohorts.

Testing for *AIP* variants should include sequencing of all the exons and exon-intron junctions by direct sequencing or next-generation sequencing. If no variant is identified by sequencing, deletion/duplication analysis should be done next; this is usually accomplished using multiplex ligation-dependent probe amplification [108]. Findings in the genetic screening should be carefully analyzed for their potential clinical significance. The PitNET-associated *AIP* variants so far reported in the literature are included in Table 6.1; their pathogenicity likelihood has been assessed taking into consideration clinical, histopathological, as well as in vitro and in vivo experimental data and classified according to the American College of Medical Genetics and Genomics guidelines [109]. The results of in silico prediction might be equivocal for some variants, and in those cases functional assays might be required. Since there is not a single functional assay that can confidently predict by itself the pathogenicity of all the *AIP* variants detected in patients, the use of multiple methods, combined with a careful interpretation of the clinical data is recommended [62]. Further testing and follow-up is currently recommended only for carriers of likely pathogenic and pathogenic variants. It is recommendable to contact research centers for guidance on how to proceed in cases where variants of uncertain significance are reported.

Table 6.1 Germline *AIP* variants reported in pituitary neuroendocrine tumor patients and their functional effects.

HGVS nomenclature DNA, protein Reference sequence: NM_003977.4	dbSNP ID	Location in gene	Location in protein	Variant type	Frequency in gnomAD (%)	Experimental evidence	References for clinical reports	ACMG classification
c.-1213_279 + 578del (ex1_ex2del)	Not in dbSNP	5′UTR and exons 1–2	N-terminus	Large genomic deletion	Not in gnomAD	N/A	[11,25,31]	Pathogenic
c.-270_269delinsAA, p.?	rs267606588	5′UTR	N/A	Promoter	Not in gnomAD	LOH in PitNET tissue [13]; Reduced promoter activity by luciferase assay [11].	[11,13,16,88,110]	Pathogenic
c.-262G > A, p.?	Not in dbSNP	Promoter	n/a	Promoter	Not in gnomAD	N/A	[13]	VUS
c.-220G > A, p.?	rs267606540	5′UTR	n/a	Promoter	Not in gnomAD	LOH in PitNET tissue [13]. No effect on promoter activity by luciferase assay [11].	[11,13,110]	Benign
c.1-?_993 + ?del (whole gene deletion)	Not in dbSNP	Exons 1–6	Absence of the whole protein	Whole gene deletion	Not in gnomAD	Full KO is lethal in mice and fruit flies [81,82,87]. Pituitary hyperplasia and tumors have been described hypomorphic *Aip* mice [83,84]. Pituitary-specific *Aip* mice develop pituitary tumors [86].	[11,16,88]	Pathogenic
c.?-50_99 + 1_100-1del, p.? (exon 1 deletion)	Not in dbSNP	Exon 1	Absence of the whole protein	Large genomic deletion	Not in gnomAD	N/A	[88]	Pathogenic
c.2T > C, p.?	rs267606546	Exon 1	N-terminus	Start codon	Not in gnomAD	Decreased AIP protein expression in patient's fibroblasts [79].	[95,111]	Pathogenic
c.3G > A, p.?	rs886037871	Exon 1	N-terminus	Start codon	Not in gnomAD	N/A	[16,35,88]	Pathogenic
c.3_4insC, p.A2Rfs*43	rs267606547	Exon 1	N-terminus	Frameshift	Not in gnomAD	N/A	[8]	Pathogenic
c.4del, p.A2Rfs*16	Not in dbSNP	Exon 1	N-terminus	Frameshift	Not in gnomAD	N/A	[91]	Pathogenic
c.25C > T, p.R9W	rs1057523115	Exon 1	N-terminus	Missense	N/A	N/A	[112]	VUS

(Continued)

Table 6.1 Germline *AIP* variants reported in pituitary neuroendocrine tumor patients and their functional effects. *Continued*

HGVS nomenclature DNA, protein Reference sequence: NM_003977.4	dbSNP ID	Location in gene	Location in protein	Variant type	Frequency in gnomAD (%)	Experimental evidence	References for clinical reports	ACMG classification
c.26G > A, p.R9Q	rs139459091	Exon 1	N-terminus	Missense	Exomes: 0.02625 Genomes: 0.009558 Combined: 0.0244	Stable protein with normal half-life [113]. Increased cAMP-driven transcription when overexpressed in forskolin-stimulated GH3 cells [113].	[19,20,111,113]	VUS
c.36G > A, p.G12 =	rs79662690	Exon 1	N-terminus	Synonymous	Exomes: 0.0521 Genomes: 0.03823 Combined: 0.05056	N/A	[13,21,96,114]	Nonpathogenic
c.38T > A, p.I13N	rs376913545	Exon 1	N-terminus	Missense	Exomes: 0.0003977	LOH in PitNET tissue [115]. Does not rescue lethality in a KO *Drosophila melanogaster* model [87].	[115]	Likely pathogenic
c.40C > T, p.Q14*	rs104894194	Exon 1	N-terminus	Nonsense	Exomes: 0.002784	LOH in PitNET tissue [5,66].	[5,16,66,88, 116,117]	Pathogenic
c.47G > A, p.R16H	rs145047094	Exon 1	N-terminus	Missense	Exomes: 0.1973 Genomes: 0.2962 Combined: 0.2082	Stable protein with normal half-life [41,113]. Rescues lethality in a KO *D. melanogaster* model [87]. Does not affect AIP interaction with PDE4A5 by 2-H/β-gal [11,118]. Increased cAMP-driven transcription and cAMP and GH concentration when overexpressed in forskolin-stimulated GH3 cells [113].	[12,16,19,21, 23,29,67,111, 116,119−128]	Benign

Variant	dbSNP	Exon/Intron	Domain	Type	gnomAD frequency	Notes	References	Classification
c.58G > T, p.E20*	Not in dbSNP	Exon 1	N-terminus	Nonsense	Not in gnomAD	N/A	[24]	Pathogenic
c.64C > T, p.R22*	rs121908357	Exon 1	N-terminus	Nonsense	Not in gnomAD	LOH in PitNET tissue [18].	[18,23,24]	Pathogenic
c.66_71del, p.G23_E24del	rs267606567	Exon 1	N-terminus	In-frame deletion	Not in gnomAD	LOH in PitNET tissue [116].	[116]	VUS
c.68G > A, p.G23E	rs116940576|	Exon 1	N-terminus	Missense	Exomes: 0.1138 Genomes: 0.1656 Combined: 0.1195	N/A	[21]	Benign
c.70G > T, p.E24*	rs267606568	Exon 1	N-terminus	Nonsense	Not in gnomAD	LOH in PitNET tissue [13].	[2,7,13,16,88,129]	Pathogenic
c.70_89del, p.E24Wfs*14	Not in dbSNP	Exon 1	N-terminus	Frameshift	Not in gnomAD	Homozygous in PitNET tumor, but germline DNA was not analyzed [66].	[66]	Pathogenic
c.74_81delins, p.L25Pfs*131	Not in dbSNP	Exon 1	N-terminus	Frameshift	Not in gnomAD	N/A	[11,16,88,130]	Pathogenic
c.88_89del, p.D30Wfs*14	rs267606584	Exon 1	N-terminus	Frameshift	Not in gnomAD	N/A	[8]	Pathogenic
c.90T > G, p.D30E	Not in dbSNP	Exon 1	N-terminus	Missense	Not in gnomAD	LOH in PitNET tissue [131].	[131]	VUS
c.100−1025_279 + 357del, p.A34_K93del (exon 2 deletion)	Not in dbSNP	Exon 2	PPIase domain	Large genomic deletion	Not in gnomAD	N/A	[16,31,88]	Pathogenic
c.100-6C > A, p.?	rs539027647	Intron 1	N/A	Intronic	Exomes: 0.03184 Genomes: 0.003187 Combined: 0.02866	N/A	[111]	Benign
c.100−18C > T, p.?	rs202156895	Intron 1	N/A	Intronic	Exomes: 0.3234 Genomes: 0.2454 Combined: 0.3147	N/A	[13,16,20,23, 35,111,116,124]	VUS
c.132C > T, p.D44 =	rs11822907	Exon 2	PPIase domain	Missense	Exomes: 0.6040 Genomes: 2.355 Combined: 0.7984	N/A	[16,21,132,133]	Benign

(Continued)

Table 6.1 Germline *AIP* variants reported in pituitary neuroendocrine tumor patients and their functional effects. *Continued*

HGVS nomenclature DNA, protein Reference sequence: NM_003977.4	dbSNP ID	Location in gene	Location in protein	Variant type	Frequency in gnomAD (%)	Experimental evidence	References for clinical reports	ACMG classification
c.135C > T, p.D45 =	rs181969066	Exon 2	PPIase domain	Synonymous	Exomes: 0.006367	N/A	[134]	Benign
c.140_163del, p.G47_R54del	rs267606537	Exon 2	PPIase domain	In-frame deletion	Exomes: 0.0003979	N/A	[12,88,135,136]	VUS
c.144C > T, p.T48 =	rs772658134	Exon 2	PPIase domain	Synonymous	Exomes: 0.006366	Does not affect splicing by minigene assay [137].	[16,131,132, 137–139]	Benign
c.145G > A, p.V49M	rs1063385	Exon 2	PPIase domain	Missense	Exomes: 0.02745	Probably involved in interactions by in silico prediction. Unstable protein with short half-life; normal half-life in a different study [41,113]. Does not affect AIP interaction with PDE4A5 by 2-H/β-gal [11,118]. Increased cAMP-driven transcription and GH concentration when overexpressed in forskolin-stimulated GH3 cells [113].	[134]	Benign
c.166C > T, p.R56C	rs267606538	Exon 2	PPIase domain	Missense	Exomes: 0.001989	N/A	[23]	Likely benign
c.174G > C, p.K58N	rs267606539	Exon 2	PPIase domain	Missense	Exomes: 0.0007955	N/A	[19,23,24]	VUS
c.208C > A, p.L70M	Not in dbSNP	Exon 2	PPIase domain	Missense	Not in gnomAD	N/A	[8]	VUS
c.217T > C, p.W73R	Not in dbSNP	Exon 2	PPIase domain	Missense	Not in gnomAD	Does not rescue lethality in a KO *D. melanogaster* model [87].	[122]	Likely pathogenic

Variant	dbSNP	Exon/Intron	Domain	Type	gnomAD	Functional data	References	Classification
c.240_241delinsTG, p.M80_R81delinsIG	Not in dbSNP	Exon 2	PPIase domain	In-frame insertion/deletion	Not in gnomAD	N/A	[88,107]	VUS
c.241C>T, p.R81*	rs267606541	Exon 2	PPIase domain	Nonsense	Not in gnomAD	LOH in PitNET tissue [13,89,140]. Abolishes AIP interaction with PDE4A5 by 2-H/β-gal [13,118].	[13,16,88,92,93,110,140,141]	Pathogenic
c.245_249del, p.E82Gfs*8	rs267606542	Exon 2	PPIase domain	Frameshift	Not in gnomAD	N/A	[23]	Pathogenic
c.249G>T, p.G83Afs*15	rs104895072	Exon 2	PPIase domain	Splice site	Not in gnomAD	Cryptic splice site generates an alternatively spliced mRNA [11].	[11,16,88]	Pathogenic
c.250G>A, p.E84K	rs267606543	Exon 2	PPIase domain	Missense	Not in gnomAD	N/A	[23]	VUS
c.280-81G>A	Not in dbSNP	Intron 2	PPIase domain	Intronic	Not in gnomAD	N/A	[21]	VUS
c.280-1G>C	Not in dbSNP	Intron 2	PPIase domain	Splice site	Not in gnomAD	N/A	[116]	Likely pathogenic
c.286_287del, p.V96Pfs*33	rs267606545	Exon 3	PPIase domain	Frameshift	Not in gnomAD	LOH in PitNET tissue [134].	[134,142]	Pathogenic
c.308A>G, p.K103R	rs267606548	Exon 3	PPIase domain	Missense	Exomes: 0.0004034	Does not affect AIP interaction with PDE4A5 by 2-H/β-gal (although it showed reduced interaction in one study) [11,118]. Stable protein with normal half-life [113]. No effect on cAMP-driven transcription and cAMP and GH concentration when overexpressed in forskolin-stimulated GH3 cells [113].	[22]	VUS
c.316C>T p.R106C	rs369414668	Exon 3	PPIase domain	Missense	Exomes: 0.004032	N/A	[143]	Likely benign
c.333del, p.K112Rfs*44	Not in dbSNP	Exon 3	PPIase domain	Frameshift	Not in gnomAD	N/A	[88,107]	Pathogenic
c.338_341dup, p.L115Pfs*16	Not in dbSNP	Exon 3	PPIase domain	Frameshift	Not in gnomAD	N/A	[16,22,88]	Pathogenic

(Continued)

Table 6.1 Germline *AIP* variants reported in pituitary neuroendocrine tumor patients and their functional effects. *Continued*

HGVS nomenclature DNA, protein Reference sequence: NM_003977.4	dbSNP ID	Location in gene	Location in protein	Variant type	Frequency in gnomAD (%)	Experimental evidence	References for clinical reports	ACMG classification
c.344del (p.L115Rfs*41)	Not in dbSNP	Exon 3	PPIase domain	Frameshift	Not in gnomAD	N/A	[88]	Pathogenic
c.350del, p.E117Afs*39	rs267606549	Exon 3	PPIase domain	Frameshift	Not in gnomAD	Decreased AIP protein expression and increased *AHRR* in patient's fibroblasts [79].	[19,23,24,79]	Pathogenic
c.355C>T, p.R119W	67256813	Exon 3	PPIase domain	Missense	Exomes: 0.001604	Removes an α-helix structure at the amino acids 188-192 binding site by 3-D modeling [131].	[131]	Likely pathogenic
c.376_377del, p.Q126Dfs*3	Not in dbSNP	Exon 3	Between PPIase and TPR1 domains	Frameshift	Not in gnomAD	N/A	[88,107]	Pathogenic
c.382C>T, p.R128C	rs140530307	Exon 3	Between PPIase and TPR1 domains	Missense	Exomes: 0.01321 Genomes: 0.04783 Combined: 0.01707	N/A	[111,128]	VUS
c.383G>A, p.R128H	rs267606550l	Exon 3	Between PPIase and TPR1 domains	Missense	Exomes: 0.001600	N/A	[23,60]	Likely benign
c.404del, p.H135Lfs*21	rs267606551	Exon 3	Between PPIase and TPR1 domains	Frameshift	Not in gnomAD	N/A	[21]	Pathogenic
c.424C>T, p.Q142*	rs267606552	Exon 3	Between PPIase and TPR1 domains	Nonsense	Not in gnomAD	N/A	[12]	Pathogenic

Variant	dbSNP	Location	Domain	Type	gnomAD		References	Classification
c.427C > T, p.QI143*	Not in dbSNP	Exon 3	Between PPIase and TPR1 domains	Nonsense	Not in gnomAD	N/A	[16,88]	Pathogenic
c.429G > A, p.QI143 =	rs267606553	Exon 3	Between PPIase and TPR1 domains	Synonymous	Not in gnomAD	N/A	[21]	VUS
c.455T > G, p.M152R	Not in dbSNP	Exon 3	Between PPIase and TPR1 domains	Missense	Not in gnomAD	N/A	[96]	VUS
c.468G > A, p.K156 =	rs1162771114	Exon 3	Between PPIase and TPR1 domains	Synonymous	Exomes: 0.0008173	N/A	[143]	VUS
c.468 + 1G > A, p.?	rs267606554	Intron 3	N/A	Intronic	Not in gnomAD	N/A	[126,127,137]	Likely pathogenic
c.468 + 15C > T	rs267607274	Intron 3	N/A	Intronic	Exomes: 0.005375 Genomes: 0.006378 Combined: 0.005490	N/A	[23,142,144,145]	Benign
c.468 + 16G > T	rs267607273	Intron 3	N/A	Intronic	Not in gnomAD	N/A	[23]	VUS
c.468 + 111C > T	rs4084113	Intron 3	N/A	Intronic	Exomes: 35.77 Genomes: 30.82 Combined: 35	N/A	[21,131]	Benign
c.469-54G > A, p.?	Not in dbSNP	Intron 3	N/A	Intronic	Not in gnomAD	N/A	[146]	VUS
c.469-17T > C, p.?	rs886037872	Intron 3	N/A	Intronic	Exomes: 0.001452 Genomes: 0.003187 Combined: 0.001681	N/A	[35]	VUS

(Continued)

Table 6.1 Germline *AIP* variants reported in pituitary neuroendocrine tumor patients and their functional effects. *Continued*

HGVS nomenclature DNA, protein Reference sequence: NM_003977.4	dbSNP ID	Location in gene	Location in protein	Variant type	Frequency in gnomAD (%)	Experimental evidence	References for clinical reports	ACMG classification
c.469−2A > G, p. E158_Q184del	rs267606556	Intron 3	Between PPIase and TPR1 domains to TPR1 domain	Splice site	Not in gnomAD	Cryptic splice site generates an alternatively spliced mRNA [147].	[16,19,21,88,148]	Likely pathogenic
c.469-1G > A	rs267606555	Intron 3	N/A	Intronic	Not in gnomAD	N/A	[5]	Likely pathogenic
c.490C > T, p.Q164*	rs104895073	Exon 4	Between PPIase and TPR1 domains	Nonsense	Not in gnomAD	Nonfunctional protein by in silico prediction. Abolishes AIP interaction with PDE4A5 by 2-H/β-gal [11,118].	[11,16,25,88]	Pathogenic
c.491A > G, p.Q164R	Not in dbSNP	Exon 4	Between PPIase and TPR1 domains	Missense	Not in gnomAD	Rescues lethality in a KO *D. melanogaster* model [87].	[87]	Likely benign
c.500del, p.P167Hfs*4	rs267606557	Exon 4	Between PPIase and TPR1 domains	Frameshift	Not in gnomAD	N/A	[149]	Pathogenic
c.504G > A, p.W168*	Not in dbSNP	Exon 4	Between PPIase and TPR1 domains	Nonsense	Not in gnomAD	N/A	[88,150]	Pathogenic
c.509T > C, p.M170T	Not in dbSNP	Exon 4	Between PPIase and TPR1 domains	Missense	Not in gnomAD	Stable protein with normal half-life [41].	[19]	VUS
c.512C > T, p.T171I	rs1287485768	Exon 4	Between PPIase and TPR1 domains	Missense	Not in gnomAD	LOH in PitNET tissue [151]. Failed to inhibit cell proliferation in 293FT and GH3 cells; increased GH secretion, promoted invasion increased the expression of Stat3-Tyr705, and reduced the expression of *Sstr2* and *Zac1* in GH3 cells [151].	[151]	Likely pathogenic

Variant	rsID	Exon	Domain	Type	gnomAD frequency	Functional/notes	References	Classification
c.516C > T, p.D172 =	rs2276020	Exon 4	Between PPIase and TPR1 domains	Synonymous	Exomes: 3.384 Genomes: 3.801 Combined: 3.431	N/A	[11,18,21,96,119, 123,131–134, 146,152]	Benign
c.521_525del, p.E174Gfs*47	rs267606558	Exon 4	Between PPIase and TPR1 domains	Frameshift	Not in gnomAD	N/A	[12,14]	Pathogenic
c.543del, p.I182Sfs*14	rs267606559	Exon 4	TPR1 domain	Frameshift	Not in gnomAD	N/A	[116,153]	Pathogenic
c.550C > T, p.Q184*	rs267606560	Exon 4	TPR1 domain	Nonsense	Not in gnomAD	Nonfunctional protein by 3-D modeling [40].	[23]	Pathogenic
c.562C > T, p.R188W	rs577617733	Exon 4	TPR1 domain	Missense	Exomes: 0.003202 Genomes: 0.003184 Combined: 0.0032	Very unstable protein with very short half-life [41].	[41,88]	Likely pathogenic
c.562del, p. Arg188Glyfs*8	Not in dbSNP	Exon 4	TPR1 domain	Frameshift	Not in gnomAD	LOH in PitNET tissue [135].	[135]	Pathogenic
c.563G > A, p.R188Q	rs866556486	Exon 4	TPR1 domain	Missense	Not in gnomAD	N/A	[19]	Likely benign
c.570C > G, p.Y190*	Not in dbSNP	Exon 4	TPR1 domain	Nonsense	Not in gnomAD	N/A	[16,88]	Pathogenic
c.573C > T, p.R191 =	rs781545373	Exon 4	TPR1 domain	Synonymous	Exomes: 0.005199 Genomes: 0.003185 Combined: 0.004975	N/A	[16]	Likely benign
c.579G > T, p.G193 =	rs1194122725	Exon 4	TPR1 domain	Synonymous	Not in gnomAD	N/A	[16]	VUS
c.584T > C, p.V195A	rs267606561	Exon 4	TPR1 domain	Missense	Exomes: 0.001599 Genomes: 0.003187 Combined: 0.001776	N/A	[23,60,154]	VUS
c.591G > A, p.E197 =	rs202006716	Exon 4	TPR1 domain	Synonymous	Exomes: 0.007194 Genomes: 0.006371 Combined: 0.007102	N/A	[23,111]	Benign

(Continued)

Table 6.1 Germline *AIP* variants reported in pituitary neuroendocrine tumor patients and their functional effects. *Continued*

HGVS nomenclature DNA, protein Reference sequence: NM_003977.4	dbSNP ID	Location in gene	Location in protein	Variant type	Frequency in gnomAD (%)	Experimental evidence	References for clinical reports	ACMG classification
c.601A > T, p.K201*	rs267606563	Exon 4	TPR1 domain	Nonsense	Not in gnomAD	Nonfunctional protein by 3-D modeling [40].	[21]	Pathogenic
c.605A > G, p.Y202C	Not in dbSNP	Exon 4	TPR1 domain	Missense	Not in gnomAD	N/A	[88,107]	VUS
c.630del, p.N211Tfs*4	Not in dbSNP	Exon 4	TPR1 domain	Frameshift	Not in gnomAD	N/A	[24]	Pathogenic
c.634C > T, p.L212 =	rs773865566	Exon 4	TPR1 domain	Synonymous	Exomes: 0.0008 Genomes: 0.003188 Combined: 0.001066	N/A	[132]	VUS
c.645 + 1G > C, p.?	Not in dbSNP	Intron 4	N/A	Splice site	Not in gnomAD	N/A	[88,107]	Likely pathogenic
c.645 + 37G > A	rs760363160	Intron 4	N/A	Intronic	Exomes: 0.000805	N/A	[21]	VUS
c.645 + 41dup	rs1171589678	Intron 4	N/A	Intronic	Not in gnomAD	N/A	[131]	VUS
c.646G > T, p.E216*	rs267606565	Exon 5	Between TPR1 and TPR2 domains	Nonsense	Not in gnomAD	LOH in PitNET tissue [90]. Nonfunctional protein by 3-D modeling [40].	[90,111]	Pathogenic
c.649C > T, p.Q217*	rs267606566	Exon 5	Between TPR1 and TPR2 domains	Nonsense	Not in gnomAD	Nonfunctional protein by 3-D modeling [40]. Abolishes AIP interaction with PDE4A5 by 2-H/β-gal [13,118].	[12,131,155]	Pathogenic
c.662dup, p.E222*	rs104895075	Exon 5	Between TPR1 and TPR2 domains	Nonsense	Not in gnomAD	Nonfunctional protein by 3-D modeling [40].	[11,16,88]	Pathogenic

Variant	dbSNP	Exon	Domain	Type	gnomAD frequency	Functional effect	References	Classification
c.682A > C, p.K228Q[a]	rs641081	Exon 5	Between TPR1 and TPR2 domains	Missense	Exomes: 3.78 Genomes: 14.85 Combined: 5.02	N/A	[16,18,21,96, 123,145,156]	Benign
c.685C > T, p.Q229*	Not in dbSNP	Exon 5	Between TPR1 and TPR2 domains	Nonsense	Not in gnomAD	N/A	[157]	Pathogenic
c.687G > A, p.Q229 =	Not in dbSNP	Exon 5	Between TPR1 and TPR2 domains	Synonymous	Not in gnomAD	N/A	[131]	VUS
c.696G > C, p.P232 =	Not in dbSNP	Exon 5	TPR2 domain	Synonymous	Not in gnomAD	N/A	[116]	VUS
c.707_716delinsGGC, p. N236Rfs*65	Not in dbSNP	Exon 5	TPR2 domain	Frameshift	Not in gnomAD	N/A	[96]	Pathogenic
c.713G > A, p.C238Y	rs267606569	Exon 5	TPR2 domain	Missense	Exomes: 0.0004025	LOH in PitNET tissue [13]. Disrupts packing of hydrophobic core by 3-D modeling [40]. Failed to inhibit cells proliferation in human cell lines [13]. Very unstable protein with very short half-life [41]. Does not rescue lethality in a KO D. melanogaster model [87]. Abolishes AIP interaction with PDE4A5 by 2-H/β-gal [11,13,118].	[2,13,16,88]	Pathogenic
c.715C > T, p.Q239*	rs267606571	Exon 5	TPR2 domain	Nonsense	Not in gnomAD	Nonfunctional protein by 3-D modeling [40].	[12]	Pathogenic
c.715_721delinsTCAACTAC, p.Q239Sfs*49	Not in dbSNP	Exon 5	TPR2 domain	Frameshift	Not in gnomAD	N/A	[96]	Pathogenic
c.718T > C, p.C240R	Not in dbSNP	Exon 5	TPR2 domain	Missense	Not in gnomAD	Disrupts packing of hydrophobic core by 3-D modeling [40].	[158]	VUS
c.720C > T, p.C240 =	rs267606572	Exon 5	TPR2 domain	Synonymous	Exomes: 0.0004026	N/A	[21]	VUS

(Continued)

Table 6.1 Germline *AIP* variants reported in pituitary neuroendocrine tumor patients and their functional effects. *Continued*

HGVS nomenclature DNA, protein Reference sequence: NM_003977.4	dbSNP ID	Location in gene	Location in protein	Variant type	Frequency in gnomAD (%)	Experimental evidence	References for clinical reports	ACMG classification
c.721A > G, p.K241E	rs267606573	Exon 5	TPR2 domain	Missense	Exomes: 0.003622	Disrupts hydrogen bonding to E246 by 3-D modeling [40]. Reduces AIP interaction with PDE4A5 by 2-H/β-gal [11,118].	[12]	VUS
c.721A > T, p.K241*	rs267606573	Exon 5	TPR2 domain	Nonsense	Not in gnomAD	Nonfunctional protein by 3-D modeling [40].	[22]	Pathogenic
c.733G > A, p.E245K	rs150645662	Exon 5	TPR2 domain	Missense	Exomes: 0.004434	N/A	[131]	Likely benign
c.736_738del, p.E246del	Not in dbSNP	Exon 5	TPR2 domain	In-frame deletion	Not in gnomAD	N/A	[8,24]	VUS
c.742_744del, p.Y248del	rs267606574	Exon 5	TPR2 domain	In-frame deletion	Not in gnomAD	LOH in PitNET tissue [159]. Abolishes AIP interaction with PDE4A5 by 2-H/β-gal [118].	[159]	Pathogenic
c.752del, p.L251Rfs*52	Not in dbSNP	Exon 5	TPR2 domain	Frameshift	Not in gnomAD	N/A	[19]	Pathogenic
c.753G > A, p.L251 =	rs147351993	Exon 5	TPR2 domain	Synonymous	Exomes: 0.01859 Genomes: 0.00956 Combined: 0.01757	N/A	[111]	VUS
c.760T > C, p.C254R	Not in dbSNP	Exon 5	TPR2 domain	Missense	Not in gnomAD	Very unstable protein with very short half-life [41].	[41,88]	Likely pathogenic
c.762C > G, p.C254W	Not in dbSNP	Exon 5	TPR2 domain	Missense	Not in gnomAD	Very unstable protein with very short half-life [41].	[41,88]	Likely pathogenic
c.769A > G, p.I257V	rs267606575	Exon 5	TPR2 domain	Missense	Not in gnomAD	Disrupts packing of hydrophobic core by 3-D modeling [40]. Unstable protein with short half-life [41]. Does not affect AIP interaction with PDE4A5 by 2-H/β-gal [11,118].	[10]	Likely pathogenic

Variant	dbSNP	Exon	Domain	Type	gnomAD	Functional effect	References	Classification
c.773T > G, p.L258R	Not in dbSNP	Exon 5	TPR2 domain	Missense	Not in gnomAD	N/A	[88]	VUS
c.772C > T, p.L258F	rs779315334	Exon 5	TPR2 domain	Missense	Exomes: 0.0004052	N/A	[111]	VUS
c.779del, p.K260Sfs*43	Not in dbSNP	Exon 5	TPR2 domain	Frameshift	Nor in gnomAD	N/A	[88]	Pathogenic
c.783C > T, p.Y =	rs267606576	Exon 5	TPR2 domain	Synonymous	Exomes: 0.0004074	N/A	[21]	VUS
c.783C > G, p.Y261*	Not in dbSNP	Exon 5	TPR2 domain	Nonsense	Not in gnomAD	LOH in PitNET tissue [140]. Probable effect on ligand binding by 3-D modeling [40].	[16,19,24, 88,140,160]	Pathogenic
c.784G > A, p.D262N	rs758918509	Exon 5	TPR2 domain	Missense	Exomes: 0.002037	LOH in PitNET tissue [131].	[131]	VUS
c.787 + 9C > T	rs749392143	Intron 5	N/A	Intronic	Exomes: 0.004515 Genomes: 0.006375 Combined: 0.004727	N/A	[16]	Likely benign
c.787 + 24C > T	rs373727233	Intron 5	N/A	Intronic	Exomes: 0.007053 Genomes: 0.001275 Combined: 0.007709	N/A	[20,21]	Likely benign
c.788-2A > C	Not in dbSNP	Intron 5	N/A	Splice site	Not in gnomAD	Affects the canonical acceptor splice site of intron 5 by in silico prediction [155].	[155]	Likely pathogenic
c.803A > G, p.Y268C	rs267606577	Exon 6	TPR3 domain	Missense	Exomes: 0.0004018	Disrupts packing of hydrophobic core by 3-D modeling [40].	[23]	Likely pathogenic
c.804C > A, p.Y268*	rs121908356	Exon 6	TPR3 domain	Nonsense	Not in gnomAD	Nonfunctional protein by 3-D modeling [40].	[7,16,67,88, 128,161,162]	Pathogenic
c.805_825dup, p. F269_H275dup	rs267606578	Exon 6	TPR3 domain	In-frame insertion	Not in gnomAD	Abolishes interaction with HSP90 by CO-IP. Very unstable protein, with very short half-life [36]. Decreased AIP protein expression and increased AHRR in patient's fibroblasts [79].	[13,16,19,36, 79,88,110,163]	Pathogenic

(Continued)

Table 6.1 Germline *AIP* variants reported in pituitary neuroendocrine tumor patients and their functional effects. *Continued*

HGVS nomenclature DNA, protein Reference sequence: NM_003977.4	dbSNP ID	Location in gene	Location in protein	Variant type	Frequency in gnomAD (%)	Experimental evidence	References for clinical reports	ACMG classification
c.807C > T, p.F269 =	rs139407567	Exon 6	TPR3 domain	Splice site	Exomes: 0.05586 Genomes: 0.04777 Combined: 0.05495	Reduced transcript level by minigene assay [11].	[11,16,20,21, 23,124,143, 159,164]	Pathogenic
c.811C > T, p.R271W	rs267606579	Exon 6	TPR3 domain	Missense	Exomes: 0.0004017	Involved in packing with numerous polar interactions by 3-D modeling [40]. Very unstable protein with very short half-life [41]. Abolishes AIP interaction with PDE4A5 by 2-H/β-gal [11,13,118].	[11,12,16,23, 88,143,165]	Pathogenic
c.815G > A, p.G272D	rs779831121	Exon 6	TPR3 domain	Missense	Exomes: 0.0004015	Does not rescue lethality in a KO *D. melanogaster* model [87].	[35,88,152]	Likely pathogenic
c.816del, p.K273Rfs*30	Not in dbSNP	Exon 6	TPR3 domain	Frameshift	Not in gnomAD	N/A	[16,88,128]	Pathogenic
c.824dup, p.H275Qfs*13	rs267606580	Exon 6	TPR3 domain	Frameshift	Not in gnomAD	LOH in PitNET tissue [116].	[116]	Pathogenic
c.825_845del, p. H275_A281del	Not in dbSNP	Exon 6	TPR3 domain	In-frame deletion	Not in gnomAD	N/A	[166]	VUS
c.827C > T, p.A276V	rs61741147	Exon 6	TPR3 domain	Missense	Exomes: 0.00281 Genomes: 0.01911 Combined: 0.004634	Very unstable protein with very short half-life [41].	[132]	VUS
c.829G > C, p.A277P	rs267606581	Exon 6	TPR3 domain	Missense	Not in gnomAD	LOH in PitNET tissue [60]. Disrupts hydrophobic packing against Tyr 247 by 3-D modeling [40].	[23,60,155]	VUS
c.831C > T, p.A277 =	rs531331351	Exon 6	TPR3 domain	Synonymous	Exomes: 0.001605	N/A	[16]	Likely benign

Variant	dbSNP	Exon	Domain	Type	gnomAD	Effect	Ref	Classification
c.836G > A, p.W279*	Not in dbSNP	Exon 6	TPR3 domain	Nonsense	Not in gnomAD	N/A	[167]	Pathogenic
c.844C > T, p.Q282*	Not in dbSNP	Exon 6	TPR3 domain	Nonsense	Not in gnomAD	N/A	[25]	Pathogenic
c.853C > T, p.Q285*	Not in dbSNP	Exon 6	TPR3 domain	Nonsense	Not in gnomAD	N/A	[8]	Pathogenic
c.854_857del, p. Q285Lfs*17	rs267606582	Exon 6	TPR3 domain	Frameshift	Not in gnomAD	N/A	[12]	Pathogenic
c.863_864del (p.F288Cfs*?)	Not in dbSNP	Exon 6	TPR3 domain	Frameshift	Not in gnomAD	N/A	[88]	Pathogenic
c.868A > T, p.K290*	Not in dbSNP	Exon 6	TPR3 domain	Nonsense	Not in gnomAD	Premature stop codon, loss of 40 amino acids at the C-terminus [27].	[16,27,88]	Pathogenic
c.871G > A, p.V291M	Not in dbSNP	Exon 6	TPR3 domain	Missense	Not in gnomAD	Disrupts packing of hydrophobic core by 3-D modeling [40]. Very unstable protein with very short half-life [41]. Abolishes AIP interaction with PDE4A5 by 2-H/β-gal [118].	[137]	Likely pathogenic
c.872T > A, p.V291E	rs1179125850	Exon 6	TPR3 domain	Missense	Exomes: 0.0004018	N/A	[19]	VUS
c.872_877del, p. V291_L292del	Not in dbSNP	Exon 6	TPR3 domain	In-frame deletion	Not in gnomAD	May theoretically disrupt the packaging of the C-terminal 7th α-helix [27].	[16,27,88]	VUS
c.878A > G, p.E293G	Not in dbSNP	Exon 6	TPR3 domain	Missense	Not in gnomAD	N/A	[168]	VUS
c.878A > T, p.E293V	Not in dbSNP	Exon 6	TPR3 domain	Missense	Not in gnomAD	Rescues lethality in a KO D. melanogaster model [87].	[87]	Likely benign
c.878_879delinsGT, p. E293G	rs267606583	Exon 6	TPR3 domain	Missense	Not in gnomAD	LOH in PitNET tissue [116].	[116]	VUS
884_895del, D295_L298del	rs267606585	Exon 6	TPR3 domain	In-frame deletion	Not in gnomAD	N/A	[116]	VUS
c.881T > C, p.L294P	Not in dbSNP	Exon 6	TPR3 domain	Missense	Not in gnomAD	N/A	[24]	VUS
c.891C > A, p.A297 =	rs35665586	Exon 6	TPR3 domain	Synonymous	Exomes: 0.1354 Genomes: 0.5448 Combined: 0.1813	N/A	[16,21]	Benign

(Continued)

Table 6.1 Germline *AIP* variants reported in pituitary neuroendocrine tumor patients and their functional effects. *Continued*

HGVS nomenclature DNA, protein Reference sequence: NM_003977.4	dbSNP ID	Location in gene	Location in protein	Variant type	Frequency in gnomAD (%)	Experimental evidence	References for clinical reports	ACMG classification
c.896C > T, p.A299V	rs148986773	Exon 6	Between TPR3 domain and C-terminal α-helix	Missense	Exomes: 0.05605 Genomes: 0.04141 Combined: 0.05441	May disrupt at some degree packing with L292 by 3-D modeling [40]. Reduces interaction with NME1 by pull-down and co-IP assays [47]. Unstable protein with short half-life [41]. Rescues lethality in a KO *D. melanogaster* model [87]. Does not affect AIP interaction with PDE4A5 by 2-H/β-gal [11,118]. LOH in PitNET tissue [128].	[11,16,27,111, 116,128,133,143]	Benign
c.906G > A, p.V302 =	rs142912418	Exon 6	C-terminal α-helix	Synonymous	Exomes: 0.008598 Genomes: 0.009558 Combined: 0.008598	N/A	[16,116,132]	Benign
c.910C > T, p.R304*	rs104894195	Exon 6	C-terminal α-helix	Nonsense	Exomes: 0.001213 Genomes: 0.003188 Combined: 0.001436	LOH in PitNET tissue [33]. Affects PDE4A5 and AHR binding by 3-D modeling [40]. Failed to inhibit cells proliferation in human cell lines [13]. Very unstable protein with very short half-life [41]. Abolishes interaction with HSPA8, NME1, SOD1, TUBB, and TUBB2A by pull-down and co-IP assays [47]. Does not rescue lethality in a KO *D. melanogaster* model [87]. Abolishes AIP interaction with PDE4A5 by 2-H/β-gal [13,118]. Decreased AIP protein expression and increased *AHRR* in patient's fibroblasts [79].	[5,11−13,15, 16,19,21,23, 24,27,33−35, 67,79,88,101, 126,127,143, 146,155,169]	Pathogenic

Variant	dbSNP	Exon	Location	Type	Frequency	Functional studies	References	Classification
c.911G > A, p.R304Q	rs104894190	Exon 6	C-terminal α-helix	Missense	Exomes: 0.1654 Genomes: 0.08918 Combined: 0.1568	Abolishes interaction with NME1 and SOD1 by pull-down and co-IP assays [47]. Weakens PDE4A5 binding by 3-D modeling [40]. Stable protein with normal half-life [41]. Rescues lethality in a KO *D. melanogaster* model [87]. Reduces AIP interaction with PDE4A5 by 2-H/β-gal [11,118].	[11,13,16,19, 21,23–25, 111,116,124, 126–128, 137,139,168,170]	VUS
c.919dup, p.Q307Pfs*?	rs267606589	Exon 6	C-terminal α-helix	Frameshift	Not in gnomAD	N/A	[22]	Pathogenic
c.920G > A, p.R307Q[b]	rs4930199	Exon 6	C-terminal α-helix	Missense	Exomes: 0.0001 Genomes: 0.0002 Combined: 0.0001	N/A	[96,156]	Benign
c.940C > T, p.R314W	rs375740557	Exon 6	C-terminal α-helix	Missense	Exomes: 0.002035	Rescues lethality in a KO *D. melanogaster* model [87].	[122]	Likely benign
c.943C > T, p.Q315*	Not in dbSNP	Exon 6	C-terminal α-helix	Nonsense	Not in gnomAD	LOH in tumor. Disrupted ability to inhibit cell proliferation in vitro [78].	[78]	Pathogenic
c.955G > A, p.E319K	Not in dbSNP	Exon 6	C-terminal α-helix	Missense	Not in gnomAD	N/A	[131]	VUS
c.965C > T, p.A322V	rs267606586	Exon 6	C-terminal α-helix	Missense	Exomes: 0.004203 Genomes: 0.003186 Combined: 0.004085	N/A	[21]	Likely benign
c.967C > T, p.R323W	rs188965257	Exon 6	C-terminal α-helix	Missense	Exomes: 0.004657 Genomes: 0.006373 Combined: 0.004858	LOH in tumor [131].	[131]	Likely benign

(Continued)

Table 6.1 Germline *AIP* variants reported in pituitary neuroendocrine tumor patients and their functional effects. *Continued*

HGVS nomenclature DNA, protein Reference sequence: NM_003977.4	dbSNP ID	Location in gene	Location in protein	Variant type	Frequency in gnomAD (%)	Experimental evidence	References for clinical reports	ACMG classification
c.967del, p.R323Gfs*39	Not in gnomAD	Exon 6	C-terminal α-helix	Frameshift	Not in gnomAD	N/A	[16,88]	Pathogenic
c.973C > A, p.R325 =	rs765927395	Exon 6	C-terminal α-helix	Synonymous	Exomes: 0.007373	N/A	[111]	Benign
c.974G > A, p.R325Q	rs754619109	Exon 6	C-terminal α-helix	Missense	Exomes: 0.005669 Genomes: 0.00956 Combined: 0.006138	LOH in PitNET tissue [171]. One residue short of the 5-residue deletion that disrupts AhR binding by 3-D modeling [40]. Stable protein with normal half-life [41]. Rescues lethality in a KO *D. melanogaster* model [87].	[19,146,171]	Likely benign
c.976G > A, p.G326R	rs1336893319	Exon 6	C-terminal α-helix	Missense	Not in gnomAD	LOH in PitNET tissue [131].	[131]	VUS
c.976_977insC, p.G326Afs*?	Not in dbSNP	Exon 6	C-terminal α-helix	Frameshift	Not in gnomAD	Changes in the last four amino acids, crucial for protein activity in silico [27].	[16,27,88]	Pathogenic
c.978dup, p.I327Dfs*?	Not in dbSNP	Exon 6	C-terminal α-helix	Frameshift	Not in gnomAD	N/A	[16,88]	Pathogenic
c.987C > T, p.S329 =	rs267606587	Exon 6	C-terminal α-helix	Synonymous	Not in gnomAD	N/A	[21]	VUS
c.991T > C, p. *331Rext*?	Not in dbSNP	Exon 6	C-terminal α-helix	Stop loss	Not in gnomAD	N/A	[88,107]	VUS
c.*14C > A, p.?	rs142567224	3′UTR	N/A	Intronic	Exomes: 0.03344 Genomes: 0.08283 Combined: 0.04049	N/A	[111,128]	Likely benign

c.*34C > G, p.?	Not in dbSNP	3'UTR	N/A	Intronic	Not in gnomAD	N/A	[21]	VUS
c.*56C > G, p.?	Not in dbSNP	3'UTR	N/A	Intronic	Not in gnomAD	N/A	[21]	VUS
c.*60G > C, p.?	rs146014363	3'UTR	N/A	Intronic	Genomes: 0.787	N/A	[13,21,23,119, 131,134]	Benign
c.*63C > T, p.?	rs562462428	3'UTR	N/A	Intronic	Not in gnomAD	N/A	[23]	VUS
c.*64G > A, p.?	rs115346238	3'UTR	N/A	Intronic	Genomes: 0.5447	N/A	[128]	Benign
c.*70C > T, p.?	rs868173257	3'UTR	N/A	Intronic	Not in gnomAD	N/A	[137]	VUS
c.*77C > A, p.?	rs778583953	3'UTR	N/A	Intronic	Not in gnomAD	N/A	[13]	VUS

2-H/β-gal, *Two-hybrid assay/β-galactosidase assay*; ACMG, *American College of Medical Genetics and Genomics*; C-terminal, *carboxy-terminal*; co-IP, *co-immunoprecipitation*; HGVS, *Human Genome Variation Society*; N-terminus, *amino-terminus*; N/A, *not available*; PitNET, *pituitary neuroendocrine tumor*; PPIase, *peptidylprolyl isomerase*; TPR, *tetratricopeptide repeat*; UTR, *untranslated region*; VUS, *variant of uncertain significance*.

[a]*The base at this position is a C in the reference sequence, but the major allele in the general population is an A.*

[b]*The base at this position is an A in the reference sequence, but the major allele in the general population is a G.*

6.11 Clinical screening and genetic counseling of *AIP* mutation carriers

Once an *AIP* mutation has been identified in a PitNET patient, cascade screening should be done to include all the potential carriers, based on their pedigree and clinical history. Genetic testing is recommended as early as possible in potential carriers [88]. Since an early diagnosis has a positive impact on the clinical outcomes of patients with PitNETs, directed screening and follow-up of *AIP* pathogenic variant carriers, even if asymptomatic, is strongly recommended. In addition to careful anamnesis and clinical examination, baseline tests should include an MRI and measurement of prolactin, IGF-1 and, in selected cases, OGTT. In children, growth velocity and pubertal development should also be carefully followed up, and MRI can be delayed in asymptomatic individuals until age 10 years. Clinical and biochemical examinations can safely be performed on a yearly basis, and MRI can be repeated every 5 years until at least age 30 years, unless symptoms of pituitary disease appear [16,88]. Clinical screening can be stopped in older individuals, since there is a very low risk of developing *AIP* LOF-associated PitNETs after the fourth decade of life [88]. Around one-quarter of variant carriers screened this way are expected to have pituitary disease [16].

Genetic counseling of variant carriers should provide guidance on the clinical manifestations of pituitary disease and emphasize the need for clinical screening, even though the disease has a relatively low penetrance. PitNETs identified during the clinical screening of *AIP* variant carriers should be managed in accordance with the treatment guidelines established for the specific tumor type in the general population.

6.12 Conclusions

The AIP protein seems to play a crucial role in pituitary tumorigenesis, not only in the setting of *AIP* genetic defects, but also in other PitNETs. Future studies should be directed to fully explain the mechanisms leading to disease penetrance in patients with *AIP* mutations, to define the complete repertoire of signaling pathways involved in *AIP* LOF-associated pituitary tumorigenesis, to identify possible therapeutic targets in *AIP*-deficient tumors, and to characterize the physiological function of circulating AIP. This will hopefully improve the accuracy of the classification of gene variants, with a direct impact on genetic counseling, and will also provide new tools for developing tailored management strategies for the treatment of PitNETs.

References

[1] Valdes-Socin H, Poncin J, Stevens V, Stevenaert A, Beckers A. Familial isolated pituitary adenomas unrelated to MEN1 mutations: a follow-up of 27 patients. 10th Meeting of the Belgian Endocrine Society: Ann Endocrinol (Paris); 2000. p. 343.

[2] Gadelha MR, Prezant TR, Une KN, Glick RP, Moskal SF, Vaisman M, et al. Loss of heterozygosity on chromosome 11q13 in two families with acromegaly/gigantism is independent of mutations of the multiple endocrine neoplasia type I gene. J Clin Endocrinol Metab 1999;84(1):249−56.

[3] Gadelha MR, Une KN, Rohde K, Vaisman M, Kineman RD, Frohman LA. Isolated familial somatotropinomas: establishment of linkage to chromosome 11q13.1-11q13.3 and evidence for a potential second locus at chromosome 2p16-12. J Clin Endocrinol Metab 2000;85(2):707–14.

[4] Frohman LA. Isolated familial somatotropinomas: clinical and genetic considerations. Trans Am Clin Climatol Assoc 2003;114:165–77.

[5] Vierimaa O, Georgitsi M, Lehtonen R, Vahteristo P, Kokko A, Raitila A, et al. Pituitary adenoma predisposition caused by germline mutations in the *AIP* gene. Science 2006;312(5777):1228–30.

[6] Daly AF, Jaffrain-Rea ML, Ciccarelli A, Valdes-Socin H, Rohmer V, Tamburrano G, et al. Clinical characterization of familial isolated pituitary adenomas. J Clin Endocrinol Metab 2006;91(9):3316–23.

[7] Marques NV, Kasuki L, Coelho MC, Lima CHA, Wildemberg LE, Gadelha MR. Frequency of familial pituitary adenoma syndromes among patients with functioning pituitary adenomas in a reference outpatient clinic. J Endocrinol Invest 2017;.

[8] Beckers A, Aaltonen LA, Daly AF, Karhu A. Familial isolated pituitary adenomas (FIPA) and the pituitary adenoma predisposition due to mutations in the aryl hydrocarbon receptor interacting protein (AIP) gene. Endocr Rev 2013;34(2):239–77.

[9] Daly AF, Rixhon M, Adam C, Dempegioti A, Tichomirowa MA, Beckers A. High prevalence of pituitary adenomas: a cross-sectional study in the province of Liege, Belgium. J Clin Endocrinol Metab 2006;91(12):4769–75.

[10] Daly AF, Tichomirowa MA, Petrossians P, Heliovaara E, Jaffrain-Rea ML, Barlier A, et al. Clinical characteristics and therapeutic responses in patients with germ-line *AIP* mutations and pituitary adenomas: an international collaborative study. J Clin Endocrinol Metab 2010;95(11):E373–83.

[11] Igreja S, Chahal HS, King P, Bolger GB, Srirangalingam U, Guasti L, et al. Characterization of aryl hydrocarbon receptor interacting protein *(AIP)* mutations in familial isolated pituitary adenoma families. Hum Mutat 2010;31(8):950–60.

[12] Daly AF, Vanbellinghen JF, Khoo SK, Jaffrain-Rea ML, Naves LA, Guitelman MA, et al. Aryl hydrocarbon receptor-interacting protein gene mutations in familial isolated pituitary adenomas: analysis in 73 families. J Clin Endocrinol Metab 2007;92(5):1891–6.

[13] Leontiou CA, Gueorguiev M, van der Spuy J, Quinton R, Lolli F, Hassan S, et al. The role of the aryl hydrocarbon receptor-interacting protein gene in familial and sporadic pituitary adenomas. J Clin Endocrinol Metab 2008;93(6):2390–401.

[14] Naves LA, Daly AF, Vanbellinghen JF, Casulari LA, Spilioti C, Magalhaes AV, et al. Variable pathological and clinical features of a large Brazilian family harboring a mutation in the aryl hydrocarbon receptor-interacting protein gene. Eur J Endocrinol 2007;157(4):383–91.

[15] Niyazoglu M, Sayitoglu M, Firtina S, Hatipoglu E, Gazioglu N, Kadioglu P. Familial acromegaly due to aryl hydrocarbon receptor-interacting protein *(AIP)* gene mutation in a Turkish cohort. Pituitary 2013;17 (3):220–6.

[16] Hernández-Ramírez LC, Gabrovska P, Dénes J, Stals K, Trivellin G, Tilley D, et al. Landscape of familial isolated and young-onset pituitary adenomas: prospective diagnosis in AIP mutation carriers. J Clin Endocrinol Metab 2015;100(9):E1242–54.

[17] Caimari F, Korbonits M. Novel genetic causes of pituitary adenomas. Clin Cancer Res 2016;22 (20):5030–42.

[18] Barlier A, Vanbellinghen JF, Daly AF, Silvy M, Jaffrain-Rea ML, Trouillas J, et al. Mutations in the aryl hydrocarbon receptor interacting protein gene are not highly prevalent among subjects with sporadic pituitary adenomas. J Clin Endocrinol Metab 2007;92(5):1952–5.

[19] Cazabat L, Bouligand J, Salenave S, Bernier M, Gaillard S, Parker F, et al. Germline *AIP* mutations in apparently sporadic pituitary adenomas: prevalence in a prospective single-center cohort of 443 patients. J Clin Endocrinol Metab 2012;97(4):E663–70.

[20] Oriola J, Lucas T, Halperin I, Mora M, Perales MJ, Alvarez-Escola C, et al. Germline mutations of *AIP* gene in somatotropinomas resistant to somatostatin analogues. Eur J Endocrinol 2013;168(1):9−13.

[21] Cazabat L, Libe R, Perlemoine K, Rene-Corail F, Burnichon N, Gimenez-Roqueplo AP, et al. Germline inactivating mutations of the aryl hydrocarbon receptor-interacting protein gene in a large cohort of sporadic acromegaly: mutations are found in a subset of young patients with macroadenomas. Eur J Endocrinol 2007;157(1):1−8.

[22] Stratakis CA, Tichomirowa MA, Boikos S, Azevedo MF, Lodish M, Martari M, et al. The role of germline *AIP*, *MEN1*, *PRKAR1A*, *CDKN1B* and *CDKN2C* mutations in causing pituitary adenomas in a large cohort of children, adolescents, and patients with genetic syndromes. Clin Genet 2010;78(5):457−63.

[23] Tichomirowa MA, Barlier A, Daly AF, Jaffrain-Rea ML, Ronchi C, Yaneva M, et al. High prevalence of *AIP* gene mutations following focused screening in young patients with sporadic pituitary macroadenomas. Eur J Endocrinol 2011;165(4):509−15.

[24] Cuny T, Pertuit M, Sahnoun-Fathallah M, Daly AF, Occhi G, Odou MF, et al. Genetic analysis in young patients with sporadic pituitary macroadenomas: beside *AIP* don't forget *MEN1* genetic analysis. Eur J Endocrinol 2013;168(4):533−41.

[25] Schofl C, Honegger J, Droste M, Grussendorf M, Finke R, Plockinger U, et al. Frequency of *AIP* gene mutations in young patients with acromegaly: a registry-based study. J Clin Endocrinol Metab 2014;99 (12):E2789−93.

[26] Rostomyan L, Daly AF, Petrossians P, Nachev E, Lila AR, Lecoq AL, et al. Clinical and genetic characterization of pituitary gigantism: an international collaborative study in 208 patients. Endocr Relat Cancer 2015;22(5):745−57.

[27] Ramírez-Rentería C, Hernández-Ramírez LC, Portocarrero-Ortiz L, Vargas G, Melgar V, Espinosa E, et al. AIP mutations in young patients with acromegaly and the Tampico Giant: the Mexican experience. Endocrine 2016;.

[28] Iacovazzo D, Hernández-Ramírez LC, Korbonits M. Sporadic pituitary adenomas: the role of germline mutations and recommendations for genetic screening. Expert Rev Endocrinol Metab 2017;12 (2):143−53.

[29] Georgitsi M, Karhu A, Winqvist R, Visakorpi T, Waltering K, Vahteristo P, et al. Mutation analysis of aryl hydrocarbon receptor interacting protein (*AIP*) gene in colorectal, breast, and prostate cancers. Br J Cancer 2007;96(2):352−6.

[30] Nord KH, Magnusson L, Isaksson M, Nilsson J, Lilljebjorn H, Domanski HA, et al. Concomitant deletions of tumor suppressor genes *MEN1* and *AIP* are essential for the pathogenesis of the brown fat tumor hibernoma. Proc Natl Acad Sci USA 2010;107(49):21122−7.

[31] Georgitsi M, Heliovaara E, Paschke R, Kumar AV, Tischkowitz M, Vierimaa O, et al. Large genomic deletions in *AIP* in pituitary adenoma predisposition. J Clin Endocrinol Metab 2008;93(10):4146−51.

[32] Ozfirat Z, Korbonits M. *AIP* gene and familial isolated pituitary adenomas. Mol Cell Endocrinol 2010;326(1−2):71−9.

[33] Occhi G, Jaffrain-Rea ML, Trivellin G, Albiger N, Ceccato F, De ME, et al. The R304X mutation of the aryl hydrocarbon receptor interacting protein gene in familial isolated pituitary adenomas: mutational hot-spot or founder effect? J Endocrinol Invest 2010;33(11):800−5.

[34] Chahal HS, Stals K, Unterlander M, Balding DJ, Thomas MG, Kumar AV, et al. *AIP* mutation in pituitary adenomas in the 18th century and today. N Engl J Med 2011;364(1):43−50.

[35] Radian S, Diekmann Y, Gabrovska P, Holland B, Bradley L, Wallace H, et al. Increased population risk of AIP-related acromegaly and gigantism in Ireland. Hum Mutat 2017;38(1):78−85.

[36] Salvatori R, Radian S, Diekmann Y, Iacovazzo D, David A, Gabrovska P, et al. In-frame seven amino-acid duplication in AIP arose over the last 3000 years, disrupts protein interaction and stability and is associated with gigantism. Eur J Endocrinol 2017;177(3):257−66.

[37] Ma Q, Whitlock Jr. JP. A novel cytoplasmic protein that interacts with the Ah receptor, contains tetratricopeptide repeat motifs, and augments the transcriptional response to 2,3,7,8-tetrachlorodibenzo-p-dioxin. J Biol Chem 1997;272(14):8878−84.

[38] Carver LA, LaPres JJ, Jain S, Dunham EE, Bradfield CA. Characterization of the Ah receptor-associated protein, ARA9. J Biol Chem 1998;273(50):33580−7.

[39] Linnert M, Haupt K, Lin YJ, Kissing S, Paschke AK, Fischer G, et al. NMR assignments of the FKBP-type PPIase domain of the human aryl-hydrocarbon receptor-interacting protein (AIP). Biomol NMR Assign 2012;6(2):209−12.

[40] Morgan RM, Hernández-Ramírez LC, Trivellin G, Zhou L, Roe SM, Korbonits M, et al. Structure of the TPR domain of AIP: Lack of client protein interaction with the C-terminal alpha-7 helix of the TPR domain of AIP is sufficient for pituitary adenoma predisposition. PLoS One 2012;7(12):e53339.

[41] Hernández-Ramírez LC, Martucci F, Morgan RM, Trivellin G, Tilley D, Ramos-Guajardo N, et al. Rapid proteasomal degradation of mutant proteins is the primary mechanism leading to tumorigenesis in patients with missense *AIP* mutations. J Clin Endocrinol Metab 2016;101(8):3144−54.

[42] Kuzhandaivelu N, Cong YS, Inouye C, Yang WM, Seto E. XAP2, a novel hepatitis B virus X-associated protein that inhibits X transactivation. Nucleic Acids Res 1996;24(23):4741−50.

[43] Cunningham F, Amode MR, Barrell D, Beal K, Billis K, Brent S, et al. Ensembl 2015. Nucleic Acids Res 2015;43(Database issue):D662−9.

[44] Chen B, Liu P, Hujber EJ, Li Y, Jorgensen EM, Wang ZW. AIP limits neurotransmitter release by inhibiting calcium bursts from the ryanodine receptor. Nat Commun 2017;8(1):1380.

[45] Bell DR, Poland A. Binding of aryl hydrocarbon receptor (AhR) to AhR-interacting protein. The role of hsp90. J Biol Chem 2000;275(46):36407−14.

[46] Yano M, Terada K, Mori M. AIP is a mitochondrial import mediator that binds to both import receptor Tom20 and preproteins. J Cell Biol 2003;163(1):45−56.

[47] Hernández-Ramírez LC, Morgan RML, Barry S, D'Acquisto F, Prodromou C, Korbonits M. Multi-chaperone function modulation and association with cytoskeletal proteins are key features of the function of AIP in the pituitary gland. Oncotarget 2018;9(10):9177−98.

[48] Heliovaara E, Raitila A, Launonen V, Paetau A, Arola J, Lehtonen H, et al. The expression of AIP-related molecules in elucidation of cellular pathways in pituitary adenomas. Am J Pathol 2009;175(6):2501−7.

[49] Trivellin G, Korbonits M. AIP and its interacting partners. J Endocrinol 2011;210(2):137−55.

[50] Schernthaner-Reiter MH, Trivellin G, Stratakis CA. Interaction of AIP with protein kinase A (cAMP-dependent protein kinase). Hum Mol Genet 2018;27(15):2604−13.

[51] Perdew GH. Association of the Ah receptor with the 90-kDa heat shock protein. J Biol Chem 1988;263(27):13802−5.

[52] Chen HS, Perdew GH. Subunit composition of the heteromeric cytosolic aryl hydrocarbon receptor complex. J Biol Chem 1994;269(44):27554−8.

[53] Nair SC, Toran EJ, Rimerman RA, Hjermstad S, Smithgall TE, Smith DF. A pathway of multi-chaperone interactions common to diverse regulatory proteins: estrogen receptor, Fes tyrosine kinase, heat shock transcription factor Hsf1, and the aryl hydrocarbon receptor. Cell Stress Chaperones 1996;1(4):237−50.

[54] Carver LA, Bradfield CA. Ligand-dependent interaction of the aryl hydrocarbon receptor with a novel immunophilin homolog in vivo. J Biol Chem 1997;272(17):11452−6.

[55] Meyer BK, Pray-Grant MG, Vanden Heuvel JP, Perdew GH. Hepatitis B virus X-associated protein 2 is a subunit of the unliganded aryl hydrocarbon receptor core complex and exhibits transcriptional enhancer activity. Mol Cell Biol 1998;18(2):978−88.

[56] Kazlauskas A, Sundstrom S, Poellinger L, Pongratz I. The hsp90 chaperone complex regulates intracellular localization of the dioxin receptor. Mol Cell Biol 2001;21(7):2594−607.

[57] Ramadoss P, Petrulis JR, Hollingshead BD, Kusnadi A, Perdew GH. Divergent roles of hepatitis B virus X-associated protein 2 (XAP2) in human versus mouse Ah receptor complexes. Biochemistry 2004;43 (3):700−9.

[58] Sun D, Stopka-Farooqui U, Barry S, Aksoy E, Parsonage G, Vossenkamper A, et al. Aryl hydrocarbon receptor interacting protein maintains germinal center B cells through suppression of BCL6 degradation. Cell Rep 2019;27(5):1461−71 e4.

[59] Bolger GB, Peden AH, Steele MR, MacKenzie C, McEwan DG, Wallace DA, et al. Attenuation of the activity of the cAMP-specific phosphodiesterase PDE4A5 by interaction with the immunophilin XAP2. J Biol Chem 2003;278(35):33351−63.

[60] Jaffrain-Rea ML, Angelini M, Gargano D, Tichomirowa MA, Daly AF, Vanbellinghen JF, et al. Expression of aryl hydrocarbon receptor (AHR) and AHR-interacting protein in pituitary adenomas: pathological and clinical implications. Endocr Relat Cancer 2009;16(3):1029−43.

[61] Stojanovic M, Wu Z, Stiles CE, Miljic D, Soldatovic I, Pekic S, et al. Circulating aryl hydrocarbon receptor-interacting protein (AIP) is independent of GH secretion. Endocr Connect 2019;8(4):326−37.

[62] Hernández-Ramírez LC, Trivellin G, Stratakis CA. Role of phosphodiesterases on the function of aryl hydrocarbon receptor-interacting protein (AIP) in the pituitary gland and on the evaluation of AIP gene variants. Horm Metab Res 2017;49(4):286−95.

[63] Kirschner LS. *PRKAR1A* and the evolution of pituitary tumors. Mol Cell Endocrinol 2010;326 (1−2):3−7.

[64] Formosa R, Xuereb-Anastasi A, Vassallo J. Aip regulates cAMP signalling and GH secretion in GH3 cells. Endocr Relat Cancer 2013;20(4):495−505.

[65] Tuominen I, Heliovaara E, Raitila A, Rautiainen MR, Mehine M, Katainen R, et al. AIP inactivation leads to pituitary tumorigenesis through defective Galphai-cAMP signaling. Oncogene 2015;34(9):1174−84.

[66] Ritvonen E, Pitkanen E, Karppinen A, Vehkavaara S, Demir H, Paetau A, et al. Impact of AIP and inhibitory G protein alpha 2 proteins on clinical features of sporadic GH-secreting pituitary adenomas. Eur J Endocrinol 2017;176(2):243−52.

[67] Denes J, Kasuki L, Trivellin G, Colli LM, Takiya CM, Stiles CE, et al. Regulation of aryl hydrocarbon receptor interacting protein (*AIP*) protein expression by MiR-34a in sporadic somatotropinomas. PLoS One 2015;10(2):e0117107.

[68] Bogner EM, Daly AF, Gulde S, Karhu A, Irmler M, Beckers J, et al. miR-34a is upregulated in AIP-mutated somatotropinomas and promotes octreotide resistance. Int J Cancer 2020;.

[69] Gadelha MR, Kasuki L, Korbonits M. Novel pathway for somatostatin analogs in patients with acromegaly. Trends Endocrinol Metab 2013;24(5):238−46.

[70] Chahal HS, Trivellin G, Leontiou CA, Alband N, Fowkes RC, Tahir A, et al. Somatostatin analogs modulate AIP in somatotroph adenomas: the role of the ZAC1 pathway. J Clin Endocrinol Metab 2012;97(8):E1411−20.

[71] Jaffrain-Rea ML, Rotondi S, Turchi A, Occhi G, Barlier A, Peverelli E, et al. Somatostatin analogues increase AIP expression in somatotropinomas, irrespective of Gsp mutations. Endocr Relat Cancer 2013;20(5):753−66.

[72] Kasuki Jomori de PL, Vieira NL, Armondi Wildemberg LE, Gasparetto EL, Marcondes J, de Almeida NB, et al. Low aryl hydrocarbon receptor-interacting protein expression is a better marker of invasiveness in somatotropinomas than Ki-67 and p53. Neuroendocrinology 2011;94(1):39−48.

[73] Kasuki L, Neto L, Wildemberg LEA, Colli LM, de Castro M, Takiya CM, et al. AIP expression in sporadic somatotropinomas is a predictor of the response to octreotide LAR therapy independent of SSTR2 expression. Endocr Relat Cancer 2012;19(3):L25−9.

[74] Kasuki L, Colli LM, Elias PC, Castro M, Gadelha MR. Resistance to octreotide LAR in acromegalic patients with high SSTR2 expression: analysis of AIP expression. Arq Bras Endocrinol Metab 2012;56 (8):501−6.

[75] Iacovazzo D, Carlsen E, Lugli F, Chiloiro S, Piacentini S, Bianchi A, et al. Factors predicting pasireotide responsiveness in somatotroph pituitary adenomas resistant to first-generation somatostatin analogues: an immunohistochemical study. Eur J Endocrinol 2016;174(2):241−50.

[76] Pagotto U, Arzberger T, Theodoropoulou M, Grubler Y, Pantaloni C, Saeger W, et al. The expression of the antiproliferative gene ZAC is lost or highly reduced in nonfunctioning pituitary adenomas. Cancer Res 2000;60(24):6794−9.

[77] Theodoropoulou M, Zhang J, Laupheimer S, Paez-Pereda M, Erneux C, Florio T, et al. Octreotide, a somatostatin analogue, mediates its antiproliferative action in pituitary tumor cells by altering phosphatidylinositol 3-kinase signaling and inducing Zac1 expression. Cancer Res 2006;66(3):1576−82.

[78] Iwata T, Yamada S, Ito J, Inoshita N, Mizusawa N, Ono S, et al. A novel C-terminal nonsense mutation, Q315X, of the aryl hydrocarbon receptor-interacting protein gene in a Japanese familial isolated pituitary adenoma family. Endocr Pathol 2014;25(3):273−81.

[79] Lecoq AL, Viengchareun S, Hage M, Bouligand J, Young J, Boutron A, et al. AIP mutations impair AhR signaling in pituitary adenoma patients fibroblasts and in GH3 cells. Endocr Relat Cancer 2016; 23(5):433−43.

[80] Barry S, Carlsen E, Marques P, Stiles CE, Gadaleta E, Berney DM, et al. Tumor microenvironment defines the invasive phenotype of AIP-mutation-positive pituitary tumors. Oncogene 2019;38(27):5381−95.

[81] Lin BC, Sullivan R, Lee Y, Moran S, Glover E, Bradfield CA. Deletion of the aryl hydrocarbon receptor-associated protein 9 leads to cardiac malformation and embryonic lethality. J Biol Chem 2007;282(49):35924−32.

[82] Lin BC, Nguyen LP, Walisser JA, Bradfield CA. A hypomorphic allele of aryl hydrocarbon receptor-associated protein-9 produces a phenocopy of the AHR-null mouse. Mol Pharmacol 2008; 74(5):1367−71.

[83] Raitila A, Lehtonen HJ, Arola J, Heliovaara E, Ahlsten M, Georgitsi M, et al. Mice with inactivation of aryl hydrocarbon receptor-interacting protein (Aip) display complete penetrance of pituitary adenomas with aberrant ARNT expression. Am J Pathol 2010;177(4):1969−76.

[84] Lecoq AL, Zizzari P, Hage M, Decourtye L, Adam C, Viengchareun S, et al. Mild pituitary phenotype in 3- and 12-month-old Aip-deficient male mice. J Endocrinol 2016;231(1):59−69.

[85] Schernthaner-Reiter MH, Trivellin G, Roetzer T, Hainfellner JA, Starost MF, Stratakis CA. Prkar1a haploinsufficiency ameliorates the growth hormone excess phenotype in Aip-deficient mice. Hum Mol Genet 2020;29(17):2951−61.

[86] Gillam MP, Ku CR, Lee YJ, Kim J, Kim SH, Lee SJ, et al. Somatotroph-specific Aip-deficient mice display pretumorigenic alterations in cell-cycle signaling. J Endocr Soc 2017;1(2):78−95.

[87] Aflorei ED, Klapholz B, Chen C, Radian S, Dragu AN, Moderau N, et al. In vivo bioassay to test the pathogenicity of missense human AIP variants. J Med Genet 2018;55(8):522−9.

[88] Marques P, Caimari F, Hernandez-Ramirez LC, Collier D, Iacovazzo D, Ronaldson A, et al. Significant benefits of AIP testing and clinical screening in familial isolated and young-onset pituitary tumors. J Clin Endocrinol Metab 2020;105(6).

[89] Dutta P, Reddy KS, Rai A, Madugundu AK, Solanki HS, Bhansali A, et al. Surgery, octreotide, temozolomide, bevacizumab, radiotherapy, and pegvisomant treatment of an AIP mutation positive child. J Clin Endocrinol Metab 2019;104(8):3539−44.

[90] Villa C, Lagonigro MS, Magri F, Koziak M, Jaffrain-Rea ML, Brauner R, et al. Hyperplasia-adenoma sequence in pituitary tumorigenesis related to aryl hydrocarbon receptor interacting protein gene mutation. Endocr Relat Cancer 2011;18(3):347−56.

[91] Xekouki P, Mastroyiannis SA, Avgeropoulos D, de la Luz SM, Trivellin G, Gourgari EA, et al. Familial pituitary apoplexy as the only presentation of a novel *AIP* mutation. Endocr Relat Cancer 2013;20(5): L11−14.

[92] Toledo RA, Mendonca BB, Fragoso MC, Soares IC, Almeida MQ, Moraes MB, et al. Isolated familial somatotropinoma: 11q13-loh and gene/protein expression analysis suggests a possible involvement of *AIP* also in non-pituitary tumorigenesis. Clinics (Sao Paulo) 2010;65(4):407−15.

[93] Guaraldi F, Corazzini V, Gallia GL, Grottoli S, Stals K, Dalantaeva N, et al. Genetic analysis in a patient presenting with meningioma and familial isolated pituitary adenoma (FIPA) reveals selective involvement of the R81X mutation of the *AIP* gene in the pathogenesis of the pituitary tumor. Pituitary 2012;15(Suppl. 1):S61−7.

[94] Coopmans EC, Muhammad A, Daly AF, de Herder WW, van Kemenade FJ, Beckers A, et al. The role of AIP variants in pituitary adenomas and concomitant thyroid carcinomas in the Netherlands: a nation-wide pathology registry (PALGA) study. Endocrine 2020;68(3):640−9.

[95] Personnier C, Cazabat L, Bertherat J, Gaillard S, Souberbielle JC, Habrand JL, et al. Clinical features and treatment of pediatric somatotropinoma: case study of an aggressive tumor due to a new AIP mutation and extensive literature review. Horm Res Paediatr 2011;75(6):392−402.

[96] Mangupli R, Rostomyan L, Castermans E, Caberg JH, Camperos P, Krivoy J, et al. Combined treatment with octreotide LAR and pegvisomant in patients with pituitary gigantism: clinical evaluation and genetic screening. Pituitary 2016;19(5):507−14.

[97] Daly A, Rostomyan L, Betea D, Bonneville JF, Villa C, Pellegata NS, et al. AIP-mutated acromegaly resistant to first-generation somatostatin analogs: long-term control with pasireotide LAR in two patients. Endocr Connect 2019.

[98] Carty DM, Harte R, Drummond RS, Ward R, Magid K, Collier D, et al. AIP variant causing familial prolactinoma. Pituitary 2020;.

[99] Vroonen L, Jaffrain-Rea ML, Petrossians P, Tamagno G, Chanson P, Vilar L, et al. Prolactinomas resistant to standard doses of cabergoline: a multicenter study of 92 patients. Eur J Endocrinol 2012;167 (5):651−62.

[100] Salenave S, Ancelle D, Bahougne T, Raverot G, Kamenicky P, Bouligand J, et al. Macroprolactinomas in children and adolescents: factors associated with the response to treatment in 77 patients. J Clin Endocrinol Metab 2015;100(3):1177−86.

[101] Williams F, Hunter S, Bradley L, Chahal HS, Storr HL, Akker SA, et al. Clinical experience in the screening and management of a large kindred with familial isolated pituitary adenoma due to an aryl hydrocarbon receptor interacting protein (AIP) mutation. J Clin Endocrinol Metab 2014;99(4):1122−31.

[102] Lu JQ, Adam B, Jack AS, Lam A, Broad RW, Chik CL. Immune cell infiltrates in pituitary adenomas: more macrophages in larger adenomas and more T cells in growth hormone adenomas. Endocr Pathol 2015;26(3):263−72.

[103] Korbonits M, Storr H, Kumar AV. Familial pituitary adenomas - who should be tested for *AIP* mutations? Clin Endocrinol (Oxf) 2012;77(3):351−6.

[104] Lecoq AL, Kamenicky P, Guiochon-Mantel A, Chanson P. Genetic mutations in sporadic pituitary adenomas—what to screen for? Nat Rev Endocrinol 2015;11(1):43−54.

[105] Marques P, Barry S, Ronaldson A, Ogilvie A, Storr HL, Goadsby PJ, et al. Emergence of pituitary adenoma in a child during surveillance: clinical challenges and the family members' view in an aip mutation-positive family. Int J Endocrinol 2018;2018:8581626.

[106] Beckers A, Petrossians P, Hanson J, Daly AF. The causes and consequences of pituitary gigantism. Nat Rev Endocrinol 2018;14(12):705−20.

[107] Caimari F, Hernández-Ramírez LC, Dang MN, Gabrovska P, Iacovazzo D, Stals K, et al. Risk category system to identify pituitary adenoma patients with AIP mutations. J Med Genet 2018;55(4):254−60.

[108] Korbonits M, Kumar AV. AIP-related familial isolated pituitary adenomas. In: Adam MP, Ardinger HH, Pagon RA, Wallace SE, Bean LJH, Stephens K, et al., editors. GeneReviews. Seattle (WA): University of Washington; 1993. Available from: https://www.ncbi.nlm.nih.gov/books/NBK97965/.

[109] Richards S, Aziz N, Bale S, Bick D, Das S, Gastier-Foster J, et al. Standards and guidelines for the interpretation of sequence variants: a joint consensus recommendation of the American College of Medical Genetics and Genomics and the Association for Molecular Pathology. Genet Med 2015;17 (5):405−24.

[110] Soares BS, Eguchi K, Frohman LA. Tumor deletion mapping on chromosome 11q13 in eight families with isolated familial somatotropinoma and in 15 sporadic somatotropinomas. J Clin Endocrinol Metab 2005;90(12):6580−7.

[111] Lecoq AL, Bouligand J, Hage M, Cazabat L, Salenave S, Linglart A, et al. Very low frequency of germline GPR101 genetic variation and no biallelic defects with AIP in a large cohort of patients with sporadic pituitary adenomas. Eur J Endocrinol 2016;174(4):523−30.

[112] Hage C, Sabini E, Alsharhan H, Fahrner JA, Beckers A, Daly A, et al. Acromegaly in the setting of Tatton-Brown-Rahman syndrome. Pituitary 2019;.

[113] Formosa R, Vassallo J. Aryl hydrocarbon receptor-interacting protein (AIP) N-terminus gene mutations identified in pituitary adenoma patients alter protein stability and function. Horm Cancer 2017;8(3):174−84.

[114] Yu R, Bonert V, Saporta I, Raffel LJ, Melmed S. Aryl hydrocarbon receptor interacting protein variants in sporadic pituitary adenomas. J Clin Endocrinol Metab 2006;91(12):5126−9.

[115] Salvatori R, Daly AF, Quinones-Hinojosa A, Thiry A, Beckers A. A clinically novel AIP mutation in a patient with a very large, apparently sporadic somatotrope adenoma. Endocrinol Diabetes Metab Case Rep 2014;2014:140048.

[116] Georgitsi M, Raitila A, Karhu A, Tuppurainen K, Makinen MJ, Vierimaa O, et al. Molecular diagnosis of pituitary adenoma predisposition caused by aryl hydrocarbon receptor-interacting protein gene mutations. Proc Natl Acad Sci USA 2007;104(10):4101−5.

[117] Raitila A, Georgitsi M, Karhu A, Tuppurainen K, Makinen MJ, Birkenkamp-Demtroder K, et al. No evidence of somatic aryl hydrocarbon receptor interacting protein mutations in sporadic endocrine neoplasia. Endocr Relat Cancer 2007;14(3):901−6.

[118] Bolger GB, Bizzi MF, Pinheiro SV, Trivellin G, Smoot L, Accavitti MA, et al. cAMP-specific PDE4 phosphodiesterases and AIP in the pathogenesis of pituitary tumors. Endocr Relat Cancer 2016;23 (5):419−31.

[119] Buchbinder S, Bierhaus A, Zorn M, Nawroth PP, Humpert P, Schilling T. Aryl hydrocarbon receptor interacting protein gene (AIP) mutations are rare in patients with hormone secreting or non-secreting pituitary adenomas. Exp Clin Endocrinol Diabetes 2008;116(10):625−8.

[120] Raitila A, Georgitsi M, Bonora E, Vargiolu M, Tuppurainen K, Makinen MJ, et al. Aryl hydrocarbon receptor interacting protein mutations seem not to associate with familial non-medullary thyroid cancer. J Endocrinol Invest 2009;32(5):426−9.

[121] Guaraldi F, Salvatori R. Familial isolated pituitary adenomas: from genetics to therapy. Clin Transl Sci 2011;4(1):55−62.

[122] Baciu I, Capatina C, Aflorei D, Botusan I, Coculescu M, Radian S. Screening of AIP mutations in young Romanian patients with sporadic pituitary adenomas. In: E. Abstracts, editor. 15th International & 14th European Congress of Endocrinology 05−09 May 2012; Florence, Italy; 2012. p. P786.

[123] Zatelli MC, Torre ML, Rossi R, Ragonese M, Trimarchi F, degli UE, et al. Should aip gene screening be recommended in family members of FIPA patients with R16H variant? Pituitary 2013;16(2):238−44.

[124] Preda V, Korbonits M, Cudlip S, Karavitaki N, Grossman AB. Low rate of germline AIP mutations in patients with apparently sporadic pituitary adenomas before the age of 40: a single-centre adult cohort. Eur J Endocrinol 2014;171(5):659−66.

[125] Dinesen PT, Dal J, Gabrovska P, Gaustadnes M, Gravholt CH, Stals K, et al. An unusual case of an ACTH-secreting macroadenoma with a germline variant in the aryl hydrocarbon receptor-interacting protein (AIP) gene. Endocrinol Diabetes Metab Case Rep 2015;2015:140105.

[126] Ferrau F, Romeo PD, Puglisi S, Ragonese M, Torre ML, Scaroni C, et al. Analysis of GPR101 and AIP genes mutations in acromegaly: a multicentric study. Endocrine 2016;54(3):762−7.

[127] Cannavo S, Ragonese M, Puglisi S, Romeo PD, Torre ML, Alibrandi A, et al. Acromegaly is more severe in patients with AHR or AIP gene variants living in highly polluted areas. J Clin Endocrinol Metab 2016;101(4):1872−9.

[128] Araujo PB, Kasuki L, de Azeredo Lima CH, Ogino L, Camacho AHS, Chimelli L, et al. AIP mutations in Brazilian patients with sporadic pituitary adenomas: a single-center evaluation. Endocr Connect 2017;6(8):914−25.

[129] Kasuki Jomori de Pinho L, Vieira Neto L, Wildemberg LE, Moraes AB, Takiya CM, Frohman LA, et al. Familial isolated pituitary adenomas experience at a single center: clinical importance of AIP mutation screening. Arq Bras Endocrinol Metab 2010;54(8):698−704.

[130] Pestell RG, Alford FP, Best JD. Familial acromegaly. Acta Endocrinol (Copenh) 1989;121(2):286−9.

[131] Cai F, Zhang YD, Zhao X, Yang YK, Ma SH, Dai CX, et al. Screening for *AIP* gene mutations in a Han Chinese pituitary adenoma cohort followed by LOH analysis. Eur J Endocrinol 2013;169(6):867−84.

[132] Rowlands JC, Urban JD, Wikoff DS, Budinsky RA. An evaluation of single nucleotide polymorphisms in the human aryl hydrocarbon receptor-interacting protein (*AIP*) gene. Drug Metab Pharmacokinet 2011;26(4):431−9.

[133] Daly A, Cano DA, Venegas E, Petrossians P, Dios E, Castermans E, et al. AIP and MEN1 mutations and AIP immunohistochemistry in pituitary adenomas in a tertiary referral center. Endocr Connect 2019;8(4):338−48.

[134] Iwata T, Yamada S, Mizusawa N, Golam HM, Sano T, Yoshimoto K. The aryl hydrocarbon receptor-interacting protein gene is rarely mutated in sporadic GH-secreting adenomas. Clin Endocrinol (Oxf) 2007;66(4):499−502.

[135] Joshi K, Daly AF, Beckers A, Zacharin M. Resistant paediatric somatotropinomas due to AIP mutations: role of pegvisomant. Horm Res Paediatr 2018;90(3):196−202.

[136] Gummadavelli A, Dinauer C, McGuone D, Vining EM, Erson-Omay EZ, Omay SB. Large-scale second-hit AIP deletion causing a pediatric growth hormone-secreting pituitary adenoma: case report and review of literature. J Clin Neurosci 2020;78:420−2.

[137] Occhi G, Trivellin G, Ceccato F, De LP, Giorgi G, Dematte S, et al. Prevalence of *AIP* mutations in a large series of sporadic Italian acromegalic patients and evaluation of *CDKN1B* status in acromegalic patients with multiple endocrine neoplasia. Eur J Endocrinol 2010;163(3):369−76.

[138] Ceccato F, Occhi G, Albiger NM, Rizzati S, Ferasin S, Trivellin G, et al. Adrenal lesions in acromegaly: do metabolic aspects and aryl hydrocarbon receptor interacting protein gene have a role? Evaluation at baseline and after long-term follow-up. J Endocrinol Invest 2011;34(5):353−60.

[139] Mian C, Ceccato F, Barollo S, Watutantrige-Fernando S, Albiger N, Regazzo D, et al. AHR overexpression in papillary thyroid carcinoma: clinical and molecular assessments in a series of Italian acromegalic patients with a long-term follow-up. PLoS One 2014;9(7):e101560.

[140] Matsumoto R, Izawa M, Fukuoka H, Iguchi G, Odake Y, Yoshida K, et al. Genetic and clinical characteristics of Japanese patients with sporadic somatotropinoma. Endocr J 2016;63(11):953−63.

[141] Luccio-Camelo DC, Une KN, Ferreira RE, Khoo SK, Nickolov R, Bronstein MD, et al. A meiotic recombination in a new isolated familial somatotropinoma kindred. Eur J Endocrinol 2004;150(5):643−8.

[142] Yamada S, Yoshimoto K, Sano T, Takada K, Itakura M, Usui M, et al. Inactivation of the tumor suppressor gene on 11q13 in brothers with familial acrogigantism without multiple endocrine neoplasia type 1. J Clin Endocrinol Metab 1997;82(1):239−42.

[143] De Sousa SM, McCabe MJ, Wu K, Roscioli T, Gayevskiy V, Brook K, et al. Germline variants in familial pituitary tumour syndrome genes are common in young patients and families with additional endocrine tumours. Eur J Endocrinol 2017;176(5):635−44.

[144] Fajardo-Montañana C, Daly AF, Riesgo-Suarez P, Gomez-Vela J, Tichomirowa MA, Camara-Gomez R, et al. AIP mutations in familial and sporadic pituitary adenomas: local experience and review of the literature. Endocrinol Nutr 2009;56(7):369−77.

[145] Foltran RK, Amorim P, Duarte FH, Grande IPP, Freire A, Frassetto FP, et al. Study of major genetic factors involved in pituitary tumorigenesis and their impact on clinical and biological characteristics of sporadic somatotropinomas and non-functioning pituitary adenomas. Braz J Med Biol Res 2018;51(9): e7427.

[146] Ozkaya HM, Comunoglu N, Sayitoglu M, Keskin FE, Firtina S, Khodzhaev K, et al. Germline mutations of aryl hydrocarbon receptor-interacting protein (AIP) gene and somatostatin receptor 1-5 and AIP immunostaining in patients with sporadic acromegaly with poor versus good response to somatostatin analogues. Pituitary 2018;21(4):335−46.

[147] Martucci F, Trivellin G, Khoo B, Owusu-Antwi S, Stals K, Kumar A, et al. Are "in silico" predictions reliable regarding splice-site mutations? − Studies in the aryl hydrocarbon receptor-interacting protein (AIP). Endocr Abstr 2012;1442 Florence.

[148] Martucci F, Trivellin G, Korbonits M. Familial isolated pituitary adenomas: an emerging clinical entity. J Endocrinol Invest 2012;35(11):1003−14.

[149] Khoo SK, Pendek R, Nickolov R, Luccio-Camelo DC, Newton TL, Massie A, et al. Genome-wide scan identifies novel modifier loci of acromegalic phenotypes for isolated familial somatotropinoma. Endocr Relat Cancer 2009;16(3):1057−63.

[150] Rojas García W, Tovar Cortes H, Florez Romero A. Pituitary gigantism: a case series from Hospital de San Jose (Bogota, Colombia). Arch Endocrinol Metab 2019;63(4):385−93.

[151] Cai F, Hong Y, Xu J, Wu Q, Reis C, Yan W, et al. A novel mutation of aryl hydrocarbon receptor interacting protein gene associated with familial isolated pituitary adenoma mediates tumor invasion and growth hormone hypersecretion. World Neurosurg 2019;123:e45−59.

[152] Karaca Z, Taheri S, Tanriverdi F, Unluhizarci K, Kelestimur F. Prevalence of AIP mutations in a series of Turkish acromegalic patients: are synonymous AIP mutations relevant? Pituitary 2015; 18(6):831−7.

[153] Bilbao Garay I, Daly AF, Egana Zunzunegi N, Beckers A. Pituitary disease in AIP mutation-positive familial isolated pituitary adenoma (FIPA): a kindred-based overview. J Clin Med 2020;9(6).

[154] Naves LA, Jaffrain-Rea ML, Vencio SA, Jacomini CZ, Casulari LA, Daly AF, et al. Aggressive prolactinoma in a child related to germline mutation in the ARYL hydrocarbon receptor interacting protein (AIP) gene. Arq Bras Endocrinol Metab 2010;54(8):761−7.

[155] Trarbach EB, Trivellin G, Grande IPP, Duarte FHG, Jorge AAL, do Nascimento FBP, et al. Genetics, clinical features and outcomes of non-syndromic pituitary gigantism: experience of a single center from Sao Paulo, Brazil. Pituitary 2020;.

[156] Yarman S, Ogret YD, Oguz FS. Do the aryl hydrocarbon receptor interacting protein variants (Q228K and Q307R) play a role in patients with familial and sporadic hormone-secreting pituitary adenomas? Genet Test Mol Biomarkers 2015;19(7):394−8.

[157] Urbani C, Russo D, Raggi F, Lombardi M, Sardella C, Scattina I, et al. A novel germline mutation in the aryl hydrocarbon receptor-interacting protein (AIP) gene in an Italian family with gigantism. J Endocrinol Invest 2014;37(10):949−55.

[158] Nozieres C, Berlier P, Dupuis C, Raynaud-Ravni C, Morel Y, Borson-Chazot F, et al. Sporadic and genetic forms of paediatric somatotropinoma: a retrospective analysis of seven cases and a review of the literature. Orphanet J Rare Dis 2011;6(1):67.

[159] Georgitsi M, De ME, Cannavo S, Makinen MJ, Tuppurainen K, Pauletto P, et al. Aryl hydrocarbon receptor interacting protein (*AIP*) gene mutation analysis in children and adolescents with sporadic pituitary adenomas. Clin Endocrinol (Oxf) 2008;69(4):621−7.

[160] Nishizawa H, Fukuoka H, Iguchi G, Inoshita N, Yamada S, Takahashi Y. AIP mutation identified in a patient with acromegaly caused by pituitary somatotroph adenoma with neuronal choristoma. Exp Clin Endocrinol Diabetes 2013;121(5):295−9.

[161] Jorge BH, Agarwal SK, Lando VS, Salvatori R, Barbero RR, Abelin N, et al. Study of the multiple endocrine neoplasia type 1, growth hormone-releasing hormone receptor, Gs alpha, and Gi2 alpha genes in isolated familial acromegaly. J Clin Endocrinol Metab 2001;86(2):542−4.

[162] Toledo RA, Lourenco Jr. DM, Liberman B, Cunha-Neto MB, Cavalcanti MG, Moyses CB, et al. Germline mutation in the aryl hydrocarbon receptor interacting protein gene in familial somatotropinoma. J Clin Endocrinol Metab 2007;92(5):1934−7.

[163] Prescott RWG, Spruce BA, Kendall-Taylor P, Hall K, Hall R. Acromegaly and gigantism presenting in two brothers. 1st Joint Mtg Brit Endocr Soc; 1982. 1982.

[164] McCarthy MI, Noonan K, Wass JA, Monson JP. Familial acromegaly: studies in three families. Clin Endocrinol (Oxf) 1990;32(6):719−28.

[165] Jennings JE, Georgitsi M, Holdaway I, Daly AF, Tichomirowa M, Beckers A, et al. Aggressive pituitary adenomas occurring in young patients in a large Polynesian kindred with a germline R271W mutation in the *AIP* gene. Eur J Endocrinol 2009;161(5):799−804.

[166] Belar O, De la Hoz C, Pérez-Nanclares G, Castaño L, Gaztambide S. Novel mutations in *MEN1*, *CDKN1B* and *AIP* genes in patients with multiple endocrine neoplasia type 1 syndrome in Spain. Clin Endocrinol (Oxf) 2012;76(5):719−24.

[167] Cansu GB, Taskiran B, Trivellin G, Faucz FR, Stratakis CA. A novel truncating AIP mutation, p.W279*, in a familial isolated pituitary adenoma (FIPA) kindred. Hormones (Athens) 2016;15(3):441−4.

[168] Vargiolu M, Fusco D, Kurelac I, Dirnberger D, Baumeister R, Morra I, et al. The tyrosine kinase receptor RET interacts in vivo with aryl hydrocarbon receptor-interacting protein to alter survivin availability. J Clin Endocrinol Metab 2009;94(7):2571−8.

[169] de Lima DS, Martins CS, Paixao BM, Amaral FC, Colli LM, Saggioro FP, et al. SAGE analysis highlights the putative role of underexpression of ribosomal proteins in GH-secreting pituitary adenomas. Eur J Endocrinol 2012;167(6):759−68.

[170] Tuncer FN, Dogansen SC, Serbest E, Tanrikulu S, Ekici Y, Bilgic B, et al. Screening of AIP gene variations in a cohort of Turkish patients with young-onset sporadic hormone-secreting pituitary adenomas. Genet Test Mol Biomarkers 2018;22(12):702−8.

[171] García-Arnés JA, González-Molero I, Oriola J, Mazuecos N, Luque R, Castaño J, et al. Familial isolated pituitary adenoma caused by a *Aip* gene mutation not described before in a family context. Endocr Pathol 2013;24(4):234−8.

The 3PAs syndrome and succinate dehydrogenase deficiency in pituitary tumors

7

Paraskevi Xekouki[1], Vasiliki Daraki[1], Grigoria Betsi[1], Maria Chrysoulaki[1], Maria Sfakiotaki[1], Maria Mytilinaiou[1] and Constantine A. Stratakis[2]

[1]*Department of Endocrinology, Diabetes and Metabolism, University Hospital of Heraklion, School of Medicine, University of Crete, Heraklion, Greece* [2]*Section on Endocrinology and Genetics, Eunice Kennedy Shriver National Institute of Child Health and Human Development (NICHD), Bethesda, MD, United States*

7.1 Introduction

Pituitary adenomas (PAs) are benign neoplasms that are found in up to 20% of pituitaries on MRI or autopsy [1], while clinically relevant pituitary adenomas occur in approximately 1:1000 people [2]. A number of oncogenes, tumor suppressor genes and cell cycle mediators, adhesion molecules, and microRNAs have been identified to be functionally involved in the initiation and progression of pituitary adenomas [3,4]; however, the initial trigger of the tumorigenic cascade remains largely unknown. Some of the genetic defects implicated in PA development, either isolated or as part of a syndrome, are the aryl hydrocarbon receptor-interacting protein *(AIP)* gene in familial isolated pituitary adenomas, the *menin* responsible for *MEN1* syndrome, the cyclin-dependent kinase inhibitor 1B *(CDKN1B)* causing *MEN4* syndrome, Xq26.3 duplications involving the gene *GPR101* which has been demonstrated in X-linked acrogigantism, the regulatory subunit of protein kinase A (PRKAR1A) in Carney Complex, the *DICER1* gene in DICER1 syndrome, the ubiquitin-specific protease 8 *(USP8)*, and the most recently discovered *CABLES1* in Cushing's disease [4,5]. In this review, we describe the role of succinate dehydrogenase (SDH) in pituitary tumor development, and we present all the available clinical and molecular data of the 3P [pheochromocytoma (PHEO) and/or paraganglioma (PGL), and pituitary adenoma] association (3PAs).

7.2 The role of SDH in physiology and disease

Succinate dehydrogenase (SDH), also named complex II or succinate:quinone oxidoreductases located in the inner mitochondrial membrane, is positioned at the intersection of oxidative phosphorylation (OXPHO) and tricarboxylic acid cycle, the essential pathways of bioenergetics in many organisms. SDH comprises four subunits *SDHA*-D, which are localized to the inner mitochondrial and cytoplasmic membranes in eukaryotes and prokaryotes, respectively. The catalytic domains (*SDHA* and *SDHB*) are extrinsic on the matrix side, and together they form the main catalytic

Gigantism and Acromegaly. DOI: https://doi.org/10.1016/B978-0-12-814537-1.00001-4

domain that oversees the oxidation of succinate to fumarate. The anchor subunits (*SDHC* and *SDHD*) are intrinsic transmembrane proteins, allowing the transfer of the electrons from succinate in the mitochondrial matrix to ubiquinone in the inner membrane. In addition, at least two other proteins, *SDHAF1* and *SDHAF2*, support the function of SDH by flavinating the *SDHA* subunit and promoting the maturation of *SDHB*, thus enabling the assembly of the full SDH complex [6] (Fig. 7.1A).

SDHx genes were the first to be recognized as tumor suppressor genes encoding a mitochondrial enzyme [7]. The concept of aerobic glycolysis or the "Warburg effect" was first reported by Otto Warburg in 1926: it proposed that even in the presence of sufficient oxygen, malignant cells prefer to produce adenosine triphosphate via glycolysis instead of OXPHOs, and this has mainly resulted from mitochondrial dysfunction [8]. Although the theory of "aerobic glycolysis" was recently challenged by many investigators for some types of tumors, this metabolic dysregulation is in fact recognized as one of the hallmarks of cancer and has undoubtedly provided a new avenue to

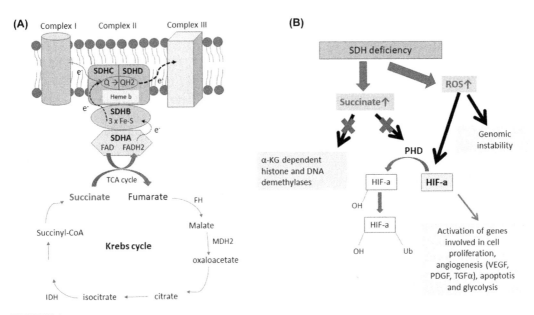

FIGURE 7.1

(A) The structure of succinate dehydrogenase (SDH) complex. The catalytic subunit A contains the flavin cofactor (FAD) that accepts electrons from succinate and passes them to Fe-S center in the B subunit. The electrons are then passed to the ubiquinone pool embedded in *SDHC* and *SDHD* subunits. Reduced Q (QH2 = ubiquinol) transfers electrons within the mitochondrial inner membrane space to complex III. (B) Mutations in any of SDH subunits could cause accumulation of succinate and the production of ROS. Both succinate and ROS could lead to the induction of hypoxic response under normoxic conditions (pseudohypoxia). In addition to pseudohypoxia, succinate might inhibit PDH1–3, which plays a role in the degradation of HIF-1 under normoxic conditions. HIF-1, when stabilized, induces the transcription of nuclear genes involved in glucose metabolism, angiogenesis, and apoptosis. ROS production inhibits the activity of PHDs, which facilitates HIF catabolism while the accumulated ROS results in oxidative damage to DNA and genomic instability. *HIF*, Hypoxia inducible factor; *PDGF*, platelet-derived growth factor; *PDH*, prolyl hydroxylase; *ROS*, reactive oxygen species; *VEGF*, vascular endothelial growth factor.

oncologic research and clinical medicine, such as the use of PET/CT as a diagnostic method in clinical oncology [9].

It has been 20 years since the first mutations in the *SDHD* and *SDHB* genes in patients with PHEO and/or PGL were identified, leading to a new era regarding the pathogenesis of these two types of neuroendocrine tumors [10,11]. To date, germline and somatic mutations in more than 17 genes have been identified in PHEOs and PGLs, and research has shown that ~40% of patients carry a causal germline mutation [12]. The list of SDH-mutant tumors has slowly expanded in the past 17 years and now include, except from SDH-deficient PGLs/PHEOs, renal cell carcinoma (RCC) [13], gastrointestinal stromal tumor (GIST) (either sporadic cases or in the context of Carney triad and/or Carney−Stratakis syndrome or dyad) [14,15], and pituitary adenomas (PAs) [16] as we describe in details in the next section.

In addition to GISTs, RCCs, and PAs, other tumors have been reported to be associated with PHEO/PGL in *SDHx* germline mutation carriers such as in thyroid cancer [17,18] although this was argued later [19], in pancreatic neuroendocrine tumor [20], in patients with Cowden-like syndrome [21] and in lymphoid malignancies (this needs further confirmation) [22].

The pathogenetic mechanism by which SDH deficiency leads to tumorigenesis is related to the accumulation of succinate in the tumor cells [23], which is the hallmark of tumorigenesis of these tumors. This accumulation inhibits several α-ketoglutarate dioxygenases and leads to the stabilization of hypoxia inducible factors (HIFs) even in the presence of oxygen (pseudohypoxic response) and activation of genes containing HIF-response elements, many of them involved in glycolysis, angiogenesis, and epithelial-to-mesenchymal transition [24−27]. Moreover, SDH deficiency or loss may lead to overproduction of reactive oxygen species (ROS), which further stabilizes HIFs and causes genomic instability [28,29] and an hypermethylator phenotype that plays an important role in furthering tumorigenesis [30,31] (Fig. 7.1B).

Loss of SDH complex activity can be detected by the corresponding loss of *SDHB* immunostaining. This was shown to significantly correlate with a germline mutation in *SDH* or *SDHAF2*; hence, this immunohistochemical test can be used as a screening method to guide genetic testing [32,33].

7.3 *SDHx* mutations in pituitary adenomas and a new syndromic association (3PAs)

In 2012 we described a family with multiple PGLs and PHEOs caused by a germline *SDHD* mutation; the index case also had an aggressive growth hormone (GH)−secreting pituitary adenoma (PA) [16]. We found loss of heterozygosity (LOH) at the *SDHD* locus in the pituitary tumor, along with increased HIF-1α levels at the pituitary tumor tissue, indicating that the *SDHD* defect was most likely implicated in the development of the pituitary tumor and that pseudohypoxia pathways were activated, as shown in PHEOs/PGLs with *SDHx* mutations [24,34]. We then studied the prevalence of *SDHx* mutations in 146 patients with sporadic and 22 patients with familial PA. We showed that *SDHx* mutations were rare (1.8% of the studied cases) in sporadic PAs. However, among the 22 familial cases included in the cohort, 4 were found to also have a PHEO and/or PGL, and 3 of the 4 were found to harbor *SDHx* defects (75%). We named this coexistence 3PAs from the acronyms of pituitary adenoma, PHEO, and PGL. Interestingly, in our data, when 3PAs was found in the sporadic setting, no *SDHx* defects were identified [35].

The very low frequency of *SDHx* mutations in PAs was also confirmed by Gill et al. who found only one *SDHA*-negative PA as the result of a double *SDHA* somatic mutation among 309 pituitary adenomas studied [36] and Papathomas et al. [19] who did not find the loss of *SDHB* protein in any of 41 PAs from patients with no family history.

Our data regarding 3PAs association were reproduced in a different cohort of patients with PAs by Dénes et al., who studied 39 patients with sporadic or familial PHEO/PGL and/or PAs. They identified nine patients with *SDHx* mutations or variants (including the assembly factor *SDHAF2)* within an international cohort of 19 patients with both tumors, which accounted for 47.4% of the affected subjects. Similar to our data, all cases were found within families with familial PHEOs/PGLs [37]. Interestingly, in addition to mutations in SDH subunits, they identified two cases who presented with PHEO/PGLs and PAs and harbored variants in the *VHL* or *MEN1* genes [37]. The patient with the *VHL* mutation (the first with 3PAs reported in the literature) had an invasive GH- and prolactin (PRL)-positive PA and was also found to harbor an *SDHA* variant. Considering the frequency with which patients with Von Hippel-Lindau (VHL) syndrome undergo regular surveillance imaging of the brain, this sole case probably does not represent a true predisposition for PA or that this *VHL* variant together with the *SDHA* variant may have increased susceptibility to PA development. On the other hand, the LOH of *MEN1* locus identified in the PHEO tissue of the patient who presented with components of *MEN1* syndrome (GH and PRL secreting PA, hyperparathyroidism, and carcinoid tumor) indicates that PHEOs and/or PGLs can be part of the *MEN1* syndrome and genetic testing for *menin* mutations should be considered in patients with PHEO/PGL, if there are other suggestive signs of *MEN1*.

Two years ago, the 3PAs syndrome was associated with germline pathogenic variants or intragenic deletions in *MYC-associated factor X (MAX)* (one of the predisposing to PHEO genes) in five patients who presented with PAs (three prolactinomas and two somatotropinomas) and bilateral PHEOs [38–40]. Immunohistochemical staining of the PHEOs demonstrated the loss of nuclear staining for *MAX;* however, the PA tissue (where available) was not tested for LOH of *MAX* locus. Daly et al. recommended that multiplex ligation-dependent probe amplification (MLPA) should be considered in such syndromic cases and in PHEO cases where a genetic cause is sought, as individual *MAX* exon deletions are generally not identifiable on Sanger sequencing [38]. Most recently, Guerrero-Pérez et al. [41] studied retrospectively 10 cases of 3PAs; 6 of them were new and unreported, whereas 4 patients were previously reported [42,43]. In accordance with the previously presented data, all familial cases of 3PAs had an *SDHx* mutation, whereas, among the sporadic cases, two were found to have a *RET* variant of unknown significance (VUS) (germline or somatic), which may indicate that *RET* may be another genetic suspect in 3PAs context.

In Tables 7.1 and 7.2, we summarize all 3PAs cases (90 in total) reported in the literature (with and without genetic mutations/variants, respectively) since 1952. Thirty-seven (41.1%) of these cases harbored germline mutations in predisposing PHEOs/PGLs or PA genes; one case had a *RET* somatic VUS. Twenty-six patients (28.9%) of all 3PAs cases had a personal or family history suggestive of a hereditary endocrine syndrome, whereas forty-one of all cases (45.6%) were isolated; for the rest 25.5%, no information regarding family history was available. Among the 90 3PAs cases described so far, 20.9% had both identified genetic mutations and family history. This coexistence was much higher (50%) if we consider only the cases with a genetic defect. Regarding the frequency of the identified genetic defects, it is obvious that most cases carry *SDHx* defects (21 out of 37 cases, 50%), whereas *MEN1* and *MAX* are the second and third most common genes involved (13.51% and 16.21%, respectively).

Table 7.1 Patients with 3PAs with identified germline genetic mutations.

Patient no	Sex	Pituitary				PHEO/PGL				Mutation	LOH/IHC in PA or PHEO	References
		Type	Size	Treatment	Age	Type	Treatment	Age	Family history			
1	F	PRL	NK	NK	27	PHEO	NK	NK	No	SDHA c.91C>T p.Arg31Ter, VHL c.589G>A p.Asp197Asn	Not performed	Dénes et al. [37]
2	F	PRL	Macro	NK	49	PGLs	NK	49	NK	SDHA p.Arg31* c.91C>T	SDHA/SDHB negative staining	Niemeijer et al. [20]
3	M	GH	Macro	SSA	84	PGL	No	84	No	SDHAF2 c.−52T>C	Not performed	Dénes et al. [37]
4	M	PRL	Macro	DA, TSS	33	PGL	Surgery	33	Mother: PRL, brother: PGL	SDHB c.298T>C p.Ser100Pro	LOH at SDBH locus, intracytoplasmic vacuoles	Dénes et al. [37]
5	F	NFPA	Macro	TSS ×3, RT	53	PGL	RT	28	Sister: glioma	SDHB c.587G>A p.Cys196Tyr	LOH at SDBH locus/SDHB staining: diffuse/intracytoplasmic vacuoles	Dénes et al. [37]
6	F	PRL	Macro	DA	38	PGLs carotid and mediastinal	Carotid: surgery, mediastinal: inoperable	35	Brother: PGLs	SDHB mutation, (actual genetic defect not available)	Not performed	Gorospe et al. [44]
7	F	PRL	Macro	DA, RT	60	PGL	RT	60	NK	SDHB c.423+1G>A	Not performed	Dénes et al. [37]
8	F	NFPA	Micro	No	50	PHEO	Surgery	50	NK	SDHB c.770dupT p.Asn258GlufsTer17	Not performed	Dénes et al. [37]
9	M	GH	NK	SSA	72	PGL	No	70	Brother and niece: PA, sister: bilateral HNPGL	SDHB c.689G>A p.Arg230His	Not performed	Xekouki et al. [35]
10	F	PRL	Micro	NK	50	PGL	NK	47	Brother: HNPGL, grandmother: GIST	SDHB c.642+1G>A, splice site alteration	Not performed	Xekouki et al. [35]
11	F	PRL	Micro	DA	33	PGL	Surgery	33	Brain tumor	SDHB c.18C>A p.Ala6Ala[a] 3 PTEN polymorphisms	Not performed	Efstathiadou et al. [45]
12	F	PRL	Macro	DA	38	PGL	SSA	38	Brother index case: PGL. Mother and sister: positive for region of ex. 1 of SDHB	Deletion affecting ex. 1 of SDHB	Not performed	Guerrero-Pérez et al. [43]

(Continued)

Table 7.1 Patients with 3PAs with identified germline genetic mutations. *Continued*

Patient no	Sex	Pituitary				PHEO/PGL				Mutation	LOH/IHC in PA or PHEO	References
		Type	Size	Treatment	Age	Type	Treatment	Age	Family history			
13	M	PRL	Macro	DA	53	PGL	Surgery	38	Cousin: PA, brother: PGL	SDHC c.380A > G p.His127Arg	Not performed	Dénes et al. [37]
14	F	PRL	Macro	NK	60	PGL	NK	60	No	SDHC c.256-257insTTTpPhe85dup	Not performed	López-Jiménez et al. [46]
15	F	PRL	Macro	TSS, DA	23	PGL	Surgery	32	Sister, aunt and grandmother: PA; sister: bilateral HNPGL	SDHD c.242C > T, p.Pro81Leu	Not performed	Xekouki et al. [35]
16	M	PRL	Macro	DA, TSS	60	PGL, PHEO	Surgery (PHEO)	62	NK	SDHD c.274G > T pAsp92Tyr	LOH at SDHD locus/SDHB positive, IHC SDHA IHC positive	Papathomas et al. [19]
17	F	GH	Macro	TSS, SSA	56	PGL	NK	56	Father and two sisters: HNPGL; sister: GIST	SDHD c.274G > T p.Asp92Tyr	LOH at SDHD locus/SDHB positive, IHC SDHA IHC positive	Papathomas et al. [19]
18	F	PRL	Macro	DA, TSS	33	PGL	Surgery × 2	39	Aunt, uncle, brother: HNPGL	SDHD c.242C > T p.Pro81Leu	Not performed	Varsavsky et al. [42]
19	M	GH	Macro	SSA, TSS	37	PGL, PHEO	Surgery	37	Sister and paternal uncle: neck PGLs HNPGL	SDHD c.298_30del, premature stop at codon 133 AIP and CDKN1B polymorphism	PA: LOH at SDHD locus, reduced SDHD protein/patchy SDHB staining	Xekouki et al. [16]
20	M	NFPA	Macro	TSS	45	Zuckerkandl organ PGL	Surgical resection	40	Twin brother died due to a metastatic PHEO	SDHB mutation (c.166-170delCCTCA)	Not performed	Guerrero-Pérez et al. [41]
21	F	PRL	Macro	DA	36	PGL	Surgery	50	Brother diagnosed with cervical PGL	SDHD p.P81L exon 3	Not performed	Guerrero-Pérez et al. [41]
22	M	GH/PRL	Macro	TSS, RT, DA	27	PHEO	Surgery	31	No	MEN1 c.1452delG p.Thr557Ter	Menin staining of the PHEO: no menin positive cells	Dénes et al. [37]

Table 7.1 Patients with 3PAs with identified germline genetic mutations. *Continued*

Patient no	Sex	Pituitary				PHEO/PGL			Family history	Mutation	LOH/IHC in PA or PHEO	References
		Type	Size	Treatment	Age	Type	Treatment	Age				
23	F	NFPA	Macro	Surveillance	45	PGL	NK	45	No	*MEN1* c.196_200dupAG-CCC frameshift (pathogenic), polymorphism C423T no amino acid change	Not performed	Jeong et al. [47]
24	M	PRL	NK	NK	41	PHEO	Surgery	48	NK	*MEN1* K119X, *RET* WT	Not performed	Langer et al. [48]
25	NK	NK	NK	NK	NK	PHEO	NK	NK	Hyperparathyroidism and pancreatic islet cell tumor	*MEN1* c.320del2	Not performed	Dackiw et al. [49]
26	NK	NK	NK	NK	NK	PHEO	NK	NK	Pancreatic islet cell tumor and rectal leiomyoma	*MEN1* 1325insA	Not performed	Dackiw et al. [49]
27	M	PRL	NK	NK	29	PHEO	NK	32	MEN1	Not performed, other family members *MEN1* mutation	Not performed	Carty et al. [50]
28	M	GH	Macro	TSS, SSA	54	Bilateral PHEO	Bilateral adrenalectomy	54	Yes	MEN1 germline VUS (c.1618C > T; p.Pro540Ser)	Not performed	Guerrero-Pérez et al. [41]
29	F	PRL	Macro	DA	49	PHEO	Bilateral adrenalectomy	49	No	*MAX* c.296G > T, Neg for *MEN1, VHL, SDHB, SDHC, SDHD, SDHAF2,* or *TMEM127* genes.	Not performed in PA/LOH in PHEO	Roszko et al. [39]
30	M	PRL	Micro	DA	49	PHEO	Surgery	32	No	Germinal heterozygous deletion of exon 3 of *MAX* (detected by MLPA). Neg for *RET, VHL, SDHx, CDKN1B, AIP, MENI*	Not performed in PA/LOH in PHEO	Daly et al. [38]
31	F	GH	Macro	SSA, DA, PGV, RT	26	Bilateral PHEOs	Bilateral adrenalectomy	35	No	Germinal heterozygous deletion of exons 1–3 and intron 3 of *MAX*. Neg for *RET, VHL, SDHx, CDKN1B, AIP, MENI*	Not performed in PA/LOH in PHEO	Daly et al. [38]

(Continued)

Table 7.1 Patients with 3PAs with identified germline genetic mutations. *Continued*

Patient no	Sex	Pituitary				PHEO/PGL						References
		Type	Size	Treatment	Age	Type	Treatment	Age	Family history	Mutation	LOH/IHC in PA or PHEO	
32	M	GH	Macro	TSS, RT	16	Bilateral PHEOs	Bilateral adrenalectomy	22	No	Germinal heterozygous deletion of exon 3 of MAX (detected by MLPA). Neg for RET, VHL, SDHx, CDKN1B, AIP, MEN1	Not performed in PA/LOH in PHEO	Daly et al. [38]
33	F	PRL	Micro	DA	33	Bilateral PHEO	Bilateral adrenalectomy	39	No	MAX c.171 + 2T > A		Kobza et al. [40]
34	M	GH	Macro	TSS	62	PHEO	Surgery	62	No	RET p.Cys618Ser	Not performed	Heinlen et al. [51]
35	M	ACTH	Micro	TSS × 2	48	PHEO	Surgery	66	Son: HPTH	RET c.1900T > C, p.Cys634Arg. Neg for MEN1 mutations	Not performed	Naziat et al. [52]
36	M	GH	Macro	TSS, SSA, DA	73	PGL (neck)	None	84	No	RET germline VUS: c.2556C > G, p.Ile852Met (*)	SDHB positive	Guerrero-Pérez et al. [41]
37	F	GH	Micro	TSS	59	Right PHEO	Right adrenalectomy	56	No	RET M918T somatic mutation No germline findings	Not performed	Guerrero-Pérez et al. [41]

A, Single nucleotide polymorphism with a minor allele frequency of 0.2% and a genotype frequency of 0.5% (1000 Genomes Project Consortium, 2012); ACTH, adrenocorticotropic hormone; DA, dopamine agonist; F, female; GH, growth hormone; GIST, gastrointestinal stromal tumor; HNPGL, head and neck paraganglioma; HPTH, hyperparathyroidism; IHC, immunohistochemistry; LOH, loss of heterozygosity; M, male; Macro, macroadenoma; MEN1, multiple endocrine neoplasia type 1; Micro, microadenoma; Neg, negative; NFPA, nonfunctional pituitary adenoma; NK, not known; PGL, paraganglioma; PGV, pegvisomant; PHEO, pheochromocytoma; PRL, prolactinoma; PTEN, phosphatase and tensin homolog; RT, radiotherapy; SSA, somatostatin analog; TSS, transsphenoidal surgery.

Table 7.2 Patients with three PAs without identified germline genetic mutations.

Patient no	Sex	Pituitary				PHEO/PGL				Mutation	Other info (when reported)	References
		Type	Size	Treatment	Age	Type	Treatment	Age	Family history			
1	M	GH	Micro	TSS	60	PHEO	Right adrenalectomy	56	No	Neg for RET, VHL, NF1, MAX, TMEM127, SDHA-D, SDHAF2, MDH2, FH, EPAS1, HRAS	None	Guerrero-Pérez et al. [41]
2	F	GH	Micro	TSS	54	PHEO	Left adrenalectomy	63	No	Neg for RET, VHL, NF1, MAX, TMEM127, SDHA-D, SDHAF2, MDH2, FH, EPAS1, HRAS, KIF1B, MEN1, EGLN1, AIP, CDKN1, EGLN	None	Guerrero-Pérez et al. [41]
3	F	GH	No sellar lesion on MRI	SSA, DA, PGV	58	PHEO	Left adrenalectomy	58	No	Neg for RET, VHL, NF1, MAX, TMEM127, SDHA-D, SDHAF2, MDH2, FH, EPAS1, HRAS, KIF1B, MEN1, EGLN1, AIP, CDKN1, EGLN	Multinodular goiter	Guerrero-Pérez et al. [41]
4	F	PRL	Micro	DA	NK	Vagal PGL	Surgery	NK	No	Neg for RET, VHL, NF1, MAX, TMEM127, SDHA-D, SDHAF2, MDH2, FH, EPAS1, HRAS, KIF1B, MEN1, EGLN1, AIP, CDKN1, EGLN	None	Guerrero-Pérez et al. [41]
5	M	PRL	Macro	DA	38	PHEO/PGL	α-adrenergic blockade	NK	NK	Neg for MEN	Catecholamine-mediated cardiac toxicity	Koshy et al. [53]
6	F	ACTH	Micro	TSS	61	PGL	Surveillance	61	No	Negative for SDHA-D, MEN1, RET, AIP	Bilateral HNPGL	Xekouki et al. [35]
7	F	PRL	Macro	DA, TSS	35	PHEO	Surgery	55	No	Negative for SDHA-D, MEN1, RET, AIP	Bilateral PHEO	Xekouki et al. [35]
8	F	GH	Macro	TSS	35	PGL	Surgery	58	No	Negative for SDHA-D, MEN1, RET, AIP	Bladder PGL	Xekouki et al. [35]

(*Continued*)

Table 7.2 Patients with three PAs without identified germline genetic mutations. *Continued*

Patient no	Pituitary					PHEO/PGL			Family history	Mutation	Other info (when reported)	References
	Sex	Type	Size	Treatment	Age	Type	Treatment	Age				
9	F	NFPA	Macro	TSS	39	PHEO	Surgery	34	No	Negative for SDHA-D, MEN1, RET, AIP		Xekouki et al. [35]
10	F	GH	Macro	TSS, RT, DA, SSA	56	PHEO	Surgery	66	No	Negative for SDHA-D, AF2, MEN1, RET, AIP, VHL, TMEM127, MAX, FH, CDKN1B	GIST, thyroid follicular adenoma	Boguszewski et al. [54], Dénes et al. [37]
11	M	NFPA	Macro	TSS	53	PGL	Surgery	50	Father: PA	SDHA c.969C>T p.Gly323Gly[a] SDHB-D, AF2, MEN1, RET, AIP, VHL, TMEM127, MAX, FH, CDKN1B all normal	Abdominal PGL Wilms' tumor, liposarcoma, renal oncocytoma; PA: no LOH at SDHA locus, intracytoplasmic vacuoles, SDHA and B staining preserved	Dénes et al. [37]
12	F	GH	Macro	TSS, RT, SSA	39	PHEO	Surgery	20	No	Negative for SDHA-D, AF2, MEN1, RET, AIP, VHL, TMEM127, MAX, FH, CDKN1B		Dénes et al. [37]
13	F	NFPA	Macro	TSS, RT	73	PGL	RT	73	No	Negative for SDHA-D, AF2, MEN1, RET, AIP, VHL, TMEM127, MAX, FH, CDKN1B	HNPGL	Dénes et al. [37]
14	M	GH	Macro	Infracted	16	PHEO	NK	16	No	Negative for SDHA-D, AF2, MEN1, RET, AIP, VHL, TMEM127, MAX, FH, CDKN1B		Dénes et al. [37]
15	M	PRL	Macro	TSS	40	PGL	NK	52	No	Negative for SDHA-D, AF2, MEN1, RET, AIP, VHL, TMEM127, MAX, FH, CDKN1B	HNPGL	Dénes et al. [37]
16	F	PRL	NK	NK	27	PHEO	NK	41	No	Negative for SDHA-D, AF2, MEN1, RET, AIP, VHL, TMEM127, MAX, FH, CDKN1B		Dénes et al. [37]
17	M	NK	NK	NK	NK	PHEO/PGL	NK	NK	No	Negative for SDHA-D, AF2, MEN1, RET, AIP, VHL, TMEM127, MAX, FH, CDKN1B		Dénes et al. [37]

Table 7.2 Patients with three PAs without identified germline genetic mutations. *Continued*

Patient no	Sex	Pituitary				PHEO/PGL			Family history	Mutation	Other info (when reported)	References
		Type	Size	Treatment	Age	Type	Treatment	Age				
18	F	PRL	Micro	DA	40	PHEO	Surgery	38	No	Negative for SDHA-D, AF2, MEN1, RET, AIP, VHL, TMEM127, MAX, FH, CDKN1B		Dénes et al. [37]
19	M	PRL	Micro	DA	56	PHEO	Surgery	56	No	Negative for SDHA-D, AF2, MEN1, RET, AIP, VHL, TMEM127, MAX, FH, CDKN1B		Dénes et al. [37]
20	F	PRL	Macro	DA	61	PHEO	Surgery	61	No	Negative for SDHA-D, AF2, MEN1, RET, AIP, VHL, TMEM127, MAX, FH, CDKN1B		Dénes et al. [37]
21	F	PRL	Macro	NK	60	PGL	RT	60	No	Negative for SDHB		Parghane et al. [55]
22	F	NFPA	Macro	No	52	PHEO	Surgery	52	No	Negative for SDHA-D, AF2, RET, MAX, TMEM127, VHL	GHRH-secreting PHEO	Mumby et al. [56]
23	M	GH	Macro	TSS	29	PHEO	Surgery	29	No	Not performed	Bilateral PHEO lipoma, metastatic PTC	Sisson et al. [57]
24	NK	GH	NK	NK	NK	PHEO	NK	NK	MEN1	Not performed	Bilateral PHEO HPTH, pNET Clinical features NF1	Gatta-Cherifi et al. [58]
25	M	GH	Macro	TSS	45	PGL, PHEO	Surgery × 3	54	Father HNPGL, sister: adrenal abnormality	Not performed	Abdominal, HN, cardiac PGLs	Zhang et al. [59]
26	M	NFPA	Micro	No	64	PHEO	Surgery	64	No	Not performed	High cortisol (cured postadrenalectomy)	Yaylali et al. [60]
27	M	NFPA	Macro	TSS	59	PHEO	Surgery	59	No	Not performed		Breckenridge et al. [61]
28	M	NFPA	Macro	TSS	56	PHEO	Nil	56	No	Not performed		Dünser et al. [62]
29	F	GH	Macro	TSS	57	PHEO	Surgery	57	No	Not performed		Sleilati et al. [63]

(Continued)

Table 7.2 Patients with three PAs without identified germline genetic mutations. *Continued*

Patient no	Sex	Pituitary				PHEO/PGL			Family history	Mutation	Other info (when reported)	References
		Type	Size	Treatment	Age	Type	Treatment	Age				
30	M	NK	Micro	No	43	PHEO	Surgery	43	No	Negative for RET	Lipoma, pectus excavatum, pleomorphic parotid adenoma GH levels responded to OGTT postadrenalectomy	Baughan et al. [64]
31	F	NK	Micro	No	44	PHEO	Surgery	44	No	Not performed	Cushing's (cured postadrenalectomy)	Khalil et al. [65]
32	M	GH	Macro	TSS	41	PHEO, PGLs	Surgery	20	NK	Negative for RET		Teh et al. [66]
33	M	PRL	Macro	TSS	20	PGL	Surgery	20	NK	Not performed	HNPGL	Azzarelli et al. [67]
34	M	PRL	Macro	DA	26	PHEO	Surgery	26	Father: metastatic MTC and probably PHEO	Not performed		Bertrand et al. [68]
35	F	NFPA	NK	NK	70	PGL	NK	70	Daughter and granddaughter: PA, bilateral HNPGL	Not performed	HPTH, PTC, gastric leiomyoma, amyloidosis	Larraza-Hernandez et al. [69]
36	F	PRL	Micro	NK	35	PHEO	NK	35	NK	Not performed		Meyers [70]
37	M	NFPA	Macro	TSS	66	PGL	Surgery	63	NK	Not performed		Blumenkopf and Boekelheide [71]
38	F	GH	NK	TSS, RT	53	PHEO	Surgery	53	Brother: hypertension	Not performed		Anderson et al. [72]
39	F	GH	NK	RT	33	PHEO	None	33	No	Not performed	Multinodular goiter	Anderson et al. [72]
40	F	GH	Macro	NK	53	PHEO	NK	53	NK	Not performed	HPTH	Myers and Eversman [73]
41	F	PRL	NK	NK	23	PHEO	NK	23	NK	Not performed	HPTH, gastrinoma, adrenal adenoma	Alberts et al. [74]
42	F	NK	Macro	NK	22	PHEO	NK	22	Granddaughter: unilateral PHEO	Not performed	Islet cell tumor/renal adenoma	Janson et al. [75]

Table 7.2 Patients with three PAs without identified germline genetic mutations. *Continued*

Patient no	Sex	Pituitary				PHEO/PGL				Mutation	Other info (when reported)	References
		Type	Size	Treatment	Age	Type	Treatment	Age	Family history			
43	F	GH	Macro	NK	15	PHEO	NK	15	NK	Not performed	HPTH	Manger and Glifford [76]
44	F	NFPA	NK	NK	49	PHEO	None	49	NK	Not performed	Papillary carcinoma of thyroid	Melicow [77]
45	M	GH	NK	NK	21	PHEO	NK	44	NK	Not performed		Kadowaki et al. [78]
46	F	GH	NK	NK	19	PGL	NK	19	NK	Not performed	PGL (HN, pelvis), HPTH	Farhi et al. [79]
47	F	GH	NK	NK	36	PGL	NK	36	NK	Not performed	HNPGL, HPTH, hyperplasia of antral and duodenal gastrin cells	Berg et al. [80]
48	F	NK	Micro	NK	43	PHEO	NK	43	NK	Not performed	MTC	Wolf et al. [81]
49	F	GH	NK	RT	36	PHEO	Surgery	36	NK	Not performed	Toxic nodular, goiter, endometriosis, and diabetes, DM	Miller and Wynn [82]
50	M	ACTH	NK	NK	NK	PHEO	Surgery	41	VI generations with MEN	Not performed	MTC	Steiner et al. [83]
51	M	GH	Macro	RT	23	PHEO	None	41	NK	Not performed		Kahn and Mullon [84]
52	M	GH	NK	NK	NK	PHEO	NK	NK	NK	Not performed		German and Flanigan [85]
53	M	GH	NK	NK	44	PHEO	NK	44	NK	Not performed		Iversen [86]

A, Single nucleotide polymorphism with a frequency of 3.5% (Bayley et al. [87]); ACTH, adrenocorticotropic hormone; DA, dopamine agonist; DM, diabetes mellitus; F, female; GH, growth hormone; GHRH, growth hormone-releasing hormone; GIST, gastrointestinal stromal tumor; HNPGL, head and neck paraganglioma; HPTH, hyperparathyroidism; LOH, loss of heterozygosity; M, male; Macro, macroadenoma; MEN1, multiple endocrine neoplasia type 1; Micro, microadenoma; MTC, medullary thyroid carcinoma; NF1, neurofibromatosis type 1; NFPA, nonfunctional pituitary adenoma; NK, not known; OGTT, oral glucose tolerance test; PGL, paraganglioma; PGV, pegvisomant; PHEO, pheochromocytoma; PHPT, primary hyperparathyroidism; pNET, pancreatic neuroendocrine tumor; PRL, prolactinoma; PTC, papillary thyroid cancer; RT, radiotherapy; SSA, somatostatin analogue; TSS, transsphenoidal surgery.

Interestingly, in four 3PAs cases, *RET* variants were identified. Of course, due to the retrospective nature of this review and the lack of genetic screening in the majority of 3PAs cases, generalizations regarding the overall frequency of the implicated genes should be avoided. However, it is obvious that there is a significant overlap between the different MEN types and an indication that reclassification of MEN syndromes may be considered.

The majority of 53 cases of 3PAs with no known genetic defect (Table 7.2) is prior to the genetic testing era. Therefore we cannot exclude the possibility that they harbor a genetic defect in any of the implicated genes; for some of them and based on the clinical presentation and family history, the genetic defect is obvious. However, in the rest of the cases and particularly where genetic screening was negative, the coexistence of a PA and a PHEO/PGL may indeed represent an extremely rare coincidence due to mutations in cosegregating genetic defects, epigenetic changes, mosaicism, pituitary hyperplasia due to ectopic hypothalamic hormone secretion, or another gene different from the ones reported could be implicated. We also must keep in mind that sequencing screening might have missed promoter, other intronic or genomic defects, or larger deletions that require additional genetic techniques such as MLPA.

Regarding the reported variations in sporadic PAs in Table 7.3, most of them are variants of unknown significance and some of them are predicted to be damaging using the available prediction tools. However, one should perform functional studies to prove their deleterious effect.

7.4 Clinical and pathology characteristics of *SDHx*-deficient pituitary adenomas

We reviewed all reported cases of PAs with *SDHx* mutations/variants in order to identify whether these have different clinical characteristics and course compared to PAs with no mutations, as previously described [19,35,37,92,93]. Unfortunately, not all clinical data were available to make a safe distinction between the PAs as part of 3PAs with and without germline mutations. However, based on the available data from the reported cases, we looked for any difference in phenotypic characteristics between PAs in the context of 3PAs with (Tables 7.1 and 7.4) and without genetic mutations (Tables 7.2 and 7.4) and the isolated PAs with *SDHx* mutations/variants (Tables 7.3 and 7.4). We also attempted to see whether the PAs in 3PAs with *MAX* mutation show any common phenotypic characteristics (although the number is very small).

Overall, PAs in 3PAs with genetic defects were significantly more common among familial cases ($P = .001776$), more frequently macroadenomas secreting PRL or GH, they often led to multiple phenotypes within the same family (somatotropinomas, prolactinomas, and nonfunctioning adenomas) and required more than one modes of treatment ($P = .011872$) (Tables 7.1 and 7.4). Patients with isolated PAs and *SDHx* mutations/variants (Table 7.3) were younger compared to the ones in cases with 3PAs regardless of the presence of a genetic mutation, and they all required surgery (regardless of the size of their tumors) but did not require multiple treatments (Table 7.4). A subanalysis between the isolated PAs with *SDHx* mutations/variants and their counterparts in 3PAs revealed that the latter was found in older patients ($P = .041399$) with positive family history of PA or PHEO/PGL, were more frequently macroadenomas, and required more than one treatment modality, which may suggest that the presence of PHEOs/PGLs may have contributed to the increased size and treatment resistance. Interestingly, in our first report, the GH receptor was found

Table 7.3 *SDHx* variants/mutations in pituitary adenomas.

Patient no	Gene	Sex	Age	Type	Size/ treatment	IHC/LOH	Other tumors/ conditions	Family history	Genetics tested/ prediction	References
1	*SDHB*	F	49	PRL	Micro/NK	*SDHB* positive	PHPT	Aunt: ACTH PA	c.5C > T; p.A2V (deleterious[a]/benign[b])	de Sousa et al. [88]
2	*SDHB*	M	70	PRL	NK	Not studied	None	Daughter 36 years old: prolactinoma	c.24C > T; p.S8S-(rs148738139) (likely benign[a]/possible effect on splice site[b])	de Sousa et al. [88]
3	*SDHC*	F	34	PRL	Micro/NK	*SDHB* positive	None	Brother: PRL PA	c.403G > C; p.E110Q (deleterious[a]/possibly damaging[b])	de Sousa et al. [88]
4	*SDHC*	F	63	GH	NK	*SDHB* positive	Pituitary gangliocytoma, PHPT	No	p.E110Q (deleterious[a]/ possibly damaging[b])	de Sousa et al. [88]
5	*SDHA*	M	62	NFPA	Macro/ TSS	Loss of staining for both *SDHA* and *SDHB*	None	No	c.725_736del/ c.989_990insTA (double hit)	Gill et al. [89]
6	*SDHB*	F	35	PRL	Macro/ TSS	intracytoplasmic vacuoles	None	Mother: prolactinoma, brother: PGL	c.298T > C (damaging mutation)	Dénes et al. [37]
7	*SDHA*	M	30	NFPA	Macro/ TSS	IHC negative for *SDHA* and *SDHB*	None	Mother carotid body paraganglioma	c.1873C > T p.His625Tyr	Dwight et al. [90]
8	*SDHB*	NK	15	NK	NK	NK	NK	NK	ex. 7c.761insC p.254fsX255	Benn et al. [91]
9	*SDHD*	M	12	ACTH	Micro/ TSS	Not studied	Thyroid	No	ex. 2c.53C > T, p. Ala18Val (very rare) SNP (allele frequency <0.001) Benign (PSIC 0.004) (by Polyphen) Deleterious (Score 0.01) (by SHIFT)	Xekouki et al. [35]
10	*SDHB*	F	31	PRL	Macro/ TSS	LOH at *SDHB* locus in the PA/ *SDHB* staining: loss of expression of *SDHB*	None	Grandmother's first cousin PGL	del ex. 6 to 8	Dénes et al. [37]

(Continued)

Table 7.3 *SDHx* variants/mutations in pituitary adenomas. *Continued*

Patient no	Gene	Sex	Age	Type	Size/ treatment	IHC/LOH	Other tumors/ conditions	Family history	Genetics tested/ prediction	References
11	*SDHD*	F	16	ACTH	Micro/ TSS	Not studied	None	No	ex. 2c.149A > G/p. His50Arg (probably damaging (PSIC 0.993) (polyphen) Benign (score 0.48) (SHIFT)	Xekouki et al. [35]
12	*SDHB*	F	14	ACTH	Micro/ TSS	Not studied	None	No	ex. 5c.487T > C, p. Ser163Pro [(reported as potentially pathogenic by Ni et al. (2012)]	Xekouki et al. [35]
13	*SDHB*	M	10	ACTH	Micro/ TSS	Not studied	Brachydactyly/ dysmorphic features	No	ex. 5c.487T > C, p. Ser163Pro [(reported as potentially pathogenic by Ni et al. (2012)]	Xekouki et al. [35]

ACTH, *adrenocorticotropic hormone*; F, *female*; GH, *growth hormone*; IHC, *immunohistochemistry*; LOH, *loss of heterozygosity*; M, *male*; Macro, *macroadenoma*; Micro, *microadenoma*; NFPA, *nonfunctional pituitary adenoma*; NK, *not known*; PA, *pituitary adenoma*; PGL, *paraganglioma*; PHPT, *primary hyperparathyroidism*; PRL, *prolactinoma*; PSIC, *position-specific independent counts*; SHIFT, *Sorting Intolerant from Tolerant*; TSS, *transsphenoidal surgery*.
[a]*ClinVar prediction tool.*
[b]*Mutation Taster.*

Table 7.4 Phenotypic characteristics of the pituitary adenomas with and without mutations in the context of 3PAs or as isolated cases (only cases with the relevant available data are included in the analysis.

	3PAs with relevant gene mutations	3PAs with no genetic mutations	3PAs with SDHx mutations	Isolated PA with SDHx mutations/variants	3PAs with MAX mutations	3Pas with MEN1 mutations
Mean age of PA patients at diagnosis	45.74 ± 15.37	43.29 ± 15.36	47.14 ± 15.2*	33.9 ± 21	34.6 ± 14.46	41.33 ± 11.23
Mean age of PHEO/PGL patients at diagnosis	47.51 ± 15.23	45.42 ± 15.87	48.05 ± 14.49		35.4 ± 9.86	44.66 ± 11.37
Type of adenoma	GH:10 PRL: 19 PRL/GH: 1 NFPA: 4 ACTH: 1	GH:23 PRL: 13 NFPA:10 ACTH: 2	GH: 4 PRL: 14 NFPA:3	GH: 1 PRL: 5 NFPA: 2 ACTH: 4	GH: 2 PRL:3	GH: 1 PRL: 2 PRL/GH: 1 NFPA: 1
Size (%)	Micro: 20 Macro: 80	Micro: 28.29 Macro: 70.27	Micro:15.78 Macro: 84.21	Micro: 60 Macro: 40	Micro: 40 Macro: 60	Macro: 100
No of cases required surgery for PA	15 (51.72%)	20 (57.14%)	7 (41.17%)	8 (100%)	1 (20%)	
>1 treatment for PA	14 (48.27%)	6 (17.14%)**	8 (47.05%)	None	2 (40%)	
Type of mutation detected	Missense: 57.14% Insertion/duplication: 17.14% Deletion: 22.85% Frameshift: 2.85%		Missense: 65% Insertion/duplication: 15% Deletion: 15% Frameshift: 5%	Missense: 76% Insertion/duplication: 7.69% Deletion: 15.38	Deletion: 60% Insertion: 20% Missense: 20%	
Family history	19 (59.37%)	8 (22.22%)***	14 (82.35%)	6 (50%)	0	4 (66.66%)

GH, Growth hormone; NFPA, nonfunctional pituitary adenoma; No, number; PA, pituitary adenoma; PGL, paraganglioma; PHEO, pheochromocytoma; PRL, prolactinoma.

*P = .041399 compared to 3PAs with SDHx mutations.

**P = .011872.

***P = .001776 compared to 3PAs with mutations.

to be expressed in PHEO samples from our patient with the *SDHD* mutation, as well as in tumor samples from other patients harboring *SDHB* or *SDHD* mutation [16]. Based on this finding and taking into account the three-fold decrease of plasma and urinary metanephrines after pituitary adenoma resection (greater than the one noted following bilateral adrenalectomy for the patient's PHEOs) [16], we assumed that normalization of GH levels after TSS contributed significantly to such biochemical changes. This was the first and only study so far reporting the expression of GHR in PHEOs, whereas there are reports of the differential expression of ghrelin and GH-releasing hormone receptors in various adrenal tumors, including PHEOs [94–96]. These two observations indicate that both PA and PHEOs/PGLs may have a mutual effect on their biochemical findings and treatment response.

The *MAX*-mutated PAs in the 3PAs context show some very interesting characteristics, although the number is small to make definite conclusions. They develop at young age (mean age 34.6 ± 14.46 years), are macroadenomas, usually invasive, can have variable secretory characteristics [acromegaly ($n = 2$) or prolactinoma ($n = 3$)] and may either precede or follow the appearance of a PHEO. In all reported cases there is no family history, and, in most cases, the genetic defect was a large deletion that required MLPA for detection (Tables 7.1 and 7.4).

At this point, there does not appear to exist an apparent phenotype–genotype correlation as it was recently reported by Bayley et al. [97] mainly due to the limited number of cases and the lack of prospective studies. However, based on the available cases (Tables 7.1 and 7.3), the age at diagnosis of SDH-deficient PAs ranged from as early as 15 years to as late as 72 years old. PHEOs or PGLs in these patients were bilateral and/or multiple, with a tendency to recur as it was found in *SDHx*-related PHEOs/PGLs [98].

A distinctive pathologic feature of *SDHx*-mutated pituitary adenomas described by Dénes et al. [37] is an extensive vacuolization of the cytoplasm; similar finding was detected in the PA tissue of the original case with the *SDHD* mutation (Fig. 7.2A) [16]. Analogous vacuoles were found in SDH-deficient renal carcinoma and GISTs [89,99] and in mouse models of SDH-deficient pituitary adenomas as we will describe later. Although electron microscopy was not used to identify the exact nature of the vacuoles, there is a possibility that they represent hypoxia-related autophagic bodies as seen in many cancers [100,101].

Traditionally, LOH has been used in oncology to confirm the causative association between a tumor and the loss of a tumor suppressor gene [103]. Loss of *SDHB* immunochemistry has been shown to be an excellent indicator of germline or somatic mutations in the *SDHx* genes [33,104]. In most of the *SDHx*-related PAs reported so far, completely absent or weak diffuse staining of *SDHB* was found (in those cases that this was performed) (Tables 7.1 and 7.3). Therefore as in PHEOs/PGLs, *SDHB* staining may be used as an additional diagnostic tool to screen for *SDHx* mutations. As shown in Table 7.1, in some cases, no LOH studies were performed, whereas in few cases, no LOH of the *SDHx*-mutated locus was identified. Does this mean that the absence of consistent LOH in PAs indicates a lack of association with the identified *SDHx* variant? The presence of LOH in SDH-deficient tumors is not always present as it was shown in bilateral adrenal medullary hyperplasia associated with a germline *SDHB* mutation, which showed the retention of heterozygosity [105], and PHEOs without loss of the normal *SDHD* allele have been shown in patients with pathogenic *SDHD* mutations [106]. Additionally, cases of "paradoxical" loss of the mutant *SDHx* allele have been shown [90,107], suggesting that the *SDHx* defects may not always lead to tumorigenesis in the classical tumor suppressor gene way. Finally, epigenetic alterations such as

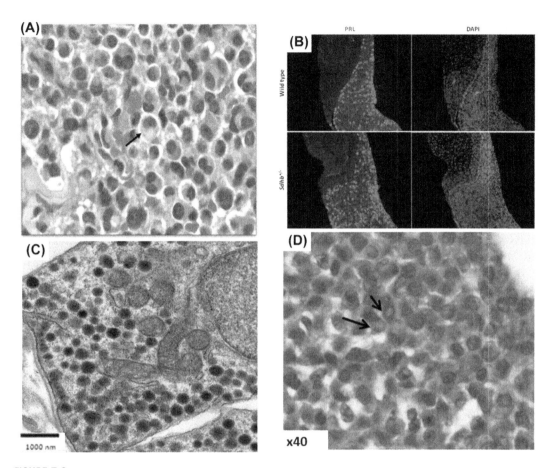

FIGURE 7.2

(A) H&E staining of the pituitary adenoma of the *SDHD* patient reported by Xekouki et al. [102] showing cytoplasmic vacuoles (arrow) (×40). (B) Pituitaries of 12-month-old *Sdhb*$^{+/-}$ mice (lower panel) stained for prolactin with Alexa Fluor 555 conjugated secondary antibody appeared hyperplastic compared to wt mice (upper panel); nuclei were stained with 4′,6′-diamino-2-phenylindole appeared hyperplastic (C) with abnormal enlarged mitochondria with marked swelling and destructive internal cristae in electron microscopy (D) and intranuclear inclusions of unknown nature.

somatic *SDHC* promoter methylation and postzygotic somatic mosaicism could provide another explanation for those cases negative for germline mutations [108,109].

Metabolomics, or global metabolite profiling, which is a new technology of functional genomics, can be used for investigating metabolite changes associated with some gene mutations [110]. As shown by Richter et al. [111], profiling of Krebs cycle metabolites with the use of mass spectrometry (MS), the ratio of succinate:fumarate was significantly higher in SDH-deficient PHEOs/PGLs, and this may be another tool to identify patients for testing of *SDHx* mutations and to assess functionality associated with *SDHx* VUS in PGLs. Although this method has not been used in any of the 3PAs cases, it would be interesting to see whether it could predict the presence of *SDHx* mutations or distinguish damaging mutations from nonfunctional polymorphisms in pituitary adenomas in the context of 3PAs as in PGLs.

7.5 Possible pathogenetic mechanism for pituitary adenomas development due to *SDHx* deficiency

We attempted to identify the pathophysiological mechanism responsible for PA development in the presence of SDH deficiency. Utilizing the $Sdhb^{+/-}$ mouse model, we showed that the pituitary of 12-month-old $Sdhb^{+/-}$ mice was hypercellular mainly due to the increased number of prolactin (PRL)−secreting cells, and to a lesser extent, GH-secreting cells (Fig. 7.2B) [35]. Although no gross pituitary tumors were detected, this observation indicates that pituitary hyperplasia in SDH-deficient cell could be the first step to neoplasm development as it was previously reported in Aip-deficient male mice [112] as well as in those lacking p19 (arf) or p27 (kip−1), and those overexpressing the beta-catenin or the pituitary tumor-transforming gene (CTNNB1 and PTTG, respectively) [112−116]. In humans, pituitary hyperplasia has been reported in patients with germline mutations in the *PRKAR1A* or *AIP* tumor suppressor genes, and in those with gigantism and chromosome Xq26 duplications [117−119]. Additional interesting findings in the $Sdhb^{+/-}$ pituitary cells were the strong cytoplasmic (and occasionally nuclear) HIF1a expression, the abnormal mitochondria (Fig. 7.2C), the nuclear inclusions of unknown nature (Fig. 7.2D), and the abnormal and poorly defined heterochromatin forming a ring-shaped pattern rather than being centrally located as in pituitaries of the normal control animals [35].

Increased expression of HIFs is a well-known characteristic of pseudohypoxic PHEO/PGL cluster [24,25,120]. Furthermore, fragmented and defective mitochondria associated with tumor progression or resistance to therapy have been found in many different cancers as well as in tumor samples from patients with Carney triad and in *SDHC* and *SDHD*-deficient PGLs [121−123]. The altered chromatin pattern that we also observed in $Sdhb^{+/-}$ mice, like the one observed in mice overexpressing PTTG and in hypoxic cells [114,124], is another indication that chronic activation of pseudohypoxia in *SDHx*-deficient pituitary cells can drive genetic instability and eventually lead to tumor formation. Although electron microscopy failed to identify the actual nature of the nuclear inclusions seen in $Sdhb^{+/-}$ pituitary cells with light microscopy, it has revealed that these appeared to be fused to the nucleus rather than being intranuclear. These inclusions resembled the vacuoles described by Dénes et al. in *SDHx*-deficient pituitary tumors and may represent late autophagic vacuoles previously filled with digested abnormal mitochondria.

7.6 Recommendation for diagnosis, follow-up, and screening in 3PAs with genetic alterations

Based on the clinical characteristics of PAs in the context of 3PAs described previously, we recommend that a detailed baseline medical and family history is taken from all patients along with careful physical examination to detect signs of other tumors/conditions associated with genetic defects in the relevant genes: GISTs in *SDHx* defects, hyperparathyroidism (HPTH), insulinoma in *MEN1* mutations, and HPTH in 3PAs with *RET* mutations. Unless there are other signs and symptoms suggestive of the presence of *MEN1* or *MEN2*, patients presenting with 3PAs should be initially screened for *SDHx* mutations, particularly if there is a strong family history of PHEOs/PGLs in other family members. If there is no family history for PHEOs/PGLs, screening for *SDHx*, *MEN1*, and *MAX* mutations may be considered (Fig. 7.3A). In case other family members of a patient with 3PAs found to carry the same mutation, attention should be paid to any symptoms that would indicate GH or PRL hypersecretion and visual disturbance, as most of the pituitary adenomas in the context of 3PAs are PRL or GH-secreting macroadenomas or NFPAs. If there are no suspicious symptoms or findings and biochemistry is normal, annual biochemical surveillance should include testing for PHEOs/PGLs, as per the most recent recommendations [125,126]. If at any time, clinical findings or biochemistry indicate hormonal hypersecretion, a pituitary MRI should be performed. Treatment of PAs due to *SDHx* mutations should not differ from the recommended treatment for sporadic tumors [127–133]. However, it is possible that more than one treatment modalities will be required, as these tumors are often more aggressive and tend to recur more frequently than their sporadic counterparts (although this impression may change as more cases are now identified by screening and are followed prospectively). In all *MAX* mutation carriers (regardless of the presence of a family history of PHEO/PGL), careful medical history and physical examination should be performed for the presence of other tumors related to *MAX* mutations such as PA and renal cancer as recently described [134] (Fig. 7.3B).

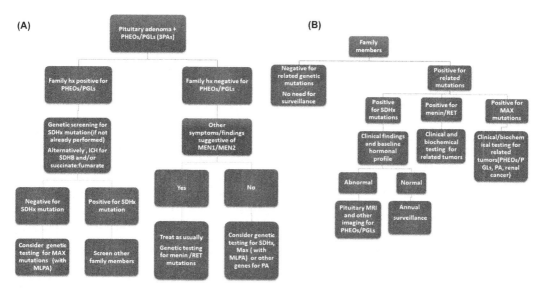

FIGURE 7.3

Recommended clinical investigation and follow-up in patients with 3PAs and in family members. *PAs*, Pituitary adenomas; *PGL*, paraganglioma; *PHEO*, pheochromocytoma.

We do not recommend systematic genetic screening for *SDHx* mutations in patients with sporadic isolated PA or with PAs, as these genetic defects are known to have low penetrance in affected families. However, a genetic study should be indicated for patients with PA and coexistent PHEO/PGL or family history of PHEO/PGL (the latter may not only include first-degree relatives).

Alternatively, and/or if genetic screening is not readily available, *SDHB* staining and succinate: fumarate ratio may be another way to look for *SDHx* defects. The presence of intracytoplasmic vacuoles is another pathology findings strongly suggestive of *SDHx* gene defect.

If screening for *SDHx* mutations is negative or pathology and biochemistry are not suggestive of *SDHx* defects, screening for *MAX* mutations with the use of MLPA may be considered. Screening for *menin* or *RET* mutations should be based on the presence of clinical features and/or tumors that would indicate the presence of the *MEN1* or *MEN2* syndromes. Screening for *MEN4* is not considered routine at this point, as is the case for other genes known to predispose to PA development [135].

7.7 Future therapeutic implications and final remarks

The importance of aberrant tricarboxylic acid cycle (TCA) cycle function in tumorigenesis and the potentials of applying small molecule inhibitors to perturb the enhanced cycle function for cancer treatment start to evolve. Profiling of gene expression and methylation can serve as a powerful tool for characterizing disease clusters and for guiding targeted therapy, an important basis for developing personalized therapy. In PHEO/PGL the initial stages of tumor development are associated with hypoxia due to excessive growth and/or high metabolic activity and an insufficient oxygen supply. Thus modifying/interrupting the HIF signaling pathway seems to be a promising therapeutic target. Emerging therapies consist predominantly of peptide receptor radionuclide therapy, vascular endothelial growth factor receptor-targeted therapy, tyrosine kinase inhibitors, mammalian target of rapamycin inhibitors, and more recently, immune checkpoint inhibitors [136]. These therapies are also under investigation for aggressive PAs resistant to temozolomide treatment [137].

Although rare, we anticipate that more 3PAs cases will be recognized and new genetic causes may be identified. This will allow for better characterization of PAs in the context of this association, genotype−phenotype correlations, and may help to develop new directions for diagnostic, therapeutic options, follow-up, and genetic screening for detecting the carriers and offering them appropriate management and counseling.

References

[1] Ezzat S, Asa SL, Couldwell WT, Barr CE, Dodge WE, Vance ML, et al. The prevalence of pituitary adenomas: a systematic review. Cancer 2004.

[2] Mete O, Cintosun A, Pressman I, Asa SL. Epidemiology and biomarker profile of pituitary adenohypophysial tumors. Mod Pathol 2018.

[3] Lim CT, Korbonits M. Update on the clinicopathology of pituitary adenomas. Endocr Pract 2018.

[4] Vandeva S, Daly AF, Petrossians P, Zacharieva S, Beckers A. Somatic and germline mutations in the pathogenesis of pituitary adenomas. Eur J Endocrinol 2019;181:R235−54.

[5] Caimari F, Korbonits M. Novel genetic causes of pituitary adenomas. Clin Cancer Res 2016.

[6] Moosavi B, Berry EA, Zhu XL, Yang WC, Yang GF. The assembly of succinate dehydrogenase: a key enzyme in bioenergetics. Cell Mol Life Sci 2019.

[7] Moosavi B, Zhu X, Yang W-C, Yang G-F. Molecular pathogenesis of tumorigenesis caused by succinate dehydrogenase defect. Eur J Cell Biol 2019.

[8] Warburg O. Origin of cancer cells. Science 1956.

[9] Fu Y, Liu S, Yin S, Niu W, Xiong W, Tan M, et al. The reverse Warburg effect is likely to be an Achilles' heel of cancer that can be exploited for cancer therapy. Oncotarget 2017.

[10] Baysal BE, Ferrell RE, Willett-Brozick JE, Lawrence EC, Myssiorek D, Bosch A, et al. Mutations in *SDHD*, a mitochondrial complex II gene, in hereditary paraganglioma. Science 2000.

[11] Astuti D, Latif F, Dallol A, Dahia PLM, Douglas F, George E, et al. Gene mutations in the succinate dehydrogenase subunit *SDHB* cause susceptibility to familial pheochromocytoma and to familial paraganglioma. Am J Hum Genet 2001.

[12] Nölting S, Ullrich M, Pietzsch J, Ziegler CG, Eisenhofer G, Grossman A, et al. Current management of pheochromocytoma/paraganglioma: a guide for the practicing clinician in the era of precision medicine. Cancers (Basel) 2019.

[13] Vanharanta S, Buchta M, McWhinney SR, Virta SK, Pęczkowska M, Morrison CD, et al. Early-onset renal cell carcinoma as a novel extraparaganglial component of *SDHB*-associated heritable paraganglioma. Am J Hum Genet 2004.

[14] McWhinney SR, Pasini B, Stratakis CA. Familial gastrointestinal stromal tumors and germ-line mutations. N Engl J Med 2007.

[15] Janeway KA, Kim SY, Lodish M, Nosé V, Rustin P, Gaal J, et al. Defects in succinate dehydrogenase in gastrointestinal stromal tumors lacking KIT and PDGFRA mutations. Proc Natl Acad Sci USA 2011.

[16] Xekouki P, Pacak K, Almeida M, Wassif CA, Rustin P, Nesterova M, et al. Succinate dehydrogenase (SDH) D subunit (*SDHD*) inactivation in a growth-hormone-producing pituitary tumor: a new association for SDH? J Clin Endocrinol Metab 2012.

[17] Neumann HPH, Pawlu C, Pęczkowska M, Bausch B, McWhinney SR, Muresan M, et al. Distinct clinical features of paraganglioma syndromes associated with *SDHB* and *SDHD* and gene mutations. J Am Med Assoc 2004.

[18] Ni Y, Seballos S, Ganapathi S, Gurin D, Fletcher B, Ngeow J, et al. Germline and somatic *SDHx* alterations in apparently sporadic differentiated thyroid cancer. Endocr Relat Cancer 2015.

[19] Papathomas TG, Gaal J, Corssmit EPM, Oudijk L, Korpershoek E, Heimdal K, et al. Non-pheochromocytoma (PCC)/paraganglioma (PGL) tumors in patients with succinate dehydrogenase-related PCC-PGL syndromes: a clinicopathological and molecular analysis. Eur J Endocrinol 2014.

[20] Niemeijer ND, Papathomas TG, Korpershoek E, De Krijger RR, Oudijk L, Morreau H, et al. Succinate dehydrogenase (SDH)-deficient pancreatic neuroendocrine tumor expands the SDH-related tumor spectrum. J Clin Endocrinol Metab 2015.

[21] Ni Y, Zbuk KM, Sadler T, Patocs A, Lobo G, Edelman E, et al. Germline mutations and variants in the succinate dehydrogenase genes in Cowden and Cowden-like syndromes. Am J Hum Genet 2008.

[22] Renella R, Carnevale J, Schneider KA, Hornick JL, Rana HQ, Janeway KA. Exploring the association of succinate dehydrogenase complex mutations with lymphoid malignancies. Fam Cancer 2014.

[23] Selak MA, Armour SM, MacKenzie ED, Boulahbel H, Watson DG, Mansfield KD, et al. Succinate links TCA cycle dysfunction to oncogenesis by inhibiting HIF-α prolyl hydroxylase. Cancer Cell 2005.

[24] Pollard PJ, Brière JJ, Alam NA, Barwell J, Barclay E, Wortham NC, et al. Accumulation of Krebs cycle intermediates and over-expression of HIF1α in tumours which result from germline FH and SDH mutations. Hum Mol Genet 2005.

[25] Weber A, Klocker H, Oberacher H, Gnaiger E, Neuwirt H, Sampson N, et al. Succinate accumulation is associated with a shift of mitochondrial respiratory control and HIF-1α upregulation in PTEN negative prostate cancer cells. Int J Mol Sci 2018.

[26] Loriot C, Burnichon N, Gadessaud N, Vescovo L, Amar L, Libé R, et al. Epithelial to mesenchymal transition is activated in metastatic pheochromocytomas and paragangliomas caused by *SDHB* gene mutations. J Clin Endocrinol Metab 2012.

[27] Sciacovelli M, Gonçalves E, Johnson TI, Zecchini VR, Da Costa ASH, Gaude E, et al. Fumarate is an epigenetic modifier that elicits epithelial-to-mesenchymal transition. Nature 2016.

[28] Tretter L, Patocs A, Chinopoulos C. Succinate, an intermediate in metabolism, signal transduction, ROS, hypoxia, and tumorigenesis. Biochim Biophys Acta 2016.

[29] Kluckova K, Tennant DA. Metabolic implications of hypoxia and pseudohypoxia in pheochromocytoma and paraganglioma. Cell Tissue Res 2018.

[30] Letouzé E, Martinelli C, Loriot C, Burnichon N, Abermil N, Ottolenghi C, et al. SDH mutations establish a hypermethylator phenotype in paraganglioma. Cancer Cell 2013.

[31] Killian JK, Kim SY, Miettinen M, Smith C, Merino M, Tsokos M, et al. Succinate dehydrogenase mutation underlies global epigenomic divergence in gastrointestinal stromal tumor. Cancer Discov 2013.

[32] Gill AJ. Succinate dehydrogenase (SDH)-deficient neoplasia. Histopathology 2018.

[33] van Nederveen FH, Gaal J, Favier J, Korpershoek E, Oldenburg RA, de Bruyn EM, et al. An immuno-histochemical procedure to detect patients with paraganglioma and pheochromocytoma with germline *SDHB*, *SDHC*, or *SDHD* gene mutations: a retrospective and prospective analysis. Lancet Oncol 2009.

[34] Brière JJ, Favier J, Bénit P, El Ghouzzi V, Lorenzato A, Rabier D, et al. Mitochondrial succinate is instrumental for HIF1α nuclear translocation in *SDHA*-mutant fibroblasts under normoxic conditions. Hum Mol Genet 2005.

[35] Xekouki P, Szarek E, Bullova P, Giubellino A, Quezado M, Mastroyannis SA, et al. Pituitary adenoma with paraganglioma/pheochromocytoma (3PAs) and succinate dehydrogenase defects in humans and mice. J Clin Endocrinol Metab 2015.

[36] Gill AJ, Toon CW, Clarkson A, Sioson L, Chou A, Winship I, et al. Succinate dehydrogenase deficiency is rare in pituitary adenomas. Am J Surg Pathol 2014.

[37] Dénes J, Swords F, Rattenberry E, Stals K, Owens M, Cranston T, et al. Heterogeneous genetic background of the association of pheochromocytoma/paraganglioma and pituitary adenoma: results from a large patient cohort. J Clin Endocrinol Metab 2015.

[38] Daly AF, Castermans E, Oudijk L, Guitelman MA, Beckers P, Potorac I, et al. Pheochromocytomas and pituitary adenomas in three patients with *MAX* exon deletions. Endocr Relat Cancer 2018.

[39] Roszko KL, Blouch E, Blake M, Powers JF, Tischler AS, Hodin R, et al. Case report of a prolactinoma in a patient with a novel *MAX* mutation and bilateral pheochromocytomas. J Endocr Soc 2017.

[40] Kobza AO, Dizon S, Arnaout A. Case report of bilateral pheochromocytomas due to a novel *max* mutation in a patient known to have a pituitary prolactinoma. AACE Clin Case Rep 2018.

[41] Guerrero-Pérez F, Fajardo C, Torres Vela E, Giménez-Palop O, Lisbona Gil A, Martín T, et al. 3P association (3PAs): pituitary adenoma and pheochromocytoma/paraganglioma. A heterogeneous clinical syndrome associated with different gene mutations. Eur J Intern Med 2019.

[42] Varsavsky M, Sebastián-Ochoa A, Torres Vela E. Coexistence of a pituitary macroadenoma and multicentric paraganglioma: a strange coincidence. Endocrinol Nutr 2013.

[43] Guerrero Pérez F, Lisbona Gil A, Robledo M, Iglesias P, Villabona Artero C. Pituitary adenoma associated with pheochromocytoma/paraganglioma: a new form of multiple endocrine neoplasia. Endocrinol Nutr (English Ed.) 2016.

[44] Gorospe L, Cabañero-Sánchez A, Muñoz-Molina GM, Pacios-Blanco RE, Ureña Vacas A, García-Santana E. An unusual case of mediastinal paraganglioma and pituitary adenoma. Surgery 2017.

[45] Efstathiadou ZA, Sapranidis M, Anagnostis P, Kita MD, Rosenthal EL. Unusual case of Cowden-like syndrome, neck paraganglioma, and pituitary adenoma. Head Neck 2014.

[46] Lopez-Jimenez E, De Campos JM, Kusak EM, Landa I, Leskelä S, Montero-Conde C, et al. *SDHC* mutation in an elderly patient without familial antecedents. Clin Endocrinol 2008.

[47] Jeong YJ, Oh HK, Bong JG. Multiple endocrine neoplasia type 1 associated with breast cancer: a case report and review of the literature. Oncol Lett 2014.

[48] Langer P, Cupisti K, Bartsch DK, Nies C, Goretzki PE, Rothmund M, et al. Adrenal involvement in multiple endocrine neoplasia type 1. World J Surg 2002.

[49] Dackiw AP, Cote GJ, Fleming JB, Schultz PN, Stanford P, Vassilopoulou-Sellin R, et al. Screening for *MEN1* mutations in patients with atypical endocrine neoplasia. Surgery 1999.

[50] Carty SA, Helm AK, Amico JA, Clarke MR, Foley TP, Watson CG, et al. The variable penetrance and spectrum of manifestations of multiple endocrine neoplasia type 1. Surgery 1998.

[51] Heinlen JE, Buethe DD, Culkin DJ, Slobodov G. Multiple endocrine neoplasia 2a presenting with pheochromocytoma and pituitary macroadenoma. Oncol 2011.

[52] Naziat A, Karavitaki N, Thakker R, Ansorge O, Sadler G, Gleeson F, et al. Confusing genes: a patient with MEN2A and Cushing's disease. Clin Endocrinol 2013.

[53] Koshy AN, Cheng VE, Sajeev JK, Venkataraman P, Profitis K, et al. Bilateral pheochromocytoma and paraganglioma: a rare cause of cardiomyopathy. Eur Heart J Cardiovasc Imag 2016.

[54] Boguszewski CL, Fighera TM, Bornschein A, Marques FM, Dénes J, Rattenbery E, et al. Genetic studies in a coexistence of acromegaly, pheochromocytoma, gastrointestinal stromal tumor (GIST) and thyroid follicular adenoma. Arq Bras Endocrinol Metabol 2012.

[55] Parghane RV, Agrawal K, Mittal BR, Shukla J, Bhattacharya A, Mukherjee KK. 68Ga DOTATATE PET/CT in a rare coexistence of pituitary macroadenoma and multiple paragangliomas. Clin Nuc Med 2014.

[56] Mumby C, Davis JRE, Trouillas J, Higham CE. Phaeochromocytoma and acromegaly: a unifying diagnosis. Endocrinol Diabetes Metab Case Rep 2014.

[57] Sisson JC, Giordano TJ, Avram AM. Three endocrine neoplasms: an unusual combination of pheochromocytoma, pituitary adenoma, and papillary thyroid carcinoma. Thyroid 2012.

[58] Gatta-Cherifi B, Chabre O, Murat A, Niccoli P, Cardot-Bauters C, Rohmer V, et al. Adrenal involvement in MEN1. Analysis of 715 cases from the Groupe d'étude des Tumeurs Endocrines database. Eur J Endocrinol 2012.

[59] Zhang C, Ma G, Liu X, Zhang H, Deng H, Nowell J, et al. Primary cardiac pheochromocytoma with multiple endocrine neoplasia. J Cancer Res Clin Oncol 2011.

[60] Yaylali GF, Akin F, Bastemir M, Yaylali YT, Ozden A. Phaeochromocytoma combined with subclinical Cushing's syndrome and pituitary microadenoma. Clin Invest Med 2008.

[61] Breckenridge SM, Hamrahian AH, Faiman C, Suh J, Prayson R, Mayberg M. Coexistence of a pituitary macroadenoma and pheochromocytoma − a case report and review of the literature. Pituitary 2003.

[62] Dünser MW, Mayr AJ, Gasser R, Rieger M, Friesenecker B, Hasibeder WR. Cardiac failure and multiple organ dysfunction syndrome in a patient with endocrine adenomatosis. Acta Anaesthesiol Scand 2002.

[63] Sleilati GG, Kovacs KT, Honasoge M. Acromegaly and pheochromocytoma: report of a rare coexistence. Endocr Pract 2002.

[64] Baughan J, De Gara C, Morrish D. A rare association between acromegaly and pheochromocytoma. Am J Surg 2001.

[65] Khalil WKA, Vadasz J, Rigo E, Kardos L, Tiszlavicz L, Gaspar L. Pheochromocytoma combined with unusual form of Cushing's syndrome and pituitary microadenoma. Eur J Endocrinol 1999.

[66] Teh BT, Hansen J, Svensson PJ, Hartley L. Bilateral recurrent phaeochromocytoma associated with a growth hormone-secreting pituitary tumour. Br J Surg 1996.

[67] Azzarelli B, Felten S, Muller J, Miyamoto R, Purvin V. Dopamine in paragangliomas of the glomus jugulare. Laryngoscope 1988.

[68] Bertrand JH, Ritz P, Reznik Y, Grollier G, Potier JC, Evrad C, et al. Sipple's syndrome associated with a large prolactinoma. Clin Endocrinol 1987.

[69] Larraza-Hernandez O, Albores-Saavedra J, Benavides G, Krause LG, Perez-Merizaldi JC, Ginzo A. Multiple endocrine neoplasia. Pituitary adenoma, multicentric papillary thyroid carcinoma, bilateral carotid body paraganglioma, parathyroid hyperplasia, gastric leiomyoma, and systemic amyloidosis. Am J Clin Pathol 1982.

[70] Meyers DH. Association of phaeochromocytoma and prolactinoma. MJA 1982.

[71] Blumenkopf B, Boekelheide K. Neck paraganglioma with a pituitary adenoma. Case report. J Neurosurg 1982.

[72] Anderson RJ, Lufkin EG, Sizemore GW, Carney JA, Sheps SG, Silliman YE. Acromegaly and pituitary adenoma with phaeochromocytoma: a variant of multiple endocrine neoplasia. Clin Endocrinol 1981.

[73] Myers JH, Eversman JJ. Acromegaly, hyperparathyroidism, and pheochromocytoma in the same patient. A multiple endocrine disorder. Arch Intern Med 1981.

[74] Alberts WM, McMeekin JO, George JM. Mixed multiple endocrine neoplasia syndromes. JAMA 1980.

[75] Janson KL, Roberts JA, Varela M. Multiple endocrine adenomatosis: in support of the common origin theories. J Urol 1978.

[76] Manger WM, Gifford RW. Studies on 38 patients with pheochromocytoma in whom catecholamines were determined. Pheochromocytoma. Springer; 1977.

[77] Melicow MM. One hundred cases of pheochromocytoma (107 tumors) at the Columbia-Presbyterian Medical Center, 1926−1976: a clinicopathological analysis. Cancer 1977.

[78] Kadowaki S, Baba Y, Kakita T, Yamamoto H, Fukase M, Goto Y. A case of acromegaly associated with pheochromocytoma. Saishin Igaku 1976.

[79] Farhi F, Dikman SH, Lawson W, Cobin RH, Zak FG. Paragangliomatosis associated with multiple endocrine adenomas. Arch Pathol Lab Med 1976.

[80] Berg B, Biörklund A, Grimelius L, Ingemansson S, Larsson LI, Stenram U, et al. A new pattern of multiple endocrine adenomatosis: chemodectoma, bronchial carcinoid, GH-producing pituitary adenoma, and hyperplasia of the parathyroid glands, and antral and duodenal gastrin cells. Acta Med Scand 1976.

[81] Wolf LM, Duduisson M, Schrub JC, Metayer J, Laumonier R. Sipple's syndrome associated with pituitary and parathyroid adenomas. Ann Endocrinol 1972.

[82] Miller GL, Wynn J. Acromegaly, pheochromocytoma, toxic goiter, diabetes mellitus, and endometriosis. Arch Intern Med 1971.

[83] Steiner A, Goodman A, Powers S. Study of a kindred with pheochromocytoma, medullary thyroid carcinoma, hyperparathyroidism and Cushing's disease: multiple endocrine neoplasia, type 2. Medicine 1968.

[84] Kahn MT, Mullon DA. Phoechromocytoma without hypertension. Report of a patient with acromegaly. JAMA 1964.

[85] German WJ, Flanigan S. Pituitary adenomas: a follow-up study of the Cushing series. Clin Neurosurg 1964.

[86] Iversen K. Acromegaly associated with phaeochromocytoma. Acta Med Scand 1952.

[87] Bayley JP, Devilee P, Taschner PEM. The SDH mutation database: an online resource for succinate dehydrogenase sequence variants involved in pheochromocytoma, paraganglioma and mitochondrial complex II deficiency. BMC Med Genet 2005.

[88] De Sousa SM, McCabe MJ, Wu K, Roscioli T, Gayevskiy V, Brook K, et al. Germline variants in familial pituitary tumour syndrome genes are common in young patients and families with additional endocrine tumours. Eur J Endocrinol 2017.

[89] Gill AJ, Hes O, Papathomas T, Šedivcová M, Tan PH, Agaimy A, et al. Succinate dehydrogenase (SDH)-deficient renal carcinoma: a morphologically distinct entity: a clinicopathologic series of 36 tumors from 27 patients. Am J Surg Pathol 2014.

[90] Dwight T, Mann K, Benn DE, Robinson BG, McKelvie P, Gill AJ, et al. Familial *SDHA* mutation associated with pituitary adenoma and pheochromocytoma/paraganglioma. J Clin Endocrinol Metab 2013.

[91] Benn DE, Gimenez-Roqueplo A-P, Reilly JR, Bertherat J, Burgess J, Byth K, et al. Clinical presentation and penetrance of pheochromocytoma/paraganglioma syndromes. J Clin Endocrinol Metab 2006.

[92] Vandeva S, Daly AF, Petrossians P, Zacharieva S, Beckers A. Genetics in endocrinology: somatic and germline mutations in the pathogenesis of pituitary adenomas. Eur J Endocrinol 2019.

[93] Mannelli M, Canu L, Ercolino T, Rapizzi E, Martinelli S, Parenti G, et al. Diagnosis of endocrine disease: *SDHx* mutations: beyond pheochromocytomas and paragangliomas. Eur J Endocrinol 2018.

[94] Ueberberg B, Unger N, Sheu SY, Walz MK, Schmid KW, Saeger W, et al. Differential expression of ghrelin and its receptor (GHS-R1a) in various adrenal tumors and normal adrenal gland. Horm Metab Res 2008.

[95] Freddi S, Arnaldi G, Fazioli F, Scarpelli M, Appolloni G, Mancini T, et al. Expression of growth hormone-releasing hormone receptor splicing variants in human primary adrenocortical tumours. Clin Endocrinol (Oxf) 2005.

[96] Ziegler CG, Brown JW, Schally AV, Erler A, Gebauer L, Treszl A, et al. Expression of neuropeptide hormone receptors in human adrenal tumors and cell lines: antiproliferative effects of peptide analogues. Proc Natl Acad Sci USA 2009.

[97] Bayley JP, Bausch B, Rijken JA, Van Hulsteijn LT, Jansen JC, Ascher D, et al. Variant type is associated with disease characteristics in *SDHB*, *SDHC* and *SDHD*-linked pheochromocytoma-paraganglioma. J Med Genet 2019.

[98] Kantorovich V, King KS, Pacak K. SDH-related pheochromocytoma and paraganglioma. Best Pract Res Clin Endocrinol Metab 2010.

[99] Powers JF, Cochran B, Baleja JD, Sikes HD, Zhang X, Lomakin I, et al. A unique model for SDH-deficient GIST: an endocrine-related cancer. Endocr Relat Cancer 2018.

[100] Kardideh B, Samimi Z, Norooznezhad F, Kiani S, Mansouri K. Autophagy, cancer and angiogenesis: where is the link? Cell Biosci 2019.

[101] Zhang H, Bosch-Marce M, Shimoda LA, Yee ST, Jin HB, Wesley JB, et al. Mitochondrial autophagy is an HIF-1-dependent adaptive metabolic response to hypoxia. J Biol Chem 2008.

[102] Xekouki P, Pacak K, Almeida M, Wassif CA, Rustin P, Nesterova M, et al. Succinate dehydrogenase (SDH) D subunit (*SDHD*) inactivation in a growth-hormone-producing pituitary tumor: a new association for SDH? J Clin Endocrinol Metab 2012;97(3).

[103] Wang LH, Wu CF, Rajasekaran N, Shin YK. Loss of tumor suppressor gene function in human cancer: an overview. Cell Physiol Biochem 2019.

[104] Gill AJ, Benn DE, Chou A, Clarkson A, Muljono A, Meyer-Rochow GY, et al. Immunohistochemistry for *SDHB* triages genetic testing of *SDHB*, *SDHC*, and *SDHD* in paraganglioma-pheochromocytoma syndromes. Hum Pathol 2010.

[105] Grogan RH, Pacak K, Pasche L, Huynh TT, Greco RS. Bilateral adrenal medullary hyperplasia associated with an *SDHB* mutation. J Clin Oncol 2011.

[106] Weber A, Hoffmann MM, Neumann HPH, Erlic Z. Somatic mutation analysis of the *SDHB*, *SDHC*, *SDHD*, and RET genes in the clinical assessment of sporadic and hereditary pheochromocytoma. Horm Cancer 2012.

[107] Pasini B, McWhinney SR, Bei T, Matyakhina L, Stergiopoulos S, Muchow M, et al. Clinical and molecular genetics of patients with the Carney-Stratakis syndrome and germline mutations of the genes coding for the succinate dehydrogenase subunits *SDHB*, *SDHC*, and *SDHD*. Eur J Hum Genet 2008.

[108] Haller F, Moskalev EA, Faucz FR, Barthelmeß S, Wiemann S, Bieg M, et al. Aberrant DNA hypermethylation of *SDHC*: a novel mechanism of tumor development in Carney triad. Endocr Relat Cancer 2014.

[109] Buffet A, Smati S, Mansuy L, Ménara M, Lebras M, Heymann MF, et al. Mosaicism in HIF2A-related polycythemia-paraganglioma syndrome. J Clin Endocrinol Metab 2014.

[110] Reitman ZJ, Jin G, Karoly ED, Spasojevic I, Yang J, Kinzler KW, et al. Profiling the effects of isocitrate dehydrogenase 1 and 2 mutations on the cellular metabolome. Proc Natl Acad Sci USA 2011.

[111] Richter S, Peitzsch M, Rapizzi E, Lenders JW, Qin N, De Cubas AA, et al. Krebs cycle metabolite profiling for identification and stratification of pheochromocytomas/paragangliomas due to succinate dehydrogenase deficiency. J Clin Endocrinol Metab 2014.

[112] Lecoq AL, Zizzari P, Hage M, Decourtye L, Adam C, Viengchareun S, et al. Mild pituitary phenotype in 3- and 12-month-old Aip-deficient male mice. J Endocrinol 2016.

[113] Lloyd RV, Ruebel KH, Zhang S, Jin L. Pituitary hyperplasia in glycoprotein hormone alpha subunit-, p18INK4C-, and p27kip-1-null mice: analysis of proteins influencing p27kip-1 ubiquitin degradation. Am J Pathol 2002.

[114] Donangelo I, Gutman S, Horvath E, Kovacs K, Wawrowsky K, Mount M, et al. Pituitary tumor transforming gene overexpression facilitates pituitary tumor development. Endocrinology 2006.

[115] Bai F, Chan HL, Smith MD, Kiyokawa H, Pei X-H. p19Ink4d is a tumor suppressor and controls pituitary anterior lobe cell proliferation. Mol Cell Biol 2014.

[116] Gaston-Massuet C, Andoniadou CL, Signore M, Jayakody SA, Charolidi N, Kyeyune R, et al. Increased Wingless (Wnt) signaling in pituitary progenitor/stem cells gives rise to pituitary tumors in mice and humans. Proc Natl Acad Sci USA 2011.

[117] Stergiopoulos SG, Abu-Asab MS, Tsokos M, Stratakis CA. Pituitary pathology in Carney complex patients. Pituitary 2004.

[118] Villa C, Lagonigro MS, Magri F, Koziak M, Jaffrain-Rea ML, Brauner R, et al. Hyperplasia-adenoma sequence in pituitary tumorigenesis related to aryl hydrocarbon receptor interacting protein gene mutation. Endocr Relat Cancer 2011.

[119] Trivellin G, Daly AF, Faucz FR, Yuan B, Rostomyan L, Larco DO, et al. Gigantism and acromegaly due to Xq26 microduplications and GPR101 mutation. N Engl J Med 2014.

[120] Jochmanová I, Yang C, Zhuang Z, Pacak K. Hypoxia-inducible factor signaling in pheochromocytoma: turning the rudder in the right direction. J Natl Cancer Inst 2013.

[121] Grandemange S, Herzig S, Martinou JC. Mitochondrial dynamics and cancer. Semin Cancer Biol 2009.

[122] Szarek E, Ball ER, Imperiale A, Tsokos M, Faucz FR, Giubellino A, et al. Carney triad, SDH-deficient tumors, and $Sdhb^{+/-}$ mice share abnormal mitochondria. Endocr Relat Cancer 2015.

[123] Vyas S, Zaganjor E, Haigis MC. Mitochondria and cancer Cell [Internet] 2016;166(3):555−66. < https://linkinghub.elsevier.com/retrieve/pii/S0092867416309084 > .

[124] Bristow RG, Hill RP. Hypoxia and metabolism. Hypoxia, DNA repair and genetic instability Nat Rev Cancer [Internet] 2008;8(3):180−92. [cited 2020 Jan 13]. < http://www.ncbi.nlm.nih.gov/pubmed/18273037 > .

[125] Plouin PF, Amar L, Dekkers OM, Fassnach M, Gimenez-Roqueplo AP, Lenders JWM, et al. European Society of Endocrinology Clinical Practice Guideline for long-term follow-up of patients operated on for a pheochromocytoma or a paraganglioma. Eur J Endocrinol 2016.

[126] Lenders JWM, Duh QY, Eisenhofer G, Gimenez-Roqueplo AP, Grebe SKG, Murad MH, et al. Pheochromocytoma and paraganglioma: an endocrine society clinical practice guideline. J Clin Endocrinol Metab 2014.

[127] Melmed S, Casanueva FF, Hoffman AR, Kleinberg DL, Montori VM, Schlechte JA, et al. Diagnosis and treatment of hyperprolactinemia: an endocrine society clinical practice guideline. J Clin Endocrinol Metab 2011.

[128] Katznelson L, Laws ER, Melmed S, Molitch ME, Murad MH, Utz A, et al. Acromegaly: an endocrine society clinical practice guideline. J Clin Endocrinol Metab 2014.

[129] Beck-Peccoz P, Lania A, Beckers A, Chatterjee K, Wemeau J-L. 2013 European Thyroid Association Guidelines for the diagnosis and treatment of thyrotropin-secreting pituitary tumors. Eur Thyroid J 2013.

[130] Nieman LK, Biller BMK, Findling JW, Murad MH, Newell-Price J, Savage MO, et al. Treatment of Cushing's syndrome: an endocrine society clinical practice guideline. J Clin Endocrinol Metab 2015.

[131] Chanson P, Raverot G, Castinetti F, Cortet-Rudelli C, Galland F, Salenave S, et al. Management of clinically non-functioning pituitary adenoma. Ann Endocrinol (Paris) 2015.

[132] Freda PU, Beckers AM, Katznelson L, Molitch ME, Montori VM, Post KD, et al. Pituitary incidentaloma: an endocrine society clinical practice guideline. J Clin Endocrinol Metab 2011.

[133] Melmed S, Bronstein MD, Chanson P, Klibanski A, Casanueva FF, Wass JAH, et al. A consensus statement on acromegaly therapeutic outcomes. Nat Rev Endocrinol 2018.

[134] Korpershoek E, Koffy D, Eussen BH, Oudijk L, Papathomas TG, Van Nederveen FH, et al. Complex *MAX* rearrangement in a family with malignant pheochromocytoma, renal oncocytoma, and erythrocytosis. J Clin Endocrinol Metab 2016.

[135] Daly AF, Beckers A. Genetic testing in pituitary adenomas: what, how, and in whom? Endocrinol Diabetes Nutr 2019.

[136] Pang Y, Liu Y, Pacak K, Yang C. Pheochromocytomas and paragangliomas: from genetic diversity to targeted therapies. Cancers (Basel) 2019.

[137] Ilie MD, Lasolle H, Raverot G. Emerging and novel treatments for pituitary tumors. J Clin Med 2019.

CDKN1B (p27) defects leading to pituitary tumors

Sebastian Gulde and Natalia S. Pellegata

Institute for Diabetes and Cancer, Helmholtz Zentrum München, Neuherberg, Germany

8.1 Introduction

Familial pituitary tumors can be divided into two groups: (1) isolated tumors, when patients develop tumors at no other organs in addition to the pituitary gland (e.g., familial isolated pituitary adenoma-FIPA); (2) syndromic tumors, when affected individuals present with tumors at other organs too. The latter group includes: multiple endocrine neoplasia type 1 (MEN1), MEN4, Carney complex, DICER1 syndrome, *SDHx* gene−associated syndromes, and neurofibromatosis type 1 syndrome. The most common hereditary type is FIPA, followed by MEN1, together accounting for 5%−7% of all pituitary tumor patients [1,2].

MEN syndromes are inherited disorders characterized by proliferative lesions (hyperplasia or adenoma) occurring in at least two endocrine organs [3]. They are inherited as autosomal-dominant traits, have high penetrance and variable clinical expression. Three MEN syndromes are currently recognized: MEN1, MEN2, and MEN4, each presenting with a different tumor spectrum and having a different causative mutation. The MEN1 and MEN2 syndromes have long been known, and extensive information is available concerning the underlying genetic mutations, their biological effects, and the associated clinical manifestations. MEN1 and MEN2 are caused by inactivating mutations in the *MEN1* tumor suppressor gene or activating mutations in the *RET* proto-oncogene, respectively. MEN1 is characterized by tumors of the parathyroid glands, the endocrine pancreas and duodenum, the anterior pituitary gland, whereas MEN2 by medullary thyroid carcinoma, bilateral pheochromocytoma, and primary hyperparathyroidism due to parathyroid adenomas (reviewed in detail in Chapter 10: GNAS, McCune−Albright Syndrome, and GH-Producing Tumors).

The MEN4 syndrome was discovered in 2006 and represents the latest and rarest addition to the MEN family [4]. MEN4 is due to germline mutations in the *CDKN1B* gene, which codes for the cell cycle inhibitor p27, and is inherited as an autosomal-dominant trait [5]. These mutations have mostly been associated with impaired protein function compatible with a tumor suppressive role of p27 in tumorigenesis [6]. *CDKN1B* mutation carriers usually have a syndromic presentation of various neuroendocrine tumors (NETs), such as parathyroid adenomas, pituitary tumors, gastro-entero-pancreatic, and bronchial NETs. However, germline *CDKN1B* mutations have also been found in presumed sporadic patients presenting with single gland parathyroid adenomas, as well as in a few patients showing only pituitary tumors [7,8].

The incidence of *CDKN1B* germline mutations in patients with a familial presentation of multiple NETs and having no mutations in *MEN1* or *RET* is difficult to estimate given their rarity, but it

Gigantism and Acromegaly. DOI: https://doi.org/10.1016/B978-0-12-814537-1.00006-3

likely ranges from 1.5% to 3.7% [9−11]. To date, a total of 18 different germline base substitutions in *CDKN1B* have been identified in 29 patients (Fig. 8.1). A subset of the *CDKN1B* mutation-positive patients developed pituitary tumors and represent the focus of this chapter. Specifically, we will discuss their phenotypic features and, where available, the functional characterization of the germline p27 mutations. By putting these genetic changes in the context of the established functions of p27, we can gather some insight into the role of *CDKN1B* in pituitary tumorigenesis.

8.2 Structure, regulation, and function of p27

8.2.1 CDKN1B and p27 structure

The *CDKN1B* gene maps to chromosome 12p13, consists of two coding and one noncoding exons resulting in a 2.4-Kb long transcript encoding the 198 amino acids long wild-type p27 protein. p27 is an ubiquitously expressed member of the KIP family of cyclin-dependent kinase (CDK) inhibitors. It is highly conserved showing a 92.9% and 91.8% sequence homology between human and mouse or rat, respectively [6].

   p27 consists of several domains mediating protein−protein interactions; it possesses sites for phosphorylation and other posttranslational modifications, as well as a nuclear localization signal and a putative nuclear exporting sequence (Fig. 8.1). p27 belongs to the intrinsically unstructured proteins that lack a stable secondary/tertiary structure and have the ability to fold upon binding to interaction partners. Indeed, several p27 protein partners have been identified to date and mediate the various roles of this protein in both CDK dependent and CDK independent processes.

   The p27 protein consists of two major regions. At the N-terminus is located the kinase inhibitory domain comprising the two regions responsible for the binding to cyclins (residues 28−37) and to CDKs (residues 60−89), as well as a linker region that connects the two domains (residues 38−59). The C-terminal region (residues 105−198) comprises several posttranslational modification sites and mediates the binding to numerous partners that are not involved in mediating that cell cycle regulatory function of p27. Among the CDK-independent interactors we find Jab1, Grb2, RhoA, stathmin, citron kinase, Hsc70, and others (reviewed in [12]) (Fig. 8.1).

   In Fig. 8.1 are also reported the germline mutations identified so far in MEN4 patients. For sake of completeness, we have included a mutation in the 5′ UTR of *CDKN1B* (c.-73G > A), which was identified in a patient hemizygous for the p27 locus due to a maternally inherited interstitial deletion of chromosome 12 [13]. This patient manifested gigantism and neurodevelopmental defects. The overgrowth phenotype might have been caused by elevated IGF-1 levels but no pituitary tumors were identified at MRI [13].

8.2.2 p27 functions

The classical and most extensively documented function of p27 is the binding to and inhibition of cyclin/CDK complexes. Specifically, p27 binds to CyclinE/CDK2 and CyclinA/CDK2 complexes in the nucleus in response to different stimuli. When bound to p27, the CDK2 kinase cannot phosphorylate its main targets that are the members of the retinoblastoma (Rb) family. Phosphorylation of Rb is essential for cell cycle progression since it allows the release of E2F transcription factors

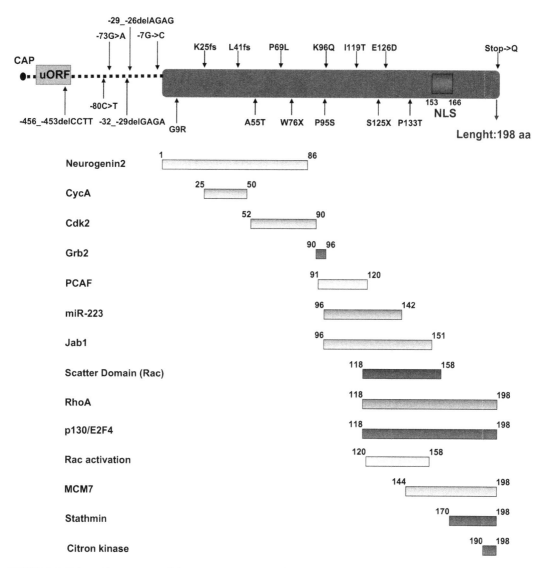

FIGURE 8.1 Schematic structure of the 5' untranslated region of the *CDKN1B* gene and of the p27 protein with indicated the position of the germline mutations identified in MEN4 patients. The regions of p27 mediating its binding to major interacting partners are reported below the protein. The germline mutations so far identified are indicated above/below the protein sequence. *CAP*, catabolite activator protein; *ORF*, open reading frame; *NLS*, nuclear localization signal.

(TFs) that in turn activate the transcription of genes responsible for the progression from G1 to S phase [14].

Furthermore, p27 binds to CyclinD/CDK4–6 complexes in the cytoplasm and represents an essential assembly factor for these complexes. Different from CyclinE/CDK2, CyclinD/CDK4–6 complexes stay catalytically active when bound to p27. The binding to CyclinD/CDK4–6 complexes prevents p27 from being transferred into the nucleus thereby hindering its interaction with CyclinE/CDK2 complexes and ultimately leading to cell cycle progression [15].

In addition to its role as critical cell cycle regulator, a number of activities not related to CDK inhibition have been ascribed to p27. Among the cell cycle-independent functions of p27, a particularly relevant one for tumorigenesis is its ability to regulate cytoskeleton assembly/disassembly, which ultimately modulates cell motility. p27 can reduce stress fibers stability and focal adhesion, thus promoting cell migration [16–19], but can also reduce cell motility depending on the cellular context, on the interaction with protein partners and on its intracellular levels [20–22]. Similarly, p27 can either promote or suppress autophagy and apoptosis in cancer cells (reviewed in [19]).

Noteworthy, it has recently been shown that p27 also plays a role as transcriptional coregulator. Chromatin immunoprecipitation-sequencing (ChIP-Seq) experiments conducted using an anti-p27 antibody led to the identification of DNA sequences that were bound by p27-containing complexes. These sequences mainly comprised chromatin regions distal from promotors or intronic regions [23–25]. Gene ontology classification of the genes located near these p27-bound regions, potentially regulated by p27, highlighted their involvement in cell adhesion, neuronal differentiation, cell signaling, and ion transport [25]. Given the fact that p27 does not possess a DNA binding domain, its role as transcriptional regulator must be mediated through the interaction with TFs. Indeed, several TFs were demonstrated to bind to p27, including p130/E2F4, Pax5, MyoD, PCAF, and others [23,25,26]. This newly recognized role of p27 as a transcriptional coregulator offers a new dimension to the role of this multifaceted protein in the tumorigenic process.

8.2.3 p27 regulation

Because of its important role as main driver of cell cycle exit in differentiated cells, the intracellular levels of p27 are tightly regulated through different mechanisms at transcriptional, translational, and posttranslational levels.

Several proteins regulate *CDKN1B* transcription, including Myc, Forkhead box O (FOXO) members, and Menin. Myc acts as an important negative regulator of *CDKN1B* transcription in two ways. On the one hand, Myc can directly bind to the *CDKN1B* promotor region and inhibit transcription, and on the other, indirectly regulates genes and miRNAs that ultimately reduce p27 levels [27]. In contrast, members of the FOXO protein family and the protein Menin promote p27 expression thereby resulting in cell cycle arrest in G1. FOXO proteins are located in the nucleus in quiescent cells, and they are able to bind to the promotor of *CDKN1B* and induce its transcription. Upon cell stimulation, FOXO proteins are phosphorylated by Akt and are then exported to the cytoplasm where they can no longer positively regulate *CDKN1B* transcription [28]. Menin, the product of the *MEN1* gene, induces p27 expression by interacting with other TFs such as MLL2 and Ash2 to form a nuclear complex that binds to the *CDKN1B* promoter and activates transcription via histone H3 methylation [27,29].

Translational regulatory elements are usually located in the 5′UTR of genes. So far three such elements have been described for the *CDKN1B* gene. The first is a G/C-rich sequence forming a hairpin loop that overlaps with a short upstream open reading frame (ORF) (-575 to -540). The second element is the ORF itself. Both these regulatory elements are essential for a correct cell cycle-dependent translation pf p27 [30]. The third element is a U-rich sequence (-575 to -420) that is essential for an efficient translation of p27 [31].

Although transcriptional and translational regulation has been reported for *CDKN1B*, the most important mechanism modulating p27 levels works at the posttranslational level and involves the phosphorylation of the protein at different residues, nuclear export mechanisms and, most importantly, ubiquitin-mediated proteasomal degradation. This posttranslational regulation allows the cell to rapidly adjust the levels of p27 in order to respond to extracellular mitogenic stimuli.

Two ubiquitin-mediated degradation pathways have been reported for p27, which are located in different cellular compartments and are required at different stages of the cell cycle. One is mediated by the Skp2-dependent SCF E2 ligase taking place in the G1-S and the G2 phase of the cell cycle in the nucleus. Following the formation of a stable complex involving p27 and CyclinE/CDK2 and upon phosphorylation of p27 at Thr 187 by CyclinE/CDK2, the Skp2 ligase recognizes p27 and signals it for degradation [15]. The other ubiquitin-mediated protein degradation takes place in the cytosol at the G0−G1 transition and involves the Kip1 ubiquitin-promoting complex (KPC). In early G1, p27 is phosphorylated at Ser 10 and shuttled to the cytosol by the carrier protein CRM1. Here it is targeted for proteasomal degradation by the KPC ubiquitin ligase [15,32].

As already hinted above, the activity of p27 depends on its subcellular localization given that both its degradation and the interaction with other proteins is strictly compartmentalized. The regulation of p27 shuttling from nucleus to cytosol and vice versa is complex and involves posttranslational modifications and various interactor proteins (reviewed in [33]). p27 localization is also tightly linked to the cell cycle status of the cell: in growth-arrested cells p27 is localized in the nucleus where it can bind to CyclinE/A-CDK2 complexes, whereas in proliferating cells it is found in the cytoplasm. Phosphorylation of p27 at Thr 157 and Thr 198 prevents its nuclear import thus leading to protein accumulation in the cytosol where it cannot interact with the above mentioned complexes [27].

Given its critical role in cell cycle control, it is no surprise that changes in the expression level or in the localization of p27 are often observed in human cancers. Specifically, downregulation and cytoplasmic localization of p27 have been associated to a more aggressive behavior and poorer prognosis [15,34−37].

8.3 p27 and cancer

8.3.1 p27: an atypical tumor suppressor

p27 was first classified as putative tumor suppressor given that its overexpression suppressed cell proliferation [38]. Subsequently, p27 deficient mice were reported to develop pituitary adenomas and other tumors further supporting its role as tumor suppressor [39−41]. However, due to the extreme rarity of somatic *CDKN1B* mutations identified in human cancers, p27 was until recently classified as "atypical" tumor suppressor. In fact, up to a few years ago, only a few somatic *CDKN1B* mutations had been reported, including a nonsense mutations in an adult T-cell leukemia/

lymphoma patient (p.W76X) [42] and in a breast cancer patient (p.Q104X) [43] as well as a missense sequence alteration in a case of unclassified myeloproliferative syndrome (p.I119T) [44]. Moreover, although the loss of one allele of *CDKN1B* was reported in a few cancer types, the remaining allele was neither mutated nor lost through loss-of-heterozygosity (LOH) [6,45]. The advent of next generation sequencing techniques with deep coverage has led to the identification of somatic *CDKN1B* mutation in a subset of human tumors. In the first such study, Francis et al. [46] discovered heterozygous somatic *CDKN1B* mutations and genomic alterations in small bowel NETs, findings validated 2 years later by two independent studies [47,48]. Importantly, somatic alterations in *CDKN1B* have not only been found in neuroendocrine cells but also in other cancer types thereby expanding the role of this gene in carcinogenesis. For instance, it has been shown that 16% of patients affected by hairy cell leukemia bear a somatic *CDKN1B* mutation thereby representing the second most common genetic alteration in these tumors [49]. Somatic *CDKN1B* mutations were also found in few breast cancer samples [50]. This shows that p27 somatic mutations, although occurring predominantly in NETs, are not limited to these tumors.

p27 finally entered the realm of the *bona fide* tumor suppressors following the discovery of germline *CDKN1B* mutations in patients with multiple endocrine tumors [5]. This discovery led to the search and identification of additional patients with an MEN1-like phenotype carrying germline *CDKN1B* mutations, which culminated in the recognition of MEN4 as a novel multiple endocrine neoplasia syndrome (OMIM: 610755).

As mentioned earlier, the reduction of p27 protein levels is by far more common than genetic mutation, as they were reported to occur in over 50% of all human cancers [15]. In many of these tumors, including breast, colon, lung, head and neck, prostate and ovarian cancer, the downregulation of p27 is associated with aggressive tumor behavior and poor clinical outcome. The causes of the decrease in p27 levels in these neoplasms span from reduced mRNA expression, increased proteasomal degradation, and shuttling of p27 into the cytoplasm [15]. Noteworthy, it has been shown that the role of p27 in the cytoplasm seems to be pro-oncogenic as it is involved in cell migration and invasion, apoptosis, and autophagy [19]. Cytoplasmic p27 modifies cytoskeletal functions by associating with RhoA and thereby having a direct impact on actin dynamics and migration [51,52]. Additionally, p27 regulates cell migration also by binding to microtubules and promoting their stabilization [20].

8.3.2 The role of p27 in pituitary tumorigenesis

The involvement of p27 in pituitary tumorigenesis has been amply demonstrated in animal studies. Indeed, p27-null mice develop pituitary intermediate lobe adenomas [39−41] and heterozygous p27 + / − mice display pituitary hyperplasia [53]. In human pituitary adenomas, downregulation of p27 is frequently observed, especially in corticotropinomas [54−57] whereas somatic *CDKN1B* mutations have not been reported to date. As mentioned above, germline *CDKN1B* mutations cause MEN syndromes in both rats (MENX) and humans (MEN4) which associate with the development of pituitary adenomas [5,58].

8.3.3 Pituitary tumors in *Cdkn1b*-mutated MENX rats

The rat MENX syndrome was first described by Fritz et al. [58]. MENX-affected rats show a tumor phenotype encompassing both MEN1 and MEN2 human syndromes as they develop pituitary adenomas (100% frequency), bilateral pheochromocytoma (100%), medullary thyroid, and endocrine pancreas hyperplasias and tumors. Initially, it was reported to be inherited as a recessive trait with affected rats being homozygous for the underlying genetic mutation. Recently, however, it was demonstrated that heterozygous mutant rats develop the same tumors observed in the homozygous animals, but with a slower progression [59]. This suggests that p27 in this model works as a haploinsufficient tumor suppressor, as already demonstrated in mice [53].

Through linkage studies, the *MENX* locus was mapped to rat chromosome 4 [60] and shortly thereafter the causative mutation was identified in the *Cdkn1b* gene. Specifically, a tandem duplication of eight nucleotides (c.520−528dupTTCAGAC, RefSeq: NM_031762) in exon 2 of the *Cdkn1b* gene was observed in every mutant rat [5]. This mutation causes a frameshift and consequently the encoded p27 variant (hereafter referred to as p27fs177) is 20 amino acids longer than the wild-type protein and has a different C-terminus. In vitro studies could demonstrate that the MENX-associated *Cdkn1b* mutation has no effect on gene transcription or mRNA processing. In contrast, p27fs177 is highly unstable and rapidly degraded [11], thereby causing a drastic reduction of the protein levels in tissues of affected rats in vivo [5]. Thus the MENX-causing *Cdkn1b* change behaves as a loss-of-function mutation.

Pituitary tumors appear at 4 months of age in homozygous mutant rats and at 5−6 months in the heterozygotes, and in both genotypes are multifocal [61]. MENX-associated pituitary adenomas occasionally invade the surrounding structures (Pellegata, unpublished) but no distant metastases have been observed to date. Based on pathology, ultramicroscopy, and hormone profiling, the rat pituitary tumors resemble human gonadotroph pituitary adenomas, which are usually clinically nonfunctioning (NFPAs) [59,61]. Indeed, they express high levels of the common alpha-subunit of the gonadotropin hormones (αGSU) in all stages, whereas the beta subunits (i.e., LHβ and FSHβ) are present in the small lesions but their level decreases with tumor progression, as it occurs in NFPA in humans. Rat pituitary tumors show mitotic activity and have a relatively high Ki67 labeling index in both homozygous [61] and heterozygous mutants [59], thereby quite faithfully recapitulating human aggressive NFPAs [61].

Interestingly, rat and human gonadotroph adenomas also share common genetic signatures, and these similarities have been exploited as to gene discovery and have led to the identification of novel pathways and biomarkers involved in pituitary tumorigenesis and having potential clinical relevance [62].

Being MENX the only spontaneous model of human NFPAs with a complete penetrance of the disease, affected rats have been used for therapy response studies evaluating novel antitumor drugs for their efficacy against gonadotroph tumors. The dual PI3K/mTOR inhibitor BEZ235 was tested both on 3D organotypic cultures of rat primary pituitary tumor cells (in vitro) and in homozygous MENX rats (in vivo) [63] and found to be highly effective at reducing cell proliferation and inducing apoptosis [63].

The molecular and therapy studies conducted on MENX rats have highlighted the translational potential of this animal model in contributing to our knowledge of the molecular mechanisms driving NET progression and in providing a platform for preclinical trials of clinically relevant drugs.

8.4 *CDKN1B* mutations and pituitary tumors

The germline base substitutions so far identified in *CDKN1B*, including those associated with the onset of pituitary tumors, are illustrated in Fig. 8.1 and Table 8.1. All the identified *CDKN1B* base

Table 8.1 Summary of the germline *CDKN1B* mutations and clinical manifestations of the patients.

CDKN1B mutation	Codon	Amino acid	Predicted effect	Clinical manifestations	Age	Sex	Race	Family history	Ref.
5′ UTR									
c.-456_-453delCCTT	—	—	Reduced protein level	AcromegalyWell-differentiated nonfunctional pancreatic NET	62	F	Caucasian	Negative	[64]
c.-29_-26delAGAG	—	—	transcription	Gigantism	30	F	Caucasian	Negative	[65]
Exon 1									
c.59_77dup19	25	K25fs	N.d.	Small-cell neuroendocrine cervical carcinomaACTH-dependent Cushing syndrome1°HPTMultiple sclerosis	47	F	Caucasian	Negative	[10]
c.121_122delTT	41	L41fs	N.d.	1°HPTNonfunctioning pituitary micro- and macroadenomasACTH-dependent Cushing syndromePancreatic NET	38–67	Both	Caucasian	Positive	[66]
c.206C > T	69	P69L	No binding to CDK2, no growth suppression	Papillary thyroid carcinoma with neck lymph node metastasesBilateral multiple lung metastases from bronchial carcinoid 1°HPTNonfunctioning pituitary microadenomaSubcutaneous epigastric lipomaType 2 diabetes mellitus	79	F	Caucasian	Negative	[11]

Table 8.1 Summary of the germline *CDKN1B* mutations and clinical manifestations of the patients. *Continued*

CDKN1B mutation	Codon	Amino acid	Predicted effect	Clinical manifestations	Age	Sex	Race	Family history	Ref.
c.227G > A	76	W76X	Mislocalization, no growth suppression	Acromegaly1°HPT	48	F	Caucasian	Mutation-positive sister with renal angiomyolipoma; her son (mutation-carrier) had testicular cancer	[5]
c.286A > C	96	K96Q	No binding to GRB2	Hhyperprolactinemia, due to a suspected prolactinomaBreast cancer at the age of 41	NA	F	NA	Asymptomatic mutation-positive sister	[8]
c.356T > C	119	I119T	Altered migration in SDS gels	Somatotropinoma	NA	F	NA	The p.I119T change was found in one member of a two person homogeneous FIPA family with somatotropinomas	[8]

1°HPT, *Primary hyperthyroidism;* NA, *not available;* N.d., *not determined.*

substitutions occur in heterozygosity. Only two mutations recurred in more than one patient (Table 8.1). As most of the reported changes in *CDKN1B* are missense substitutions, classifying them as "pathogenic" is not straightforward. However, given that most base changes have not been identified in healthy individuals, and some of them alter the protein's activity, we consider them potentially pathogenic and we hereafter refer to them as "mutations."

Of all the reported *CDKN1B* mutations, six were found in patients displaying pituitary tumors: four of them were GH-secreting adenomas that led to the development of acromegaly (three patients) or gigantism (one patient). The other tumors were an ACTH-secreting adenoma, causing Cushing's disease, and a nonfunctioning pituitary tumor (Table 8.1). This suggests that mutation of *CDKN1B* can predispose different pituitary cell types to tumor formation. The phenotype of the six mutation-positive patients presenting with pituitary tumors is here sequentially discussed according to the position of the causative genetic alteration along the *CDKN1B* sequence from the 5' to the 3' end of the gene (Fig. 8.1).

By studying a cohort of 25 Italian patients with a MEN1-like phenotype, Occhi et al. [64] identified a 4 bp deletion (c.-456-453delCCTT) upstream of *CDKN1B* in a female patient (age 62) presenting with a GH-secreting adenoma and acromegaly, as well as a well-differentiated nonfunctioning pancreatic endocrine tumor. This mutation occurred in the highly conserved ORF located in the 5' UTR region of *CDKN1B* (Fig. 8.1). This mutation deletes the termination codon of the ORF thereby extending its coding sequence and decreasing the distance between the ORF and the beginning of the *CDKN1B* gene. In vitro experiments demonstrated that the shortening of the intercistronic distance between ORF and *CDKN1B* causes a decrease in the rate of translation reinitiation at the *CDKN1B* transcript, thus reducing the expression of p27 [64]. The hypothesis that this mutation decreases the levels of p27 was further supported by the observation that the pancreatic tumor of the mutation-carrier patient had reduced p27 expression when compared to tumors of the same histotype but without mutations. Importantly, this deletion was not found in 600 chromosomes of healthy control individuals. Altogether, the in vitro, ex vivo and population studies support a causative role of the c.-456-453delCCTT mutation in promoting tumor formation via the downregulation of p27 levels and the resulting impairment of its function [64].

Three of the 19 *CDKN1B* mutations so far reported occur in a small repetitive sequence located in the 5' UTR of *CDKN1B*, upstream of the translation start site of p27, which represents a sort of mutational "hot spot" for this gene (Fig. 8.1). One of them, a c.-29_-26delAGAG deletion, was discovered in an Italian female patients manifesting acromegaly at age 30 due to a GH-secreting pituitary adenoma that she originally developed in childhood, at 5 years of age [65]. In vitro assays using a construct where the 5' region of *CDKN1B* containing the wild-type sequence or the c.-29_-26delAGAG deletion, was cloned upstream of a reporter gene demonstrated that this alteration decreases the gene transcription rate [65]. The same mutation was also found in an unrelated Italian lady presenting with 1°HPT [67]. Interestingly, the parathyroid tumor of this patient showed loss of p27 nuclear positivity by immunohistochemistry [67]. Therefore both the in vitro studies [65] and the p27 staining of the tumor tissue suggest [67] that the c.-29_-26delAGAG deletion suppresses the transcription/translation rates of *CDKN1B* ultimately leading to reduced protein expression.

In a Dutch cohort of patients with a MEN1-like phenotype but no *MEN1* mutations, a 19-bp duplication (c.59-77dup19) in exon 1 was identified in a patient presenting with ACTH-secreting pituitary adenoma (Cushing's disease; age 46 years), small-cell neuroendocrine cervical carcinoma (age 45 years), and 1°HPT (age 47 years) [10]. This duplication causes a frameshift after codon 25 (p.K25fs), thus leading to a truncated p27 protein having a different amino acid sequence starting from codon 25 when compared with the wild-type protein. The tissue of the cervical carcinoma was available for molecular analyses and showed LOH at the wild-type *CDKN1B* allele and no p27 protein expression by IHC [10]. Although no additional studies were conducted to assess the effect of this variant, the drastic alteration of the protein sequence with deletion of most of p27's domains supports its pathogenic role.

In a recent study, a *CDKN1B* mutation was found to segregate with a MEN4 phenotype in a large Danish family comprising 13 carriers over two generations [66]. The mutation was a deletion causing a frameshift (c121-122delTT, p. Leu41Asnfs*83). The proband was a young female patient who at the age of 37 was diagnosed with 1°HPT followed by Cushing's disease caused by an ACTH-secreting pituitary microadenoma. Upon genetic testing, the family members carrying this mutation were tested for calcium and parathyroid hormone levels, and underwent imaging to search for pituitary and gastro-intestinal tumors. Beside the proband, three additional family members were found to have nonfunctioning pituitary adenomas. LOH of the wild-type allele was investigated in the pituitary adenoma tissue of the proband, but the results were negative [66]. No additional functional analyses on this mutation were conducted. This represents the largest so far reported MEN4 family.

By analyzing 27 Italian patients with suspected MEN1, Molatore et al. [11] found a missense mutation at codon 69 (c.678C > T, p.P69L) in a female patient diagnosed with papillary thyroid carcinoma and 1°HPT caused by a parathyroid adenoma (age 64 years), bilateral multiple lung metastases of bronchial carcinoid (67 years), and with a nonfunctioning pituitary microadenoma (79 years). Proline 69 is among the residues that mediate the interaction between wild-type p27 and CDK2. In pull-down in vitro assays, the p27P69L variant was found to bind CDK2 with much lower affinity than wild-type p27. Following ectopic expression of p27P69L in various recipient cells, it was observed that this variant is less stable than the wild-type protein in vitro and thus it is present at reduced level in the transfected cells. Moreover, p27P69L has lost the ability of wild-type p27 to suppress cell growth [11]. Immunohistochemical staining of the patient's parathyroid adenoma and bronchial carcinoid with an anti-p27 antibody showed that the levels of p27 were extremely reduced or lost when compared to tumors of the same type but without *CDKN1B* mutations. Further analyses confirmed that the bronchial carcinoid had LOH of the wild-type *CDKN1B* allele [11]. Thus this mutation affects both the amount and the function of p27.

The first germline *CDKN1B* mutation to be identified was a heterozygous nonsense mutation at codon 76 (c.227G > A; p.W76X) discovered in a patient showing a MEN1-like phenotype (i.e., acromegaly and 1°HPT) [5]. Analysis of the available relatives identified a mutation-positive sister of the proband manifesting a renal angiomyolipoma, a MEN1-associated tumor. Since the truncated p27W76X variant has lost the nuclear localization signal (Fig. 8.1), it was predicted to be exclusively localized to the cytoplasm. Indeed, when ectopically expressed in cells in vitro, the p27W76X truncated protein was mislocalized to the cytosol. Accordingly, the angiomyolipoma of the mutation-carrier sister of the proband displayed a cytoplasmic staining for p27 by immunohistochemistry [5]. p27 needs to enter the nucleus in order to interact with Cyclin-CDK complexes and inhibit cell cycle progression. As expected, the cytoplasmic p27W76X variant does not suppress the growth of GH3 pituitary adenoma cells in vitro [5].

Sequence analysis of 124 affected subjects belonging to Familial Isolated Pituitary Adenoma (FIPA) kindreds with no mutations in the *AIP* gene identified two *CDKN1B* point mutations in two female patients: c.286A > C (p.K96Q) and c.356T > C (p.I119T). The K96Q variant occurred in a homogeneous FIPA family presenting with prolactinomas, but this change did not segregate with tumor phenotype. The patient with the K96Q change had hyperprolactinemia due to a suspected prolactinoma that was treated chronically with cabergoline upon referral. Interestingly, this patient developed breast cancer at the age of 41 [8]. Proline 96 is located in the proline-rich region of p27 (amino acids 90−96), which mediates the interaction between wild-type p27 and the adaptor molecule Grb2. The binding of the two molecules ultimately promotes the degradation of p27 in the cytoplasm [68,69]. The K96Q change abolishes the binding of p27 to Grb2 in vitro [8], although the consequences of this mutation on p27's function still have to be elucidated.

The second mutation (i.e., c.356T > C; p.I119T) was found in a member of a two-person homogeneous FIPA family presenting with GH-producing pituitary tumors. The I119T variant affects a residue located in the "scatter domain" of p27 (amino acids 118−158), which is responsible for actin cytoskeletal rearrangement and cell migration, processes involved in metastatic spread of human tumors [70]. When exogenously expressed in cultured cells, the p27I119T protein was more stable than wild-type p27. Moreover, this change causes a shift in the migration of the p27 protein in SDS−PAGE gels. The unique migration pattern of p27I119T, indicative of posttranslational modifications, was not affected by multiple kinase inhibitors, suggesting that it was not due to atypical phosphorylation at this newly created threonine residue [8]. However, whether this change is indeed a pathogenic mutation is still unclear. Noteworthy, the I119T alteration had been previously reported as a somatic change in an unclassified myeloproliferative syndrome [44].

8.5 Conclusions

CDKN1B is one of the susceptibility genes causing the predisposition to pituitary tumors. The majority of the patients reported to date as having a germline *CDKN1B* mutation had a syndromic presentation of various neuroendocrine tumors including pituitary tumors, but occasionally they presented with only pituitary tumors. In MEN4 patients, pituitary adenomas represent the second most prevalent phenotypic feature, the first being 1°HPT. The most common pituitary tumors associated with *CDKN1B* alterations are GH-secreting (somatotropinomas) giving rise to acromegaly or gigantism, although the small number of cases does not allow to make definitive statements about prevalence of the various tumor subtypes. The *CDKN1B* mutations so far functionally characterized in vitro were associated with impairment of p27 function, compatible with its role as tumor suppressor. The identification of additional MEN4 patients will help developing guidelines for the diagnosis and treatment of the disease, and the clinical follow-up of the mutation-positive individuals in the largest family so far identified [66] may give insight into disease penetrance.

References

[1] Daly AF, Jaffrain-Rea M-L, Ciccarelli A, Valdes-Socin H, Rohmer V, Tamburrano G, et al. Clinical characterization of familial isolated pituitary adenomas. J Clin Endocrinol Metab 2006;91(9):3316−23.

[2] Tatsi C, Stratakis CA. The genetics of pituitary adenomas. J Clin Med 2019;9(1).

[3] Brandi ML, Gagel RF, Angeli A, Bilezikian JP, Beck-Peccoz P, Bordi C, et al. Guidelines for diagnosis and therapy of MEN type 1 and type 2. J Clin Endocrinol Metab 2001;86:5658−71.

[4] Alrezk R, Hannah-Shmouni F, Stratakis CA. MEN4 and *CDKN1B* mutations: the latest of the MEN syndromes. Endocr Relat Cancer 2017;24:T195−208.

[5] Pellegata NS, Quintanilla-Martinez L, Siggelkow H, Samson E, Bink K, Höfler H, et al. Germ-line mutations in p27Kip1 cause a multiple endocrine neoplasia syndrome in rats and humans. Proc Natl Acad Sci USA 2006;103:15558−63.

[6] Minaskan Karabid N, Pellegata NS. Multiple endocrine neoplasia-type 4 (MEN4) and other MEN1-like syndromes. In: Colao A, Jaffrain-Rea M-L, Beckers A, editors. Polyendocrine disorders and endocrine neoplastic syndromes. Cham: Springer International Publishing; 2019. p. 1−30.

[7] Costa-Guda J, Marinoni I, Molatore S, Pellegata NS, Arnold A. Somatic mutation and germline sequence abnormalities in CDKN1B, encoding p27Kip1, in sporadic parathyroid adenomas. J Clin Endocrinol Metab 2011;96:E701−6.

[8] Tichomirowa MA, Lee M, Barlier A, Daly AF, Marinoni I, Jaffrain-Rea ML, et al. Cyclin-dependent kinase inhibitor 1B (CDKN1B) gene variants in AIP mutation-negative familial isolated pituitary adenoma kindreds. Endocr Relat Cancer 2012;19:233−41.

[9] Agarwal SK, Mateo CM, Marx SJ. Rare germline mutations in cyclin-dependent kinase inhibitor genes in multiple endocrine neoplasia type 1 and related states. J Clin Endocrinol Metab 2009;94:1826−34.

[10] Georgitsi M, Raitila A, Karhu A, van der Luijt RB, Aalfs CM, Sane T, et al. Germline CDKN1B/p27Kip1 mutation in multiple endocrine neoplasia. J Clin Endocrinol Metab 2007;92:3321−5.

[11] Molatore S, Marinoni I, Lee M, Pulz E, Ambrosio MR, degli Uberti EC, et al. A novel germline CDKN1B mutation causing multiple endocrine tumors: clinical, genetic and functional characterization. Hum Mutat 2010;31:E1825−35.

[12] Cusan M, Mungo G, De Marco Zompit M, Segatto I, Belletti B, Baldassarre G. Landscape of CDKN1B mutations in luminal breast cancer and other hormone-driven human tumors. Front Endocrinol 2018;9(393).

[13] Grey W, Izatt L, Sahraoui W, Ng YM, Ogilvie C, Hulse A, et al. Deficiency of the cyclin-dependent kinase inhibitor, CDKN1B, results in overgrowth and neurodevelopmental delay. Hum Mutat 2013;34:864−8.

[14] Sherr CJ, Roberts JM. CDK inhibitors: positive and negative regulators of G1-phase progression. Genes Dev 1999;13:1501−12.

[15] Chu IM, Hengst L, Slingerland JM. The Cdk inhibitor p27 in human cancer, prognostic potential and relevance to anticancer therapy. Nat Rev Cancer 2008;8:253−67.

[16] Godin JD, Thomas N, Laguesse S, Malinouskaya L, Close P, Malaise O, et al. p27(Kip1) is a microtubule-associated protein that promotes microtubule polymerization during neuron migration. Develop Cell 2012;23:729−44.

[17] Larrea MD, Hong F, Wander SA, da Silva TG, Helfman D, Lannigan D, et al. RSK1 drives p27Kip1 phosphorylation at T198 to promote RhoA inhibition and increase cell motility. Proceedings of the National Academy of Sciences. 2009;106:9268−73.

[18] Wang HC, Lee WS. Molecular mechanisms underlying progesterone-enhanced breast cancer cell migration. Sci Rep 2016;6:31509.

[19] Bencivenga D, Caldarelli I, Stampone E, Mancini FP, Balestrieri ML, Della Ragione F, et al. p27Kip1 and human cancers: a reappraisal of a still enigmatic protein. Cancer Lett 2017;403:354−65.

[20] Baldassarre G, Belletti B, Nicoloso MS, Schiappacassi M, Vecchione A, Spessotto P, et al. p27(Kip1)-stathmin interaction influences sarcoma cell migration and invasion. Cancer Cell 2005;7:51−63.

[21] Belletti B, Pellizzari I, Berton S, Fabris L, Wolf K, Lovat F, et al. p27kip1 controls cell morphology and motility by regulating microtubule-dependent lipid raft recycling. Mol Cell Biol 2010;30(9):2229−40.

[22] Berton S, Pellizzari I, Fabris L, D'Andrea S, Segatto I, Canzonieri V, et al. Genetic characterization of p27kip1 and stathmin in controlling cell proliferation in vivo. Cell Cycle 2014;13:3100−11.

[23] Bicer A, Orlando S, Islam A, Gallastegui E, Besson A, Aligue R, et al. ChIP-Seq analysis identifies p27 (Kip1)-target genes involved in cell adhesion and cell signalling in mouse embryonic fibroblasts. PLoS One 2017;12:e0187891.

[24] Bachs O, Gallastegui E, Orlando S, Bigas A, Morante-Redolat JM, Serratosa J, et al. Role of p27(Kip1) as a transcriptional regulator. Oncotarget 2018;9:26259−78.

[25] Perearnau A, Orlando S, Islam A, Gallastegui E, Martinez J, Jordan A, et al. p27Kip1, PCAF and PAX5 cooperate in the transcriptional regulation of specific target genes. Nucl Acids Res 2017;45:5086−99.

[26] Pippa R, Espinosa L, Gundem G, García-Escudero R, Dominguez A, Orlando S, et al. p27Kip1 represses transcription by direct interaction with p130/E2F4 at the promoters of target genes. Oncogene 2012;31:4207−20.

[27] Hnit SS, Xie C, Yao M, Holst J, Bensoussan A, De Souza P, et al. p27(Kip1) signaling: transcriptional and post-translational regulation. Int J Biochem Cell Biol 2015;68:9−14.

[28] Zhang X, Tang N, Hadden TJ, Rishi AK. Akt, FoxO and regulation of apoptosis. Biochim Biophys Acta 2011;1813:1978−86.

[29] Karnik SK, Hughes CM, Gu X, Rozenblatt-Rosen O, McLean GW, Xiong Y, et al. Menin regulates pan-creatic islet growth by promoting histone methylation and expression of genes encoding p27Kip1 and p18INK4c. Proc Natl Acad Sci USA 2005;102:14659−64.

[30] Goepfert U, Kullmann M, Hengst L. Cell cycle-dependent translation of p27 involves a responsive ele-ment in its 5'-UTR that overlaps with a uORF. Hum Mol Genet 2003;12:1767−79.

[31] Millard SS, Vidal A, Markus M, Koff A. A U-rich element in the 5' untranslated region is necessary for the translation of p27 mRNA. Mol Cell Biol 2000;20:5947−59.

[32] Kamura T, Hara T, Matsumoto M, Ishida N, Okumura F, Hatakeyama S, et al. Cytoplasmic ubiquitin ligase KPC regulates proteolysis of p27(Kip1) at G1 phase. Nat Cell Biol 2004;6:1229−35.

[33] Tomoda K, Kubota Y, Arata Y, Mori S, Maeda M, Tanaka T, et al. The cytoplasmic shuttling and subse-quent degradation of p27Kip1 mediated by Jab1/CSN5 and the COP9 signalosome complex. J Biol Chem 2002;277(3):2302−10.

[34] Armengol C, Boix L, Bachs O, Sole M, Fuster J, Sala M, et al. p27(Kip1) is an independent predictor of recurrence after surgical resection in patients with small hepatocellular carcinoma. J Hepatol 2003;38:591−7.

[35] De Almeida MR, Perez-Sayans M, Suarez-Penaranda JM, Somoza-Martin JM, Garcia-Garcia A. p27 (Kip1) expression as a prognostic marker for squamous cell carcinoma of the head and neck. Oncol Lett 2015;10:2675−82.

[36] Porter PL, Barlow WE, Yeh IT, Lin MG, Yuan XP, Donato E, et al. p27(Kip1) and cyclin E expression and breast cancer survival after treatment with adjuvant chemotherapy. J Natl Cancer Inst 2006;98:1723−31.

[37] Wander SA, Zhao D, Slingerland JM. p27: a barometer of signaling deregulation and potential predictor of response to targeted therapies. Clin Cancer Res 2011;17:12−18.

[38] Polyak K, Lee MH, Erdjument-Bromage H, Koff A, Roberts JM, Tempst P, et al. Cloning of p27Kip1, a cyclin-dependent kinase inhibitor and a potential mediator of extracellular antimitogenic signals. Cell 1994;78:59−66.

[39] Fero ML, Rivkin M, Tasch M, Porter P, Carow CE, Firpo E, et al. A syndrome of multiorgan hyperpla-sia with features of gigantism, tumorigenesis, and female sterility in p27[Kip1]-deficient mice. Cell 1996;85:733−44.

[40] Kiyokawa H, Kineman RD, Manova-Todorova KO, Soares VC, Hoffman ES, Ono M, et al. Enhanced growth of mice lacking the cyclin-dependent kinase inhibitor function of p27[Kip1]. Cell 1996;85:721−32.

[41] Nakayama K, Ishida N, Shirane M, Inomata A, Inoue T, Shishido N, et al. Mice lacking p27[Kip1] display increased body size, multiple organ hyperplasia, retinal dysplasia, and pituitary tumors. Cell 1996;85::707−20.

[42] Morosetti R, Kawamata N, Gombart AF, Miller CW, Hatta Y, Hirama T, et al. Alterations of the p27KIP1 gene in non-Hodgkin's lymphomas and adult T-cell leukemia/lymphoma. Blood 1995;86:1924−30.

[43] Spirin KS, Simpson JF, Takeuchi S, Kawamata N, Miller CW, Koeffler HP. p27/Kip1 mutation found in breast cancer. Cancer Res 1996;56:2400−4.

[44] Pappa V, Papageorgiou S, Papageorgiou E, Panani A, Boutou E, Tsirigotis P, et al. A novel p27 gene mutation in a case of unclassified myeloproliferative disorder. Leuk Res 2005;29:229−31.

[45] Philipp-Staheli J, Payne SR, Kemp CJ. p27(Kip1): regulation and function of a haploinsufficient tumor suppressor and its misregulation in cancer. Exp Cell Res 2001;264:148−68.

[46] Francis JM, Kiezun A, Ramos AH, Serra S, Pedamallu CS, Qian ZR, et al. Somatic mutation of CDKN1B in small intestine neuroendocrine tumors. Nat Genet 2013;45:1483−6.

[47] Crona J, Gustavsson T, Norlen O, Edfeldt K, Akerstrom T, Westin G, et al. Somatic mutations and genetic heterogeneity at the CDKN1B locus in small intestinal neuroendocrine tumors. Ann Surg Oncol 2015;22(Suppl. 3):S1428−35.

[48] Maxwell JE, Sherman SK, Li G, Choi AB, Bellizzi AM, O'Dorisio TM, et al. Somatic alterations of CDKN1B are associated with small bowel neuroendocrine tumors. Cancer Genet 2015;S2210-7762:00184-2.

[49] Dietrich S, Hüllein J, Lee SC, Hutter B, Gonzalez D, Jayne S, et al. Recurrent CDKN1B (p27) mutations in hairy cell leukemia. Blood 2015;126:1005−8.

[50] Stephens PJ, Tarpey PS, Davies H, Van Loo P, Greenman C, Wedge DC, et al. The landscape of cancer genes and mutational processes in breast cancer. Nature 2012;486(7403):400−4.

[51] Besson A, Assoian RK, Roberts JM. Regulation of the cytoskeleton: an oncogenic function for CDK inhibitors? Nat Rev Cancer 2004;4:948−55.

[52] Besson A, Gurian-West M, Schmidt A, Hall A, Roberts JM. p27Kip1 modulates cell migration through the regulation of RhoA activation. Genes Dev 2004;18:862−76.

[53] Fero ML, Randel E, Gurley KE, Roberts JM, Kemp CJ. The murine gene p27Kip1 is haplo-insufficient for tumour suppression. Nature 1998;396:177−80.

[54] Ikeda H, Yoshimoto T, Shida N. Molecular analysis of p21 and p27 genes in human pituitary adenomas. Br J cancer 1997;76:1119−23.

[55] Jin L, Qian X, Kulig E, Sanno N, Scheithauer BW, Kovacs K, et al. Transforming growth factor-beta, transforming growth factor-beta receptor II, and p27Kip1 expression in nontumorous and neoplastic human pituitaries. Am J Pathol 1997;151:509−19.

[56] Kawamata N, Morosetti R, Miller CW, Park D, Spirin KS, Nakamaki T, et al. Molecular analysis of the cyclin-dependent kinase inhibitor gene p27/Kip1 in human malignancies. Cancer Res 1995;55:2266−9.

[57] Takeuchi S, Koeffler HP, Hinton DR, Miyoshi I, Melmed S, Shimon I. Mutation and expression analysis of the cyclin-dependent kinase inhibitor gene p27/Kip1 in pituitary tumors. J Endocrinol 1998;157 (2):337−41.

[58] Fritz A, Walch A, Piotrowska K, Rosemann M, Schäffer E, Weber K, et al. Recessive transmission of a multiple endocrine neoplasia syndrome in the rat. Cancer Res 2002;62:3048−51.

[59] Molatore S, Kügler A, Irmler M, Wiedemann T, Neff F, Feuchtinger A, et al. Characterization of neuroendocrine tumors in heterozygous mutant MENX rats: a novel model of invasive medullary thyroid carcinoma. Endocr Relat Cancer 2018;25:145−62.

[60] Piotrowska K, Pellegata SN, Rosemann M, et al. Mapping of a novel MEN-like syndrome locus to rat Chromosome 4. Mamm Genome 2004;15:135−41.

[61] Marinoni I, Lee M, Mountford S, Perren A, Bravi I, Jennen L, et al. Characterization of MENX-associated pituitary tumours. Neuropathol Appl Neurobiol 2013;39:256−69.

[62] Lee M, Marinoni I, Irmler M, Psaras T, Honegger JB, Beschorner R, et al. Transcriptome analysis of MENX-associated rat pituitary adenomas identifies novel molecular mechanisms involved in the pathogenesis of human pituitary gonadotroph adenomas. Acta Neuropathol 2013;126:137−50.

[63] Lee M, Wiedemann T, Gross C, Leinhauser I, Roncaroli F, Braren R, et al. Targeting PI3K/mTOR signaling displays potent antitumor efficacy against nonfunctioning pituitary adenomas. Clin Cancer Res 2015;21:3204−15.

[64] Occhi G, Regazzo D, Trivellin G, Boaretto F, Ciato D, Bobisse S, et al. A novel mutation in the upstream open reading frame of the CDKN1B gene causes a MEN4 phenotype. PLoS Genet 2013;9: e1003350.

[65] Sambugaro S, Di Ruvo M, Ambrosio MR, Pellegata NS, Bellio M, Guerra A, et al. Early onset acromegaly associated with a novel deletion in CDKN1B 5′UTR region. Endocrine 2015;49:58−64.

[66] Frederiksen A, Rossing M, Hermann P, Ejersted C, Thakker RV, Frost M. Clinical features of multiple endocrine neoplasia type 4: novel pathogenic variant and review of published cases. J Clin Endocrinol Metab 2019;104:3637−46.

[67] Borsari S, Pardi E, Pellegata NS, Lee M, Saponaro F, Torregrossa L, et al. Loss of p27 expression is associated with MEN1 gene mutations in sporadic parathyroid adenomas. Endocrine 2017;55:386−97.

[68] Pagano M, Tam SW, Theodoras AM, Beer-Romero P, Del Sal G, Chau V, et al. Role of the ubiquitin-proteasome pathway in regulating abundance of the cyclin-dependent kinase inhibitor p27. Science 1995;269:682−5.

[69] Vervoorts J, Luescher B. Post-translational regulation of the tumor suppressor p27(KIP1). Cell Mol Life Sci CMLS 2008;65(20):3255−64.

[70] McAllister SS, Becker-Hapak M, Pintucci G, Pagano M, Dowdy SF. Novel p27(kip1) C-terminal scatter domain mediates Rac-dependent cell migration independent of cell cycle arrest functions. Mol Cell Biol 2003;23:216−28.

Multiple endocrine neoplasia syndromes and somatotroph adenomas

Carolina R.C. Pieterman[1] and Steven G. Waguespack[2]

[1]*Section of Surgical Endocrinology, Department of Surgical Oncology, The University of Texas MD Anderson Cancer Center, Houston, TX, United States* [2]*Department of Endocrine Neoplasia & Hormonal Disorders, The University of Texas MD Anderson Cancer Center, Houston, TX, United States*

9.1 Introduction

Gigantism and acromegaly are very rare disorders caused by excessive growth hormone (GH) secretion that develops before and after epiphyseal fusion, respectively. Somatotroph tumors, especially those causing gigantism, are enriched for hereditary diseases, most of which are discussed elsewhere. These neoplasms can be associated with other endocrine tumors in the context of a multiple endocrine neoplasia (MEN) syndrome, chiefly MEN type 1 (MEN1), the initial description of which was a patient who suffered from acromegaly. This chapter provides an updated review of the history, epidemiology, clinical characteristics, and management of somatotroph adenomas occurring as a clinical manifestation of MEN1, discusses the possibility of ectopic acromegaly in MEN1, and reviews the literature regarding pituitary neoplasia in MEN type 2 (MEN2), which is not associated with somatotroph tumors.

9.2 Multiple endocrine neoplasia type 1

MEN1 (OMIM #131100) is an autosomal dominantly inherited neuroendocrine tumor syndrome characterized by the occurrence of hyperplasia and tumors in two or more specific endocrine glands, mainly the parathyroid, the pituitary, and the neuroendocrine cells of the duodenum and pancreas. With an estimated prevalence between 1:20,000 and 1:40,000 [1], MEN1 is rare and considered an orphan disease. It is caused by pathogenic variants in the *MEN1* gene, a classic tumor suppressor gene located on chromosome 11q13.1 and encoding for the menin protein [2,3] (OMIM #613733). Patients with MEN1 have a germline-inactivating mutation in one of the alleles of the *MEN1* gene. In most cases, the mutant allele is inherited from one of their parents, but in 10% an inactivating mutation can occur de novo [4]. One wild-type (normal) *MEN1* allele is sufficient to prevent tumor formation. However, during a patient's lifetime, somatic mutations (the "second hit") in a single cell disrupt the only normally functioning *MEN1* allele, thereby releasing the brake and leading to tumor formation. As these second hits often delete the entire gene, this leads to loss of heterozygosity (LOH) at the *MEN1* locus in tumor DNA. Indeed, LOH has been demonstrated in

Gigantism and Acromegaly. DOI: https://doi.org/10.1016/B978-0-12-814537-1.00008-7

many MEN1-related tumors [5–7]. Because patients with MEN1 already have one mutated copy of the *MEN1* gene in every cell, the hallmark of the syndrome is the early onset and multiplicity of tumors within a single gland/organ.

In 1903, Erdheim published a necropsy case of a patient with acromegaly who had a pituitary adenoma (PA) and enlarged parathyroid glands [8]; this is now considered the first published case of MEN1. In 1945, the first case was reported where a pituitary tumor, parathyroid adenoma, and functional pancreatic neuroendocrine tumors (panNETs) were diagnosed clinically instead of at autopsy [9]. In 1953, Underdahl [10] introduced the term "multiple endocrine adenomas" for the combined occurrence of tumors in the parathyroids, pituitary, and pancreas and recognized a possible familial occurrence. In 1954, Wermer [11] described a family with adenomatosis of the anterior pituitary, the parathyroids, and the pancreatic islets and postulated that this syndrome was caused by an autosomal dominant gene with a high degree of penetrance. In 1988, the *MEN1* gene was mapped to chromosome 11 [7] and finally cloned in 1997 (celebrating its 20th birthday in 2017 [12]); its 610-amino acid protein product was named menin [2,3].

Besides the three classic endocrine tumors associated with MEN1, known as the "three Ps" (parathyroid, pancreas, and pituitary), patients with MEN1 are at risk of developing neuroendocrine tumors of the lung and thymus, neuroendocrine tumors of the stomach, which are mostly associated with the presence of duodenal gastrinomas and the Zollinger–Ellison syndrome, adrenal tumors, meningiomas, and dermatologic manifestations (lipomas, collagenomas, and angiofibromas). Recently, female patients with MEN1 were also shown to have an increased risk of breast cancer [13,14]. Penetrance of the clinical syndrome in MEN1 is high, with almost all germline mutation carriers developing either clinical or biochemical evidence of manifestations by the age of 50 [4]. Penetrance differs for the different manifestations, being 90%–100% for primary hyperparathyroidism (pHPT), 60%–85% for duodenopancreatic neuroendocrine tumors, 40%–80% for pituitary tumors, and 26%–40% for adrenal tumors [4,15–17].

MEN1 can be diagnosed by one of three criteria: (1) a patient who has two of the three primary endocrine tumors associated with the syndrome, (2) a patient who has one of the primary MEN1-associated tumors and a first-degree relative with a clinical MEN1 diagnosis, and (3) identification of a germline (likely) pathogenic *MEN1* DNA variant, regardless of the clinical phenotype [4]. Patients who have a MEN1 diagnosis are advised to undergo regular screening to identify associated tumors at an early stage, thereby preventing later morbidity and mortality from unrecognized disease [4,18,19]. First-degree family members of patients with an identified *MEN1* germline mutation are advised to undergo genetic counseling and predictive genetic testing [4].

9.3 Tumorigenesis of pituitary adenomas in MEN1

The *MEN1* gene consists of 9000 bp of genomic DNA and has 10 exons. Menin, the gene product, is a nuclear protein that is ubiquitously expressed, has no homology to other known proteins, and is without intrinsic enzymatic activity [20]. It is involved in transcriptional regulation as a corepressor or coactivator of transcription through interaction with transcription factors such as JunD and through participation in epigenetic regulation (through histone posttranslational modification, chromatin remodeling, DNA methylation, and noncoding RNA interactions) [20,21]. Menin also has functional contributions to DNA repair, cell signaling, cytoskeletal structure, cell division, cell adhesion, and cell motility [20].

The exact role of menin in pituitary pathogenesis is still largely unknown. In keeping with its role as a tumor suppressor gene, LOH has been demonstrated at 11q13 in pituitary tumors of patients with MEN1 [5,22−24]. Menin seems to play a limited role in the tumorigenesis of sporadic PAs, in contrast to sporadic panNETs, where somatic *MEN1* mutations are found in over 40% [25,26]. In early studies, allelic loss on chromosome 11 was found in 19%−33% of sporadic PAs [22,27]. However, in a follow-up study, none of 23 cases previously shown to have LOH at chromosome 11 had somatic mutations in the *MEN1* gene and the menin transcript was readily detectable [28]. Subsequent studies investigating LOH, somatic mutations, and messenger RNA (mRNA) expression confirmed the limited involvement of *MEN1* in sporadic pituitary tumors [29−32]. However, Theodoropoulou et al. [33] suggested that the involvement of menin in tumorigenesis of sporadic PAs might be higher. They used immunohistochemistry (IHC) to assess the expression of menin in normal pituitary tissue, 68 PAs, and a pituitary carcinoma. They found that 31/68 (46%) of the sporadic PAs had weak menin staining; this did not correlate with any clinical features. In the pituitary carcinoma, the tissue from the pituitary resection had weak menin staining, while metastases (obtained at autopsy) were all immunonegative for menin. The authors suggest that posttranslational mechanisms must play an important role in menin expression and that defects in these mechanisms may be responsible for the reduced levels of menin expression seen in sporadic PAs [33]. Another study [34], on the other hand, compared menin expression as assessed by immunoblotting in 11 PAs with four normal pituitary glands obtained at autopsy and found menin expression in the PAs to be higher than in the control tissue, contradicting the findings of Theodoropoulou et al.

There are different pathways through which menin loss might drive pituitary tumorigenesis, such as transforming growth factor beta pathways [35−37] and cell cycle regulation pathways [38,39].

MicroRNAs have also been implicated in pituitary tumorigenesis. Downregulation of miR-15a and miR-16-1 has been demonstrated in PAs [40], and results in, among other effects, increased cyclin D1 (*CCND1*) expression, which contributes to tumor formation [40]. In heterozygous *Men1* mice, there is decreased expression of miR-15a and miR-16-1 and this correlates with an increase in *CCND1* mRNA and cyclin D1 protein. In cell lines, it was demonstrated that menin regulates the expression of miR-15a [40].

9.4 Epidemiology and clinical characteristics of somatotroph adenomas in MEN1

9.4.1 Somatotroph adenomas in MEN1—historical context

The case now recognized to be the first ever reported case of MEN1 and acromegaly was published by Erdheim in 1903 [8]. Autopsy in a 42-year-old patient revealed a PA and multiple enlarged parathyroid glands; no mention was made of duodenopancreatic tumors. In the early descriptions of MEN1, acromegaly was a prominent feature of the syndrome. Ballard et al. [41], reviewing the 85 published cases of MEN1 (many of which were necropsy cases) up until 1964, found that PAs were seen in 65%; most (42%) were chromophobe adenomas but one in four were somatotroph tumors. When prolactin was ultimately identified, it was soon recognized that lactotroph adenomas were in fact the PA subtype most frequently occurring in MEN1 [42]. Clinical series published in the 1980s and 1990s described a prevalence of PA in patients with MEN1 ranging from 18% to 47%, with 2%−23% of cases being somatotroph adenomas [43−48]. In an older surgical series from Mayo Clinic, 40 patients with MEN1 were described, 18 (44%) of whom had somatotropinomas by IHC, while 15 of these patients had "clinical" acromegaly [49]. In contrast, in the large Tasman 1 MEN1 kindred, no somatotroph adenomas were identified [50]. In those early days, no gigantism cases were described even though the lower limit of the age range was 13.8 years in one of the case series [45]. Based on the early data, PAs in MEN1 were considered to be similar to sporadic PAs in terms of distribution of tumor types (although some series noted an overrepresentation of lactotroph tumors), size, aggressiveness, and response to treatment [45,49,51].

9.4.2 Somatotroph adenomas in MEN1—epidemiology and clinical characteristics

In MEN1 cohorts published after the discovery of the *MEN1* gene (Table 9.1), the overall prevalence of PA ranges from 37% to 72% [15−17,19,52−55,57], and the penetrance by the age of 80 years is estimated to be 49%−58% [17,56]. In most of these studies, somatotroph adenomas made up 3.7%−10.8% of all PAs [15,19,53−55] with the outlier being a Japanese cohort in which 19% of the PAs were somatotropinomas [57]. On the other hand, somatotropinomas are notably absent in the Tasman 1 kindred and families with the MEN1Burin variant [50,61]. Pediatric somatotroph adenomas are extremely rare in MEN1. Of the five studies that report on pediatric, adolescent, or young adult MEN1 manifestations (Table 9.1), four cohorts do not report any somatotropinomas [56,59,60,62], whereas the French Groupe d'étude des Tumeurs Endocrines (GTE) reports one case of acromegaly among 55 PAs occurring below the age of 21 and two cases of GH- and prolactin-secreting adenomas [58]. In case reports of MEN1-associated somatotroph adenomas published since the identification of the *MEN1* gene (Table 9.2) [63−72], two cases of pediatric somatotropinoma associated with gigantism are described, one case of a 6-year-old girl with an invasive somatotroph macroadenoma [68], and the other case of a 5-year-old boy with an invasive mammosomatotroph macroadenoma [70].

After the identification of the *MEN1* gene, five studies (Table 9.3) have specifically reported on MEN1-related PAs: two cohort studies from the French multicenter GTE (one focusing on the clinical characteristics, published in 2002 [73], and the second on the histopathological characteristics, comparing MEN1-related PAs with sporadic cases, published in 2008 [74]); one cohort study from the national multicenter DutchMEN Study Group (DMSG) published in 2015 [56]; one single-center series from China published in 2019 [75]; and one single-center series from the Mayo Clinic [76].

In the initial French study, PAs were identified in 42% of the patients at a mean age of 38 years, 85% were macroadenomas, and there was a female preponderance (69%) (Table 9.3) [73]. In the patients who had PAs, it was the first clinical manifestation of MEN1 in 37%. Thirteen percent of the PAs (*n* = 18) were GH secreting, of which 12 were pure somatotroph tumors and 6 were cosecreting (5 prolactin, 1 alpha-subunit). The patients with tumors solely secreting GH were diagnosed at a mean age of 43.6 years, had a female predilection (9/12), and had macroadenomas (12/12), eight of which were invasive. In 9/12 of the patients with pure somatotroph adenomas, MEN1 was nonfamilial (no details on genotype available for

Table 9.1 Prevalence of PAs and SAs in MEN1 cohorts published since the discovery of the *MEN1* gene in 1997.

	Population	Prevalence of PA	SA characteristics
General cohorts			
Goudet et al. [52] (GTE) Multicenter France/Belgium	$N = 734$ MEN1	$N = 291$ (39.6%) PA 46% of females versus 30% of males Unknown % SA	N/A
Romanet et al. [17][a] Multicenter France/Belgium	$N = 1604$ MEN1 genotype positive	Penetrance PA by age 80: 49%	N/A
Soczomski et al. [53] Multicenter ($N = 2$ referral centers) Poland	$N = 79$ MEN1 genotype positive	$N = 37$ (46.8%) PA Mean age at dx 36.7 years (SD 15.4) *Subtype:* NF 37.8% PRL 43.2% GH 10.8% ACTH 8% LH/FSH 2.7% >50% microadenomas	N/A
Giusti et al. [54] (Italian MEN1 database) Multicenter ($N = 14$ referral centers) Italy	$N = 475$ MEN1 $N = 400$ (84%) genotype positive $N = 75$ (16%) genotype negative	$N = 178$ (37%) PA Female 59% Mean age at dx 33.4 years (SD 14.7) $N = 105$ (63%) microadenomas $N = 73$ (37%) macroadenomas *Subtype:* $N = 36$ (20.2%) NF $N = 120$ (67.4%) PRL $N = 12$ (6.7%) GH $N = 8$ (4.4%) ACTH $N = 2$ (2.5%) cosecreting (PRL-GH and PRL-ACTH) $N = 56$ (31.4%) PA first manifestation of MEN1 PA presenting as first manifestations more often functioning	N/A

(Continued)

Table 9.1 Prevalence of PAs and SAs in MEN1 cohorts published since the discovery of the *MEN1* gene in 1997. *Continued*

	Population	Prevalence of PA	SA characteristics
Goroshi et al. [55] Single center Western India	$N = 18$ MEN1 $N = 11$ genotype positive $N = 2$ genotype negative $N = 5$ not tested	$N = 13$ (72%) PA Female 46% Mean age at dx 26.7 years *Subtype:* $N = 2$ (15%) NF $N = 8$ (62%) PRL $N = 2$ (15%) ACTH $N = 1$ (8%) GH 46% macroadenoma	$N = 1$ SA (micro) treated with SRL for concomitant metastatic insulinoma
de Laat et al. [56] (DMSG) Multicenter (8 university medical centers) The Netherlands	$N = 323$ MEN1 91% genotype positive 9% genotype negative	$N = 123$ (38.1%) PA *Subtype:* $N = 52$ (42%) NF $N = 52$ (42%) PRL $N = 8$ (6.5%) GH $N = 4$ (3.3%) ACTH $N = 2$ (1.6%) LH/FSH $N = 5$ (4%) cosecreting	See Table 9.3
Sakurai et al. [57] (MEN Consortium, Japan) Multicenter (27 consortium members + nonmember physicians) Japan	$N = 560$ MEN1 $N = 347$ genotype positive $N = 72$ genotype negative $N = 141$ not tested	$N = 266$ (47.5%) PAs Mean dx age: Probands 46.1 years Family members 38.9 years $N = 227$ *subtype information available*: $N = 75$ (33%) NF $N = 97$ (43%) PRL $N = 43$ (19%) GH $N = 9$ cosecreting with prolactin $N = 10$ (4.4%) ACTH $N = 1$ (0.4%) TSH 58% microadenomas	Treatment SA: 65% surgical 21% nonsurgical 14% unknown
Machens et al. [16] Multicenter ($N = 6$ referral centers) Germany	$N = 258$ MEN1 genotype positive	PAs 40% Median age at dx 37 years Subtype not specified	N/A
Lourenço-Jr et al. [19] Single center Brazil	$N = 52$ MEN1 genotype positive	PAs 51.9% *Subtype*: 29.6% NF 66.7% PRL 3.7% GH	N/A

Table 9.1 Prevalence of PAs and SAs in MEN1 cohorts published since the discovery of the *MEN1* gene in 1997. *Continued*

	Population	Prevalence of PA	SA characteristics
Cohort data specific for pediatric, adolescent, and young adult patients			
de Laat et al. [56] (DMSG) Multicenter (8 university medical centers) The Netherlands	$N = 323$ MEN1 $N = 123$ (38%) PAs	$N = 11$ (9%) pediatric cases *Subtype:* $N = 7$ NF (all microadenomas) $N = 3$ PRL $N = 1$ ACTH	No SA among pediatric cases
Goudet et al. [58] (GTE) Multicenter France + Belgium	$N = 160$ MEN1 + manifestations <21 years of age $N = 55$ (34%) PAs	Female 76% Mean age at dx 17 years (SD 4) Diagnosed 6−10 years: $N = 2$ both 10 years (0% symptomatic, 0% resection) Diagnosed 11−15 years: $N = 25$ (36% symptomatic, 16% resection) Diagnosed 16−20 years: $N = 28$ (86% symptomatic, 18% resection) *Subtype* $N = 14$ (25%) NF $N = 38$ (69%) PRL $N = 2$ GH cosecretion $N = 1$ (2%) GH $N = 1$ (2%) ACTH $N = 1$ (2%) unknown *Size* $N = 27$ (49%) Hardy 1 $N = 15$ (27%) Hardy 2 $N = 7$ (13%) Hardy 3 $N = 6$ (11%) Hardy 4 PAs Hardy 3/4 in 54% of males versus 4% of females 21% PA first manifestation of MEN1	$N = 3$ GH-secreting PA (unclear if clinical or IHC), of which 2 were PRL cosecreting GH $N = 1$ clinical acromegaly

Herath et al. [62] (Tasman 1 kindred)
 	
Single center
 	
Australia
	$N = 180$ MEN1 patients born 1843−2018
	
	$N = 96$ born before 1965; retrospective cohort before prospective surveillance
	$N = 84$ born 1965−2015; prospective cohort
	
	Manifestations of disease <22 years investigated
		Retrospective cohort:
		$N = 3$ (3%) PA
		  $N = 2$ PRL
		   ($N = 1$ macroadenoma aged 21 years,
		   $N = 1$ microadenoma aged 16 years)
		$N = 1$ died of intracranial hemorrhage aged 9 years with possible pituitary apoplexy
		Prospective cohort:
		$N = 14$ (19.7%) PA
		  $N = 3$ NF
		  $N = 9$ PRL
		  $N = 2$ ACTH
			No SA

(*Continued*)

Table 9.1 Prevalence of PAs and SAs in MEN1 cohorts published since the discovery of the *MEN1* gene in 1997. *Continued*

	Population	Prevalence of PA	SA characteristics
Vannucci et al. [59] Single center Italy	N = 22 children and adolescents with MEN1 aged 6–31 years at the time of enrollment N = 22 (100%) genotype positive N = 21 presymptomatic genetic testing N = 1 index case	N = 7 PA N = 5 microPRL (4 females, 1 male), asymptomatic N = 2 macroPRL, both males (N = 1 index case) Youngest age at PA dx 13 years	No SA
Manoharan et al. [60] Multicenter (prospective databases Marburg + Heidelberg) Germany	N = 20 diagnosed with a MEN1 manifestation <18 years Screening started at 16 years or earlier in case of symptoms	N = 6 (30%) PA N = 4 microPRL N = 2 NF	No SA

ACTH, *Adrenocorticotropic hormone;* DMSG, *Dutch MEN Study Group;* dx, *diagnosis;* FSH, *follicle-stimulating hormone;* GH, *growth hormone;* GTE, *Groupe d'etude des Tumeurs Endocrines;* IHC, *immunohistochemistry;* LH, *luteinizing hormone;* MEN1, *multiple endocrine neoplasia type 1;* N/A, *not available;* NF, nonfunctioning; PA, *pituitary adenoma;* PRL, *prolactinoma;* SA, *somatotroph adenoma;* SD, *standard deviation;* SRL, *somatostatin receptor ligand;* TSH, *thyroid-stimulating hormone.*
[a]*Partial overlap in population with Goudet et al. [52].*

these patients). Comparison of MEN1-related with sporadic PAs (for the entire group, no separate comparison for somatotropinomas available) showed that age at diagnosis and adenoma subtype were not different. MEN1-related PAs were more often macroadenomas and consequently had more symptoms related to tumor size. In the follow-up study by Trouillas on the histopathological characteristics of PAs that were resected, 23 tumors stained positively for GH, 18 of which were plurihormonal based on IHC [74]. For the entire MEN1 PA group, compared with sporadic tumors, patients with MEN1 more often had double adenomas (4% vs 0.1%) or plurihormonal adenomas, especially GH-immunoreactive plurihormonal adenomas. Markers of proliferation were not different, but MEN1-related PAs were more commonly histologically invasive.

The Dutch series, published almost a decade later, consisted of a highly screened cohort [56] (Table 9.3). Screening by magnetic resonance imaging (MRI) was performed in 87% and biochemical screening with prolactin and insulin-like growth factor 1 (IGF-1) in 97% and 92%, respectively; 66% of the PAs were diagnosed by screening. The prevalence of PA was comparable to the French series (38%), and in patients with a PA, it was the first manifestation of MEN1 in 29%. In distinction to the French cohort, only 33% of the tumors were macroadenomas. PAs diagnosed before an MEN1 diagnosis were more often macroadenomas and more often functioning compared with those diagnosed by screening. Somatotropinomas were seen in eight patients (6.5% of PAs) at a mean age of 51 years: six were females and six had macroadenomas. In seven patients, a somatotroph adenoma was the earliest presentation of MEN1; only one patient was diagnosed during follow-up

Table 9.2 Case reports of MEN1-associated somatotroph adenomas, published after discovery of the *MEN1* gene in 1997.

	Sex, age (years)	Index	*MEN1* genotype	SA first MEN1 tumor	Clinical characteristics	Treatment	Pathology	Outcome
Chung et al. [63]	M, 19	Yes	Not stated[a]	Yes	Rapid progression of scoliosis IGF-1 452 ng/mL (ULN 414) OGTT not mentioned PRL 432 mIU/L (ULN 375) Macroadenoma	1. TSS	Plurihormonal PA, mainly GH +	Hypopituitarism f/u unk
Gezer et al. [64]	M, 35	Yes	Not done[b]	Yes (also pHPT; PA was the reason for seeking medical care)	Acromegaly Elevated IGF-1, GH not suppressed on OGTT Macroadenoma	1. TSS 2. SRL for persistence 3. Add DA 4. MRI 2 micro-adenomas so TSS 5. relapse: SRL	1. IHC GH + ; Ki-67 8% 2. IHC GH + PRL + ; Ki-67 4%	Unk f/u 7 years
Herrero-Ruiz et al. [65]	F, 35	Yes	Pos.	Yes (synchronous pHPT and panNET; PA was the reason for seeking medical care)	Acromegaly Diabetes with ketosis Macroadenoma, spread to cavernous sinus, dis- placement of optic chiasm	1. SRL + DA 2. TSS 3. SRL	Densely granulated SA IHC GH+ + + P53 neg. Ki-67 <1%	Biochemically controlled f/u unk
Uraki et al. [66]	M, 54	Yes	Pos.	Yes (synchronous pHPT and gastrinoma were the reasons for seeking medical care)	Hyperpigmentation Confirmed CD HyperPRL IGF-1 >2 SD (no OGTT) No acromegaly No Cushingoid features No symptoms of hyperPRL macroadenoma with bilateral cavernous sinus invasion	1. DA 2. SRL (added for metastatic gastrinoma)	Autopsy pituitary (IHC): ACTH: strong + GH + LH weak + TSH/FSH −	Biochemically controlled Died due to metastatic gastrinoma f/u 3 years
Liu et al. [67]	M, 31	Yes	Pos.	Yes (synchronous pHPT, panNET, and ADR; hyperglycemia and kidney stones were the reasons for seeking medical care)	PRL >200 ng/mL (ULN 13.13) IGF-1 922 ng/mL (ULN 307), OGTT failed to suppress GH Enlarged pituitary on imaging; no adenoma	DA with compliance issues	N/A	IGF-1 normal, OGTT improved, PRL decreased Size of pituitary decreased f/u 1 year
Nozières et al. [68]	F, 6	Unk	Pos.	Unk	Gigantism (accelerated growth) Invasive macroadenoma	1. TSS 2. SRL + multiple operations 3. Add DA 4. Proton therapy	N/A	Uncontrolled f/u 3 years

(Continued)

Table 9.2 Case reports of MEN1-associated somatotroph adenomas, published after discovery of the *MEN1* gene in 1997. *Continued*

	Sex, age (years)	Index	*MEN1* genotype	SA first MEN1 tumor	Clinical characteristics	Treatment	Pathology	Outcome
Kontogeorgos et al. [69]	M, 19	Yes	Not done[c]	Yes (synchronous pHPT and panNET; PA was the reason for seeking medical care)	Gigantism (height gain and enlargement of feet) Bitemporal hemianopsia Provoked galactorrhea PRL 80,000 mU/L (ULN 400) GH elevated; not suppressed on OGTT Macroadenoma with suprasellar extension and obstructive hydrocephalus	1. DA + TSS	Chromophobic adenoma IHC primarily PRL with a subset of GH Monosomy chromosome 11 in PA	Unk
Stratakis et al. [70]	M, 5	Yes	Pos.	Yes	Gigantism (headaches + accelerated growth + coarse facial features) PRL 1600 ng/mL (ULN 11) GH not suppressed on OGTT Macroadenoma with erosion of sella floor and compression on optic chiasm	1. DA 2. TSS 3. DA + SRL	Mammosomatotroph adenoma, IHC PRL +, GH + LOH for the *MEN1* locus	PRL decreased IGF-1 normal f/u unk
Hermans et al. [71]	F, 69	Yes	Not done[d]	No	Diagnosed by screening; no symptoms of acromegaly IGF-1 alternating elevated and normal, GH not suppressed on OGTT Pituitary microadenoma	No therapy	N/A	f/u unk
Schuppe et al. [72]	M, 36	Unk	Pos.	No	Secondary infertility PRL 1400 Initially no PA visible on imaging After DA and f/u, acromegaly developed (IGF-1 >1150 (ULN 330) Macroadenoma	1. DA 2. Add SRL 3. TSS	N/A	Biochemically controlled f/u 10 years

ACTH, Adrenocorticotropic hormone; ADR, adrenal adenoma; CD, Cushing disease; DA, dopamine agonist; F, female; FSH, follicle-stimulating hormone; IGF-1, insulin-like growth factor 1; IHC, immunohistochemistry; LH, luteinizing hormone; LOH, loss of heterozygosity; M, male; MEN1, multiple endocrine neoplasia type 1; MRI, magnetic resonance imaging; N/A, not available; neg., negative; OGTT, oral glucose tolerance test; PA, pituitary adenoma; pHPT, primary hyperparathyroidism; panNET, pancreatic neuroendocrine tumor; Pos., positive; PRL, prolactin; SA, somatotroph adenoma; SD, standard deviation; SRL, somatostatin receptor ligand; TSH, thyroid-stimulating hormone; TSS, transsphenoidal surgery; ULN, upper limit of normal; unk, unknown.

Not included in this table are cases in which genetic testing was not performed and there was a negative family history, and the clinical diagnosis of MEN1 rested solely upon pHPT and PA only. Cases with germline MEN1 variants of uncertain pathogenicity are also not included.

[a]Clinical MEN1; this patient was also diagnosed with an insulinoma.

[b]Clinical MEN1; this patient also developed a panNET.

[c]Clinical MEN1; this patient had synchronous pHPT and panNET.

[a]Clinical MEN1; this patient presented with Zollinger—Ellison syndrome/gastrinoma and multiple recurrences of pHPT.

Table 9.3 Original research on MEN1-related PAs published after the identification of the *MEN1* gene in 1997.

	Cohort	PA characteristics	SA: clinical	SA: pathology	SA: treatment
Verges et al. [73] (GTE) Multicenter retrospective	$N = 324$ Genotype positive in 175/197 tested $N = 136$ (42%) PA	Female, $N = 94$ (69%) PA more frequent in females (50% vs 31%, $P < .001$) Mean age at dx 38 ± 15.3 years 25% dx <26 years, more often after 1994 50/136 (37%) PA first manifestation of MEN1 Macroadenomas 85% $N = 20$ (15%) NF $N = 85$ (63%) PRL $N = 12$ (9%) GH $N = 6$ (4%) ACTH $N = 13$ (10%) cosecreting	$N = 18$ GH (13%) $N = 12$ GH $N = 5$ prolactin cosecreting $N = 1$ alpha-subunit cosecreting *PAs solely secreting GH:* $N = 9$ (75%) female Mean age at dx 43.6 years $N = 12$ (100%) macroadenoma $N = 8$ (67%) invasive $N = 8$ (67%) dx before other MEN1 lesions $N = 9$ (75%) nonfamilial MEN1	N/A	*PAs solely secreting GH* $(N = 12)$: $N = 8$ Sx $N = 5$ RT $N = 3$ Rx *Result:* $N = 6$ cure $N = 4$ persistent $N = 2$ hypopituitarism
Trouillas et al. [74] (GTE) Multicenter retrospective	$N = 77$ resected PA 48% genotype positive 39% genotype negative 13% not tested $N = 64$ also reported by Verges et al.	Sex ratio female/male 1.75 Mean age at surgery 41 years 72% functional PAs	N/A, unclear if all PAs' staining positive for GH were associated with clinical acromegaly/ gigantism	*$N = 71$ with IHC:* $N = 23$ (23%) GH of which $N = 18$ (78%) plurihormonal Plurihormonal cases: GH–alpha-subunit (3) GH–FSH–LH (1) GH–TSH (1) GH–prolactin–ACTH (1) GH–prolactin–alpha-subunit (3) GH–prolactin-LH–FSH (4) GH–prolactin–TSH (1) GH–prolactin (4)	All patients were surgically treated by inclusion criteria, other treatments not specified

(Continued)

Table 9.3 Original research on MEN1-related PAs published after the identification of the *MEN1* gene in 1997. *Continued*

	Cohort	PA characteristics	SA: clinical	SA: pathology	SA: treatment
de Laat et al. [56] (DMSG) Multicenter retrospective	*N* = 323 91% genotype positive, 9% genotype negative *Entire cohort:* MRI screening 87% Biochemical screening (prolactin, IGF-1) in 97% and 92% *N* = 123 (38.1%) PA	Female, *N* = 78 (63%) *N* = 36/123 (29%) PA first manifestation of MEN1 *N* = 57 (46.3%) dx before MEN1 *N* = 66 (54%) detected by screening PAs dx before MEN1 more often macroadenomas and functioning compared with screening-detected PAs Age-related penetrance of PA ca. 58% at age 80 Macroadenomas 33% *N* = 52 (42%) NF *N* = 52 (42%) PRL *N* = 8 (7%) GH *N* = 4 (3%) ACTH *N* = 2 (2%) LH/FSH *N* = 5 (4%) cosecreting	*N* = 8 (7%) GH (unclear if GH + in cosecreting tumors) *N* = 6 (75%) female Mean age at dx 51 years *N* = 6 (75%) macroadenoma *N* = 7 dx prior to MEN1 diagnosis, only *N* = 1 diagnosed during follow-up *N* = 6 genotype negative *N* = 1 genotype positive *N* = 1 unknown	*N* = 7 *Sx:* *N* = 1 (genotype positive): atypical adenoma Mitoses 3 *N* = 6 (genotype negative): *N* = 5 Ki-67 <1% *N* = 1 Ki-67 2% Necrosis was not seen	*N* = 6 hormonal control; *N* = 4 Sx + Rx *N* = 1 Sx/Rx/RT *N* = 1 Rx only *N* = 2 persistence with Rx so subsequent Sx with resultant stable hormone levels and no tumor regrowth
Wu et al. [75] Single-center retrospective	*N* = 151 MEN1 (*N* = 6 genetically tested, *N* = 4 genotype positive) *N* = 54 (36%) PA	Female, *N* = 28 (52%) Mean age at dx 53.9 years (SD 17.8) *N* = 26 (48%) NF *N* = 15 (28%) PRL *N* = 5 (9%) GH *N* = 3 (6%) ACTH *N* = 5 (9%) cosecreting of which *N* = 3 GH +	*N* = 8 (15%) GH (all acromegaly) *N* = 4 (50%) female Mean age at dx 47.5 years *N* = 1 genotype positive *N* = 7 not tested *N* = 4 (50%) ≥ Knosp III	N/A	*N* = 8 (100%) Sx *N* = 2 followed by SRL *N* = 7 initial normalization of hormonal hypersecretion *N* = 1 persistence for which gamma-knife treatment *N* = 1 relapse and second surgery

Table 9.3 Original research on MEN1-related PAs published after the identification of the *MEN1* gene in 1997. *Continued*

	Cohort	PA characteristics	SA: clinical	SA: pathology	SA: treatment
Cohen-Cohen et al. [76] Single-center retrospective	N = 268 MEN1 Genetic testing N = 101 (37.7%), positive N = 90 (89.1%) N = 139 (52%) PA	Female, N = 90 (65%) PA more frequent in females (65% vs 35%, P < .005) Median age at dx 36 years (range 5–80) Macroadenomas 27% N = 60 (43%) NF N = 60 (43%) PRL N = 6 (4%) GH N = 4 (3%) ACTH N = 9 (6%) cosecreting, all at least PRL	N = 6 (4%) GH (unclear if GH + in cosecreting tumors) N = 4 (67%) female Median age at dx 47 (range 37–69) N = 2 (33%) macroadenoma N = 1 (17%) Knosp III/IV N = 1 (17%) visual symptoms	N/A	N = 4 Sx N = 3 remission

ACTH, Adrenocorticotropic hormone; DMSG, Dutch MEN Study Group; dx, diagnosis; FSH, follicle-stimulating hormone; GH, growth hormone; GHRH, growth hormone–releasing hormone; GTE, Groupe d'etude des Tumeurs Endocrines; IGF-1, insulin-like growth factor 1; IHC, immunohistochemistry; LH, luteinizing hormone; MEN1, multiple endocrine neoplasia type 1; MRI, magnetic resonance imaging; N/A, not available; NF, nonfunctioning; PA, pituitary adenoma; PRL, prolactinoma; RT, radiotherapy; Rx, medical therapy; SD, standard deviation; SRL, somatostatin receptor ligand; Sx, surgical resection; TSH, thyroid-stimulating hormone.

after an MEN1 diagnosis. Of the eight patients with acromegaly, six had a clinical MEN1 diagnosis but negative comprehensive genetic testing. Of the seven patients who were operated, one patient (*MEN1* genotype positive) had an atypical adenoma with three mitoses; in the others (*MEN1* genotype negative), Ki-67 was <1% in five and 2% in one.

In the Chinese cohort, PAs were seen in 36% of the patients at a mean age of 53.9 years and the male/female ratio was roughly equal (52% females) [75] (Table 9.3). Somatotroph adenomas were diagnosed in eight patients (15%), three of whom had cosecreting tumors (one adrenocorticotropic hormone, two prolactin). Half of the patients with somatotropinomas were female and all had clinical acromegaly with elevated IGF-1 levels. Half of the tumors were invasive with a Knosp grade of ≥ 3.

In a recently published series from the Mayo Clinic, the prevalence of PAs (52%) was the highest of these cohorts [76] (Table 9.3). Similar to the French and Dutch series, PAs were more frequent in females. Median age at diagnosis was 36 years, and in this series, the lowest percentage of macroadenomas (27%) was seen. Only six patients (4%) had GH-producing adenomas, diagnosed at a median age of 47 years and none at a pediatric age. Four of these patients were females and two had macroadenomas. Surgical resection was performed in four, resulting in remission in three.

Combined data from these studies show that PAs in MEN1 have the same distribution of subtypes and the same female predominance as in the general population [77,78]. Somatotroph adenomas are reported to make up 11%−13% of PAs in population-based studies [77,78], and the prevalence in MEN1 is comparable or even lower. The reported mean age at diagnosis of somatotropinomas in MEN1 is 43−51 years in the large cohorts [56,73,76], which is similar to population-based studies (mean age 43 years, median 47 years) [77] and large acromegaly studies (median age 45 years) [79]. Age was significantly lower in the 10 published case reports (mean 31 years, median 33 years) (Table 9.2) [63−72], which reflects the inclusion of the rare cases of gigantism. Despite the French GTE studies demonstrating that macroadenomas were considerably more frequent in MEN1 than in sporadic tumors, recent studies report a much lower frequency of macroadenomas, probably reflecting the effect of biochemical and radiological screening [56,73,75,76]. In most series, somatotroph tumors in MEN1 are predominantly macroadenomas, consistent with the presentation of sporadic disease [78−80]. Multiple synchronous PAs and plurihormonal PAs are rare, even in MEN1, but compared with sporadic PAs, the frequency is higher in MEN1. Plurihormonal PAs associated with MEN1 are often GH immunoreactive.

For an accurate characterization of MEN1-related somatotroph adenomas, knowledge regarding the germline genotype status is very important. In patients without a positive family history who have a clinical diagnosis of MEN1 based on the combination of pHPT and PA, germline *MEN1* (likely) pathogenic variants are detected in only a minority of cases [81,82]. One study specifically focused on patients with clinical MEN1 and somatotropinoma (n = 24), and none of the 18 patients with acromegaly and pHPT who had germline genetic testing were found to have a pathogenic variant in the *MEN1* gene [83]. In addition, the clinical course of these genotype-negative, clinical MEN1 patients appears different. Recent data show that such patients rarely develop additional MEN1 manifestations and that their overall survival is more comparable to the general population than to patients with MEN1 [15,84]. This subgroup of clinical MEN1 patients also seems to have a higher proportion of somatotroph adenomas [81,82,84,85]. Given all of the published data, patients with sporadic (i.e., without a family history), genotype-negative clinical MEN1 based on a diagnosis of pHPT and PA may represent a different clinical entity, separate from genotype-positive MEN1. Of note, in the GTE pituitary cohort, two-thirds of patients with somatotroph adenomas had sporadic MEN1 and genotype status was not reported, but only

50% of the cohort was genetically tested at that time [73]. In the DMSG pituitary cohort, 75% of the somatotroph adenoma patients were genotype negative [56]. In the Chinese study, only one of the eight patients with a somatotroph adenoma had positive genetic testing. In the other patients (six with a panNET and one with pHPT), genetic testing was not performed [75].

There are no MEN1-specific data on acromegaly related comorbidity. However, in the context of MEN1, the interaction of excessive GH production with other endocrine neoplasms should be considered. For example, both pHPT and acromegaly are considered risk factors for cardiovascular disease and might have a cumulative effect on cardiovascular morbidity and mortality in MEN1. As another example, the effect of increased GH levels may ameliorate the symptoms of insulinoma, while on the other hand, glucagon-secreting panNETs may potentiate the diabetogenic effects of GH. Another important consideration is the fact that pancreatic and bronchopulmonary NETs can ectopically secrete GH or growth hormone−releasing hormone (GHRH) and thereby cause acromegaly [86]. Thus ectopic acromegaly should be considered in any MEN1 patient with acromegaly and a normal pituitary MRI or, in the case of ectopic GHRH secretion, MRI findings suggestive of pituitary hyperplasia. As pituitary somatotroph hyperplasia is not associated with MEN1, this finding on histopathology should also prompt the search for ectopic GHRH production. In the French GTE histopathology series, four cases of pituitary hyperplasia were noted among patients with MEN1, three of which were due to ectopic GHRH secretion from a panNET [74]. It is therefore important to consider ectopic acromegaly in MEN1-related acromegaly so as not to subject the patient to unnecessary pituitary surgery.

9.5 Apparently sporadic somatotroph adenomas: when to think of MEN1

Even before the discovery of the *MEN1* gene, it was recognized that MEN1 was infrequently seen among patients with acromegaly, with a reported prevalence of 0%−3.3% [87,88]. Shortly after the identification of the *MEN1* gene, Teh et al. [89] found no *MEN1* germline mutations in eight families with familial acromegaly. In a Turkish study among 20 patients from 13 familial isolated pituitary adenoma (FIPA) families, an *MEN1* mutation was detected only in 3 (15%) patients [90]. In a Spanish study of at-risk patients with PAs, no *MEN1* germline mutation was identified in 3 members of acromegaly-containing FIPA families, 1 patient with pituitary gigantism, 17 patients with somatostatin receptor ligand (SRL)−resistant somatotropinomas, and 4 patients with somatotroph macroadenomas diagnosed ≤ 30 years [91]. In a series from the national institutes of health (NIH), no *MEN1* mutation was found in three pediatric patients with isolated somatotroph adenomas and one *MEN1* mutation (this patient was the 5-year old previously mentioned [70]) was identified among four pediatric patients (25%) with somatotroph adenomas and a family history of endocrine tumors [92]. Among six adult cases of familial/syndromal somatotropinoma, no *MEN1* mutation was identified [92]. In 44 Australian patients with diagnosis of a PA ≤ 40 years of age or a personal or family history of endocrine neoplasia, no *MEN1* (likely) pathogenic variants were identified among 19 patients with somatotroph adenomas [93]. In a French study by Cuny et al. [94], among 79 patients diagnosed with a sporadic somatotroph macroadenoma before the age of 30, only 1 patient was found to have a *MEN1* likely pathogenic variant (1.3%). In patients with sporadic somatotropinomas, Cazabat et al. [95] found no *MEN1* mutation in 28 patients under 30 years old without mutations in the aryl hydrocarbon receptor−interacting protein (*AIP*) gene. Finally, in a recent series of 414 patients with acromegaly, 6.6% ($n = 24$) were found to have clinical MEN1; however, none of the 18 patients who underwent germline genetic testing was identified with a (likely) pathogenic *MEN1* variant [83].

Taken together, these data show that *MEN1* (likely) pathogenic variants are rare among patients with apparently sporadic somatotroph adenomas and that genetic analysis for MEN1 should only be considered in familial cases, especially if other components of MEN1 besides PA are diagnosed in the family. Recently published systematic reviews provide recommendations on genetic analysis in patients with sporadic PAs (not specific for somatotropinomas), and genetic testing for *MEN1* is recommended in young ($\leq$30 years at diagnosis) patients, especially those with lactotroph tumors [96,97]. The authors' approach has been to screen periodically for pHPT in young patients presenting with an apparently sporadic PA. Nevertheless, for children with somatotroph tumors presenting with gigantism, pathogenic variants in *AIP* (OMIM #605555) or *Xq26*.3 microduplications (X-linked acrogigantism) (OMIM #300942) will be much more common [96–98].

9.6 Screening and surveillance of pituitary tumors in MEN1

Because PAs are prevalent in MEN1 and cure rates (from both a tumor perspective and the normalization of hormonal hypersecretion) improve when tumors are detected earlier, it is recommended that patients with MEN1 undergo regular surveillance for PAs. Current guidelines advise annual measurement of prolactin and IGF-1 and MRI of the pituitary every 3 years if no abnormalities are detected [4]. In prepubertal children with MEN1, an annual auxological assessment to identify perturbations in linear growth and an assessment of symptoms that could suggest an underlying PA (headaches, visual changes, etc.) are also critical. If abnormalities are detected, workup according to established PA guidelines is advised, and further follow-up depends on the findings and should not differ from what is advised in sporadic PAs. Given that PAs are diagnosed in MEN1 as young as 5 years of age [70], guidelines recommend starting screening for PAs at the age of 5 [4], although the authors typically defer MRI screening until the age of 15 years in the absence of abnormal screening labs or concerning signs/symptoms.

9.7 Treatment of somatotroph adenomas in MEN1

Treatment of MEN1-related somatotroph PAs is similar to sporadic cases [4,80]. Briefly, transsphenoidal surgery (TSS) is the preferred first-line treatment, with remission rates of >85% for microadenomas and 40%–50% for macroadenomas [80]. With persistent disease, SRLs are generally used as first-line medical therapy, with dopamine agonists (DAs) and the GH receptor antagonist pegvisomant being additional treatment options [80]. Combination therapy is considered if hormonal control is not achieved. The Endocrine Society acromegaly guidelines suggest using radiotherapy, preferably stereotactic radiotherapy, in the setting of a residual mass after surgery and if medical therapy is unavailable, unsuccessful, or not tolerated [80]. It is unclear if having MEN1 increases the long-term risks of radiation therapy, such as the development of secondary tumors.

The number of MEN1 patients with acromegaly for which treatment information is available in the literature is very limited. In the French GTE pituitary study (Table 9.3), it was reported that treatment of functional PAs was less successful in MEN1 compared with sporadic PAs, with biochemical control in 42% of MEN1 cases in contrast to 90% of sporadic cases [73]. For the acromegaly patients ($n = 12$), treatment included surgery in eight, radiotherapy (not further specified) in five, and medical treatment (not further specified) in three. This led to biochemical control in 8/12 (67%) patients at the expense of hypopituitarism in 2. In the DMSG pituitary study ($n = 8$ somatotroph adenomas), five somatotroph adenomas were controlled with surgery and medication (one also had radiotherapy), one patient achieved biochemical control with medication alone, and two other patients treated with medication only had additional surgery to attain stable hormone levels (Table 9.3) [56]. In the Chinese study, all ($n = 8$) patients with somatotroph adenomas were operated, two of whom received SRLs afterwards (Table 9.3) [75]. This led to initial biochemical control in seven, one of whom relapsed and underwent a second surgery. In one patient, there was persistence for which stereotactic radiotherapy was given [75]. In the Mayo Clinic series, four patients underwent surgical resection, with remission in three [76]. In the 10 case reports (Table 9.2), one patient was not treated due to mild disease [71]. In the other nine cases, biochemical control (normal IGF-1 level) was attained in five, mostly after multiple lines of therapy [65–67,70,72], and uncontrolled in one case of pediatric gigantism treated with TSS multiple times, SRLs, DA, and proton therapy [68]. In two cases, the final outcome was not reported [64,69], and in one case with an unknown duration of follow-up, the patient had panhypopituitarism after surgery [63]. Although the numbers are too small to draw any firm conclusions, it would appear that treatment response in MEN1-related acromegaly is similar to sporadic acromegaly [80,99,100].

In the future, MEN1-specific treatments might become available. As *MEN1* is a tumor suppressor gene, restoration of *MEN1* expression might reduce tumor proliferation. Indeed, in a mouse MEN1 model, gene replacement therapy using an adenoviral vector containing *Men1* cDNA delivered directly to the pituitary showed restoration of menin expression and decreased proliferation in pituitary tumors [101].

9.8 Somatotroph adenomas in multiple endocrine neoplasia type 2

MEN2 (OMIM #171400 and #162300) is an autosomal dominantly inherited syndrome caused by gain-of-function mutations in the *RET* proto-oncogene on chromosome 10. Medullary thyroid carcinoma (MTC) is the hallmark of the disease and has near complete penentrance. Patients with MEN2 are also at risk for the development of pheochromocytomas and pHPT, in addition to other medical problems and phenotypic traits based on the underlying *RET* pathogenic variant [102]. PAs are not considered part of the tumor spectrum in MEN2. However, there have been five reported cases of PAs in patients with MEN2: three cases of Cushing disease in two adult patients with MEN2A [103,104] and one pediatric patient with MEN2B [105], one patient with MEN2A and a nonfunctioning macroadenoma [106], and one case of a somatotroph adenoma in a patient with MEN2A [107]. (The authors also care for a patient with MEN2A and a lactotroph macroadenoma [108].) The patient with MEN2A and acromegaly presented at age 40 with a high serum carcinoembryonic antigen level and was subsequently diagnosed with MTC and MEN2A. He had a diagnosis of acromegaly at age 35 that was cured by TSS. Pathology showed an adenoma with positive staining for GH and prolactin [107]. Although RET signaling pathways may play a role in somatotroph tumors [109,110], somatic *RET* mutations have not been associated with pituitary tumorogenesis [111,112]. Therefore given the published molecular studies and only one case report of MEN2 and acromegaly, it can be concluded that somatotroph adenomas are not part of the MEN2 phenotype.

9.9 Concluding remarks

GH-secreting adenomas are among the least common clinical manifestations in MEN1 and are not associated with MEN2. In more recently published MEN1 case series, the percentage of somato-troph adenomas in patients diagnosed with a PA ranges from 3.7% to 19%, and a significant number of these cases are not associated with a germline *MEN1* mutation but are instead diagnosed with clinical MEN1. Clinical presentation does not vary significantly from sporadic acromegaly, and onset of an MEN1-related somatotroph adenoma causing gigantism is rare, with other genetic causes (*AIP* mutations and *Xq26*.3 microduplications) being much more prevalent in very young patients. Although representing a small minority of cases, multifocal and plurihormonal adenomas (many of which are GH immunoreactive) seem to be more frequent in patients with MEN1-associated PAs. Few data exist regarding the optimal therapy of GH excess in MEN1, which is similar to sporadic cases, and more research may lead to a better understanding of MEN1-associated somatotroph tumors and differences between genotype-positive and genotype-negative cases. Contemporary approaches to tumor screening in genotype-positive MEN1 patients should lead to an earlier diagnosis with smaller and less-invasive tumors identified.

References

[1] Lloyd RV, Osamura RY, Kloppel G, Rosai J, editors. WHO classification of tumours of endocrine organs. 4th ed. Lyon: IARC; 2017.

[2] Chandrasekharappa SC, Guru S, Manickam P, et al. Positional cloning of the gene for multiple endocrine neoplasia-type 1. Science 1997;276:404−7.

[3] Lemmens I, Van de Ven W, Kas K, et al. Identification of the multiple endocrine neoplasia type 1 (MEN1) gene. Hum Mol Genet 1997;6:1177−83.

[4] Thakker RV, Newey PJ, Walls GV, et al. Clinical practice guidelines for multiple endocrine neoplasia type 1 (MEN1). J Clin Endocrinol Metab 2012;97:2990−3011.

[5] Dong Q, Debelenko LV, Chandrasekharappa SC, et al. Loss of heterozygosity at 11q13: analysis of pituitary tumors, lung carcinoids, lipomas, and other uncommon tumors in subjects with familial multiple endocrine neoplasia type 1. J Clin Endocrinol Metab 1997;82:1416−20.

[6] Thakker RV, Bouloux P, Wooding C, et al. Association of parathyroid tumors in multiple endocrine neoplasia type 1 with loss of alleles on chromosome 11. N Engl J Med 1989;321:218−24.

[7] Larsson C, Skogseid B, Oberg K, Nakamura Y, Nordenskjöld M. Multiple endocrine neoplasia type 1 gene maps to chromosome 11 and is lost in insulinoma. Nature 1988;85−7.

[8] Erdheim J. Zur normalen und pathologischen Histologie der Glandula thyreoidea, parathyreoidea und Hypophysis. Beitr z path Anat u z allg Path 1903;33:158−236.

[9] Shelburne S, McLaughlin Jr CW. Coincidental adenomas of islet-cells, parathyroid gland and pituitary gland. J Clin Endocrinol 1945;5:232−4.

[10] Underdahl LO, Woolner LB, Black BM. Multiple endocrine adenomas; report of 8 cases in which the parathyroids, pituitary and pancreatic islets were involved. J Clin Endocrinol Metab 1953;13:20−47.

[11] Wermer P. Genetic aspects of adenomatosis of endocrine glands. Am J Med 1954;16:363−71.

[12] Weber F, Mulligan LM. Happy 20th anniversary MEN1: from positional cloning to gene function restoration. Endocr Relat Cancer 2017;24:E7−11.

[13] Dreijerink KM, Goudet P, Burgess JR, Valk GD. Breast-cancer predisposition in multiple endocrine neoplasia type 1. N Engl J Med 2014;371:583−4.

[14] van Leeuwaarde RS, Dreijerink KM, Ausems MG, et al. MEN1-dependent breast cancer: indication for early screening? results from the Dutch MEN1 Study Group. J Clin Endocrinol Metab 2017;102:2083−90.

[15] de Laat JM, van der Luijt RB, Pieterman CR, et al. MEN1 redefined, a clinical comparison of mutation-positive and mutation-negative patients. BMC Med 2016;14:182.

[16] Machens A, Schaaf L, Karges W, et al. Age-related penetrance of endocrine tumours in multiple endocrine neoplasia type 1 (MEN1): a multicentre study of 258 gene carriers. Clin Endocrinol (Oxf) 2007;67:613−22.

[17] Romanet P, Mohamed A, Giraud S, et al. UMD-MEN1 database: an overview of the 370 MEN1 variants present in 1676 patients from the French population. J Clin Endocrinol Metab 2019;104:753−64.

[18] Pieterman CR, Schreinemakers JM, Koppeschaar HP, et al. Multiple endocrine neoplasia type 1 (MEN1): its manifestations and effect of genetic screening on clinical outcome. Clin Endocrinol 2009;70:575−81.

[19] Lourenço Jr DM, Toledo RA, Coutinho FL, et al. The impact of clinical and genetic screenings on the management of the multiple endocrine neoplasia type 1. Clinics 2007;62:465−76.

[20] Agarwal SK. The future: genetics advances in MEN1 therapeutic approaches and management strategies. Endocr Relat Cancer 2017;24:T119−34.

[21] Dreijerink KMA, Timmers HTM, Brown M. Twenty years of menin: emerging opportunities for restoration of transcriptional regulation in MEN1. Endocr Relat Cancer 2017;24:T135−45.

[22] Thakker RV, Pook MA, Wooding C, Boscaro M, Scanarini M, Clayton RN. Association of somatotrophinomas with loss of alleles on chromosome 11 and with gsp mutations. J Clin Invest 1993;91:2815−21.

[23] Weil RJ, Vortmeyer AO, Huang S, et al. 11q13 allelic loss in pituitary tumors in patients with multiple endocrine neoplasia syndrome type 1. Clin Cancer Res 1998;4:1673−8.

[24] Yoshimoto K, Iwahana H, Kubo K, Saito S, Itakura M. Allele loss on chromosome 11 in a pituitary tumor from a patient with multiple endocrine neoplasia type 1. Jpn J Cancer Res 1991;82:886−9.

[25] Jiao Y, Shi C, Edil BH, et al. DAXX/ATRX, MEN1, and mTOR pathway genes are frequently altered in pancreatic neuroendocrine tumors. Science 2011;331:1199−203.

[26] Scarpa A, Chang DK, Nones K, et al. Whole-genome landscape of pancreatic neuroendocrine tumours. Nature 2017;543:65−71.

[27] Boggild MD, Jenkinson S, Pistorello M, et al. Molecular genetic studies of sporadic pituitary tumors. J Clin Endocrinol Metab 1994;78:387−92.

[28] Farrell WE, Simpson DJ, Bicknell J, et al. Sequence analysis and transcript expression of the MEN1 gene in sporadic pituitary tumours. Br J Cancer 1999;80:44−50.

[29] Zhuang Z, Ezzat SZ, Vortmeyer AO, et al. Mutations of the MEN1 tumor suppressor gene in pituitary tumors. Cancer Res 1997;57:5446−51.

[30] Prezant TR, Levine J, Melmed S. Molecular characterization of the men1 tumor suppressor gene in sporadic pituitary tumors. J Clin Endocrinol Metab 1998;83:1388−91.

[31] Asa SL, Somers K, Ezzat S. The MEN-1 gene is rarely down-regulated in pituitary adenomas. J Clin Endocrinol Metab 1998;83:3210−12.

[32] Tanaka C, Kimura T, Yang P, et al. Analysis of loss of heterozygosity on chromosome 11 and infrequent inactivation of the MEN1 gene in sporadic pituitary adenomas. J Clin Endocrinol Metab 1998;83:2631–4.

[33] Theodoropoulou M, Cavallari I, Barzon L, et al. Differential expression of menin in sporadic pituitary adenomas. Endocr Relat Cancer 2004;11:333–44.

[34] Wrocklage C, Gold H, Hackl W, Buchfelder M, Fahlbusch R, Paulus W. Increased menin expression in sporadic pituitary adenomas. Clin Endocrinol (Oxf) 2002;56:589–94.

[35] Kaji H, Canaff L, Lebrun JJ, Goltzman D, Hendy GN. Inactivation of menin, a Smad3-interacting protein, blocks transforming growth factor type beta signaling. Proc Natl Acad Sci U S A 2001;98:3837–42.

[36] Lacerte A, Lee EH, Reynaud R, et al. Activin inhibits pituitary prolactin expression and cell growth through Smads, Pit-1 and menin. Mol Endocrinol 2004;18:1558–69.

[37] Namihira H, Sato M, Murao K, et al. The multiple endocrine neoplasia type 1 gene product, menin, inhibits the human prolactin promoter activity. J Mol Endocrinol 2002;29:297–304.

[38] Gillam MP, Nimbalkar D, Sun L, et al. MEN1 tumorigenesis in the pituitary and pancreatic islet requires Cdk4 but not Cdk2. Oncogene 2015;34:932–8.

[39] Pepe S, Korbonits M, Iacovazzo D. Germline and mosaic mutations causing pituitary tumours: genetic and molecular aspects. J Endocrinol 2019;240:R21–45.

[40] Lines KE, Newey PJ, Yates CJ, et al. MiR-15a/miR-16-1 expression inversely correlates with cyclin D1 levels in Men1 pituitary NETs. J Endocrinol 2018;240:41–50.

[41] Ballard HS, Fame B, Hartsock RJ. Familial multiple endocrine adenoma-peptic ulcer complex. Medicine 1964;43:481–516.

[42] Prosser PR, Karam JH, Townsend JJ, Forsham PH. Prolactin-secreting pituitary adenomas in multiple endocrine adenomatosis, type I. Ann Intern Med 1979;91:41–4.

[43] Marx S, Spiegel AM, Skarulis MC, Doppman JL, Collins FS, Liotta LA. Multiple endocrine neoplasia type 1: clinical and genetic topics. Ann Intern Med 1998;129:484–94.

[44] Eriksson B, Oberg K, Skogseid B. Neuroendocrine pancreatic tumors. Clinical findings in a prospective study of 84 patients. Acta Oncol 1989;28:373–7.

[45] O'Brien T, O'Riordan DS, Gharib H, Scheithauer BW, Ebersold MJ, van Heerden JA. Results of treatment of pituitary disease in multiple endocrine neoplasia, type I. Neurosurgery 1996;39:273–8 discussion 8-9.

[46] Samaan NA, Ouais S, Ordonez NG, Choksi UA, Sellin RV, Hickey RC. Multiple endocrine syndrome type I. Clinical, laboratory findings, and management in five families. Cancer 1989;64:741–52.

[47] Trump D, Farren B, Wooding C, et al. Clinical studies of multiple endocrine neoplasia type 1 (MEN1). QJM 1996;89:653–69.

[48] Vasen HF, Lamers CB, Lips CJ. Screening for the multiple endocrine neoplasia syndrome type I. A study of 11 kindreds in The Netherlands. Arch Intern Med 1989;149:2717–22.

[49] Scheithauer BW, Laws Jr ER, Kovacs K, Horvath E, Randall RV, Carney JA. Pituitary adenomas of the multiple endocrine neoplasia type I syndrome. Semin Diagn Pathol 1987;4:205–11.

[50] Burgess JR, Shepherd JJ, Parameswaran V, Hoffman L, Greenaway TM. Somatotrophinomas in multiple endocrine neoplasia type 1: a review of clinical phenotype and insulin-like growth factor-1 levels in a large multiple endocrine neoplasia type 1 kindred. Am J Med 1996;100:544–7.

[51] McCutcheon IE. Management of individual tumor syndromes. Pituitary neoplasia. Endocrinol Metab Clin North Am 1994;23:37–51.

[52] Goudet P, Bonithon-Kopp C, Murat A, et al. Gender-related differences in MEN1 lesion occurrence and diagnosis: a cohort study of 734 cases from the Groupe d'etude des Tumeurs Endocrines. Eur J Endocrinol 2011;165:97–105.

[53] Soczomski P, Jurecka-Lubieniecka B, Rogozik N, Tukiendorf A, Jarzab B, Bednarczuk T. Multiple endocrine neoplasia type 1 in Poland: a two-centre experience. Endokrynol Pol 2019;70:385–91.

[54] Giusti F, Cianferotti L, Boaretto F, et al. Multiple endocrine neoplasia syndrome type 1: institution, management, and data analysis of a nationwide multicenter patient database. Endocrine 2017;58:349−59.

[55] Goroshi M, Bandgar T, Lila AR, et al. Multiple endocrine neoplasia type 1 syndrome: single centre experience from western India. Fam Cancer 2016;15:617−24.

[56] de Laat JM, Dekkers OM, Pieterman CR, et al. Long-term natural course of pituitary tumors in patients with MEN1: results from the DutchMEN1 Study Group (DMSG). J Clin Endocrinol Metab 2015;100:3288−96.

[57] Sakurai A, Suzuki S, Kosugi S, et al. Multiple endocrine neoplasia type 1 in Japan: establishment and analysis of a multicentre database. Clin Endocrinol 2012;76:533−9.

[58] Goudet P, Dalac A, Le Bras M, et al. MEN1 disease occurring before 21 years old: a 160-patient cohort study from the Groupe d'etude des Tumeurs Endocrines. J Clin Endocrinol Metab 2015;100:1568−77.

[59] Vannucci L, Marini F, Giusti F, Ciuffi S, Tonelli F, Brandi ML. MEN1 in children and adolescents: data from patients of a regional referral center for hereditary endocrine tumors. Endocrine 2018;59:438−48.

[60] Manoharan J, Raue F, Lopez CL, et al. Is routine screening of young asymptomatic MEN1 patients necessary? World J Surg 2017;41:2026−32.

[61] Hao W, Skarulis MC, Simonds WF, et al. Multiple endocrine neoplasia type 1 variant with frequent prolactinoma and rare gastrinoma. J Clin Endocrinol Metab 2004;89:3776−84.

[62] Herath M, Parameswaran V, Thompson M, Williams M, Burgess J. Paediatric and young adult manifestations and outcomes of multiple endocrine neoplasia type 1. Clin Endocrinol (Oxf) 2019;91:633−8.

[63] Chung WH, Chiu CK, Wei Chan CY, Kwan MK. Rapid progression of scoliosis curve in a mature patient with undiagnosed pituitary macroadenoma: a rare case report. Acta Orthop Traumatol Turc 2020;54:561−4.

[64] Gezer E, Cetinarslan B, Canturk Z, Tarkun I, Sozen M, Selek A. Metastatic MEN1 syndrome treated with lutetium-177—a case report. Eur Endocrinol 2019;15:92−4.

[65] Herrero-Ruiz A, Villanueva-Alvarado HS, Corrales-Hernandez JJ, Higueruela-Minguez C, Feito-Perez J, Recio-Cordova JM. Coexistence of GH-producing pituitary macroadenoma and meningioma in a patient with multiple endocrine neoplasia type 1 with hyperglycemia and ketosis as first clinical sign. Case Rep Endocrinol 2017;2017:2390797.

[66] Uraki S, Ariyasu H, Doi A, et al. Hypersecretion of ACTH and PRL from pituitary adenoma in MEN1, adequately managed by medical therapy. Endocrinol Diabetes Metab Case Rep 2017;2017 17-0027.

[67] Liu W, Han X, Hu Z, et al. A novel germline mutation of the MEN1 gene caused multiple endocrine neoplasia type 1 in a Chinese young man and 1 year follow-up. Eur Rev Med Pharmacol Sci 2013;17:3111−16.

[68] Nozieres C, Berlier P, Dupuis C, et al. Sporadic and genetic forms of paediatric somatotropinoma: a retrospective analysis of seven cases and a review of the literature. Orphanet J Rare Dis 2011;6:67.

[69] Kontogeorgos G, Kapranos N, Tzavara I, Thalassinos N, Rologis D. Monosomy of chromosome 11 in pituitary adenoma in a patient with familial multiple endocrine neoplasia type 1. Clin Endocrinol (Oxf) 2001;54:117−20.

[70] Stratakis CA, Schussheim DH, Freedman SM, et al. Pituitary macroadenoma in a 5-year-old: an early expression of multiple endocrine neoplasia type 1. J Clin Endocrinol Metab 2000;85:4776−80.

[71] Hermans MM, Lips CJ, Bravenboer B. Growth hormone overproduction in a patient with multiple endocrine neoplasia type I. J Intern Med 2000;248:525−30.

[72] Schuppe HC, Neumann NJ, Schock-Skasa G, Hoppner W, Feldkamp J. Secondary infertility as early symptom in a man with multiple endocrine neoplasia-type 1. Hum Reprod (Oxford, Engl) 1999;14:252−4.

[73] Verges B, Boureille F, Goudet P, et al. Pituitary disease in MEN type 1 (MEN1): data from the France-Belgium MEN1 multicenter study. J Clin Endocrinol Metab 2002;87:457−65.

[74] Trouillas J, Labat-Moleur F, Sturm N, et al. Pituitary tumors and hyperplasia in multiple endocrine neoplasia type 1 syndrome (MEN1): a case-control study in a series of 77 patients versus 2509 non-MEN1 patients. Am J Surg Pathol 2008;32:534−43.

[75] Wu Y, Gao L, Guo X, et al. Pituitary adenomas in patients with multiple endocrine neoplasia type 1: a single-center experience in China. Pituitary 2019;22:113−23.

[76] Cohen-Cohen S, Brown DA, Himes BT, et al. Pituitary adenomas in the setting of multiple endocrine neoplasia type 1: a single-institution experience. J Neurosurg 2020;1−7.

[77] Daly AF, Rixhon M, Adam C, Dempegioti A, Tichomirowa MA, Beckers A. High prevalence of pituitary adenomas: a cross-sectional study in the province of Liege, Belgium. J Clin Endocrinol Metab 2006;91:4769−75.

[78] Fernandez A, Karavitaki N, Wass JA. Prevalence of pituitary adenomas: a community-based, cross-sectional study in Banbury (Oxfordshire, UK). Clin Endocrinol (Oxf) 2010;72:377−82.

[79] Petrossians P, Daly AF, Natchev E, et al. Acromegaly at diagnosis in 3173 patients from the Liege Acromegaly Survey (LAS) Database. Endocr Relat Cancer 2017;24:505−18.

[80] Katznelson L, Laws Jr ER, Melmed S, et al. Acromegaly: an endocrine society clinical practice guideline. J Clin Endocrinol Metab 2014;99:3933−51.

[81] Hai N, Aoki N, Shimatsu A, Mori T, Kosugi S. Clinical features of multiple endocrine neoplasia type 1 (MEN1) phenocopy without germline MEN1 gene mutations: analysis of 20 Japanese sporadic cases with MEN1. Clin Endocrinol (Oxf) 2000;52:509−18.

[82] Ozawa A, Agarwal SK, Mateo CM, et al. The parathyroid/pituitary variant of multiple endocrine neoplasia type 1 usually has causes other than p27Kip1 mutations. J Clin Endocrinol Metab 2007;92:1948−51.

[83] Nachtigall LB, Guarda FJ, Lines KE, et al. Clinical MEN-1 among a large cohort of patients with acromegaly. J Clin Endocrinol Metab 2020;105:e2271−81.

[84] Pieterman CRC, Hyde SM, Wu SY, et al. Understanding the clinical course of genotype-negative MEN1 patients can inform management strategies. Surgery 2021;169:175−84.

[85] Pardi E, Borsari S, Saponaro F, et al. Mutational and large deletion study of genes implicated in hereditary forms of primary hyperparathyroidism and correlation with clinical features. PLoS One 2017;12:e0186485.

[86] Akirov A, Asa SL, Amer L, Shimon I, Ezzat S. The clinicopathological spectrum of acromegaly. J Clin Med 2019;8:1962.

[87] Andersen HO, Jorgensen PE, Bardram L, Hilsted L. Screening for multiple endocrine neoplasia type 1 in patients with recognized pituitary adenoma. Clin Endocrinol (Oxf) 1990;33:771−5.

[88] Corbetta S, Pizzocaro A, Peracchi M, Beck-Peccoz P, Faglia G, Spada A. Multiple endocrine neoplasia type 1 in patients with recognized pituitary tumours of different types. Clin Endocrinol (Oxf) 1997;47:507−12.

[89] Teh BT, Kytola S, Farnebo F, et al. Mutation analysis of the MEN1 gene in multiple endocrine neoplasia type 1, familial acromegaly and familial isolated hyperparathyroidism. J Clin Endocrinol Metab 1998;83:2621−6.

[90] Yarman S, Tuncer FN, Serbest E. Three novel MEN1 variants in AIP-negative familial isolated pituitary adenoma patients. Pathobiology 2019;86:128−34.

[91] Daly A, Cano DA, Venegas E, et al. AIP and MEN1 mutations and AIP immunohistochemistry in pituitary adenomas in a tertiary referral center. Endocr Connect 2019;8:338−48.

[92] Stratakis CA, Tichomirowa MA, Boikos S, et al. The role of germline AIP, MEN1, PRKAR1A, CDKN1B and CDKN2C mutations in causing pituitary adenomas in a large cohort of children, adolescents, and patients with genetic syndromes. Clin Genet 2010;78:457−63.

[93] De Sousa SMC, McCabe MJ, Wu K, et al. Germline variants in familial pituitary tumour syndrome genes are common in young patients and families with additional endocrine tumours. Eur J Endocrinol 2017;176:635−44.

[94] Cuny T, Pertuit M, Sahnoun-Fathallah M, et al. Genetic analysis in young patients with sporadic pituitary macroadenomas: besides AIP don't forget MEN1 genetic analysis. Eur J Endocrinol 2013;168:533−41.

[95] Cazabat L, Libe R, Perlemoine K, et al. Germline inactivating mutations of the aryl hydrocarbon receptor-interacting protein gene in a large cohort of sporadic acromegaly: mutations are found in a subset of young patients with macroadenomas. Eur J Endocrinol 2007;157:1−8.

[96] van den Broek MFM, van Nesselrooij BPM, Verrijn Stuart AA, van Leeuwaarde RS, Valk GD. Clinical relevance of genetic analysis in patients with pituitary adenomas: a systematic review. Front Endocrinol (Lausanne) 2019;10:837.

[97] Iacovazzo D, Hernandez-Ramirez LC, Korbonits M. Sporadic pituitary adenomas: the role of germline mutations and recommendations for genetic screening. Expert Rev Endocrinol Metab 2017;12:143−53.

[98] Rostomyan L, Daly AF, Beckers A. Pituitary gigantism: causes and clinical characteristics. Ann Endocrinol 2015;76:643−9.

[99] Bollerslev J, Heck A, Olarescu NC. Management of endocrine disease: individualised management of acromegaly. Eur J Endocrinol 2019;181:R57−71.

[100] Coopmans EC, van Meyel SWF, van der Lely AJ, Neggers S. The position of combined medical treatment in acromegaly. Arch Endocrinol Metab 2019;63:646−52.

[101] Walls GV, Lemos MC, Javid M, et al. MEN1 gene replacement therapy reduces proliferation rates in a mouse model of pituitary adenomas. Cancer Res 2012;72:5060−8.

[102] Waguespack SG, Rich TA, Perrier ND, Jimenez C, Cote GJ. Management of medullary thyroid carcinoma and MEN2 syndromes in childhood. Nat Rev Endocrinol 2011;7:596−607.

[103] Ezzat T, Paramesawaran R, Phillips B, Sadler G. MEN 2 syndrome masquerading as MEN 1. Ann R Coll Surg Engl 2012;94:e206−7.

[104] Naziat A, Karavitaki N, Thakker R, et al. Confusing genes: a patient with MEN2A and Cushing's disease. Clin Endocrinol (Oxf) 2013;78:966−8.

[105] Kasturi K, Fernandes L, Quezado M, et al. Cushing disease in a patient with multiple endocrine neoplasia type 2B. J Clin Transl Endocrinol Case Rep 2017;4:1−4.

[106] Heinlen JE, Buethe DD, Culkin DJ, Slobodov G. Multiple endocrine neoplasia 2a presenting with pheochromocytoma and pituitary macroadenoma. ISRN Oncol 2011;2011:732452.

[107] Saito T, Miura D, Taguchi M, Takeshita A, Miyakawa M, Takeuchi Y. Coincidence of multiple endocrine neoplasia type 2A with acromegaly. Am J Med Sci 2010;340:329−31.

[108] Waguespack SG. Personal communication. A case of macroprolactinoma in multiple endocrine neoplasia, Type 2A, 2021.

[109] Canibano C, Rodriguez NL, Saez C, et al. The dependence receptor Ret induces apoptosis in somatotrophs through a Pit-1/p53 pathway, preventing tumor growth. EMBO J 2007;26:2015−28.

[110] Chenlo M, Rodriguez-Gomez IA, Serramito R, et al. Unmasking a new prognostic marker and therapeutic target from the GDNF-RET/PIT1/p14ARF/p53 pathway in acromegaly. EBioMedicine 2019;43:537−52.

[111] Komminoth P, Roth J, Muletta-Feurer S, Saremaslani P, Seelentag WK, Heitz PU. RET proto-oncogene point mutations in sporadic neuroendocrine tumors. J Clin Endocrinol Metab 1996;81:2041−6.

[112] Yoshimoto K, Tanaka C, Moritani M, et al. Infrequent detectable somatic mutations of the RET and glial cell line-derived neurotrophic factor (GDNF) genes in human pituitary adenomas. Endocr J 1999;46:199−207.

GNAS, McCune–Albright syndrome, and GH-producing tumors

10

Erika Peverelli[1], Donatella Treppiedi[1], Federica Mangili[1], Rosa Catalano[1,2] and Giovanna Mantovani[1,3]

[1]*Department of Clinical Sciences and Community Health, University of Milan, Milan, Italy* [2]*PhD Program in Endocrinological Sciences, Sapienza University of Rome, Rome, Italy* [3]*Endocrinology Unit, Fondazione IRCCS Ca' Granda Ospedale Maggiore Policlinico, Milan, Italy*

10.1 GNAS

GNAS is a complex imprinted gene that generates several gene products, including the heterotrimeric G protein α-subunit (Gsα). This protein plays a crucial role in the activation of the cAMP/PKA pathway.

10.1.1 GNAS complex locus

GNAS gene is one of the most complex eukaryotic genes. It is located on the long arm of chromosome 20 (20q13.3) in humans [1–4], while its homolog in mouse is located on chromosome 2 [5]. *GNAS* gives rise to several coding and noncoding transcripts, which are transcribed either in sense or antisense orientation.

The most abundant and best characterized gene product is the alpha subunit of the stimulatory guanine nucleotide-binding protein (Gsα), encoded by exons 1–13 [6] (Fig. 10.1).

Upstream of Gsα exon there are alternative promoters, giving rise to different mRNAs: extra-large Gsα (XLαs), neuroendocrine secretory protein 55 (NESP55), the A/B (1A in mice), and the GNAS-AS1 transcripts (Nespas in mice) [7–11] (Fig. 10.1). XLαs, NESP55, and A/B individually contain their own unique first exon that splices onto exon 2–13 of GNAS [7–10].

NESP55 belongs to the chromogranin family of proteins. In the NESP55 transcript, sequences derived from exons 2–13 are located within the 3′-untranslated region, and the coding region is limited to its specific first exon.

Conversely, exons 2–13 are located within the coding region of XLαs transcript; thus the XLαs protein is partly identical to the Gsα protein, being an isoform of Gsα with a long amino-terminal extension encoded by the XLαs-specific first exon [12,13]. In vitro experiments have shown that XLαs, like Gsα, can stimulate adenylyl cyclase (AC) activation in transfected cells lacking endogenous Gsα [14–17]. In contrast to Gsα, which is ubiquitously expressed at high levels, XLαs presents a more restricted tissue distribution in mice. Both NESP55 and XLαs are highly expressed in neuroendocrine tissues [18,19].

Gigantism and Acromegaly. DOI: https://doi.org/10.1016/B978-0-12-814537-1.00009-9

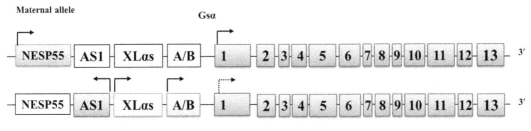

FIGURE 10.1

Genomic organization of human GNAS1 locus. The maternal and paternal alleles of *GNAS* are shown. The figure shows the alternative first exons which splice into common exon 2, generating five different transcripts: Gsα, NESP55, XLαs, A/B, and AS1 (arrows designate transcription start sites in the direction of transcription). Exon 1 is the first coding exon for Gsα. The XLαs, AS1, and A/B transcripts are expressed only from the paternal allele, while NESP55 is expressed specifically from the maternal allele. The dashed line at the paternal Gsα promoter indicated that Gsα is biallelically expressed in most tissues, but it is predominantly expressed from the maternal allele in specific endocrine tissues such as the pituitary gland, the thyroid and the gonad. The diagram is not drawn to scale.

The A/B and GNAS-AS1 transcripts are noncoding. The GNAS-AS1 transcript is derived from the antisense strand and traverses the promoter and the first exon of NESP55 [20,21]. GNAS-AS1 acts as a negative regulator, in cis, of the NESP55 expression. A/B regulates imprinting within GNAS, but some evidence suggest that it could be translated, and its product is an amino-terminally truncated form of Gsα that antagonizes the latter [22].

The GNAS locus shows a highly complex imprinted expression pattern. Genomic imprinting is the epigenetic phenomenon by which one allele (maternal or paternal), either during the embryogenesis or in the postnatal period, is subject to a partial or total loss of expression [23]. Initiation and the maintenance of imprinting are controlled by DNA methylation. Imprinted genes have regions in which CpG dinucleotides are differentially methylated between the paternal and maternal alleles. The GNAS transcripts promoters are located in differentially methylated regions and show parent-specific methylation, allowing transcription only from the unmethylated allele. Promoters giving rise to the XLαs, A/B, and GNAS-AS1 transcripts are methylated on the maternal allele and are thus transcribed exclusively from the paternal allele (Fig. 10.1). An exception is represented by GNAS-AS1, that in adrenal gland and in testes shows biallelic expression [19].

The maternal GNAS allele gives rise to transcripts encoding NESP55, since its promoter is methylated at CpG dinucleotides on the paternally inherited allele [8] (Fig. 10.1).

Studies in rodents have shown that the promoter and the first exon of Gsα lie in a CpG-rich island, but Gsα promoter is unmethylated on both parental alleles [24], explaining biallelic Gsα expression in most tissues [25], as confirmed in several human fetal tissues, bone, and adipose tissues [12,26,27]. However, in few tissues such as the renal proximal tubule, various areas of the central nervous system, brown adipose tissue, and specific endocrine tissues, including thyroid, gonads, and pituitary, Gsα is derived predominantly from the maternal allele, and the paternal allele is silenced through yet undefined mechanisms [25,28,29].

Germline or somatic alterations of *GNAS* are responsible for several disorders. Heterozygous inactivating mutations in Gsα-coding GNAS exons cause Albright's Hereditary Osteodystrophy and related diseases, including the different subtypes of pseudohypoparathyroidism type 1 [30].

As described in detail in the next paragraphs, somatic activating mutations in *GNAS* are found in 30%−40% of growth hormone (GH)−secreting pituitary tumors, and less commonly in other endocrine tumors such as toxic thyroid adenomas, hyperfunctioning adrenal tumors, and Leydigiomas. In GH-secreting pituitary tumors, *GNAS* mutations are located on the maternal allele [26−28]. When these mutations occur as an early postzygotic event, they cause the McCune−Albright syndrome (MAS).

10.1.2 Cellular actions of Gsα

As mentioned in Section 10.1.1, the most abundant and best characterized *GNAS* gene product is Gsα, that stimulates the cAMP-dependent pathway by activating AC.

10.1.2.1 Heterotrimeric G-proteins: families, structure, and function

Heterotrimeric G-proteins activate intracellular signaling cascades in response to the activation of seven-transmembrane G protein-coupled receptors (GPCRs) by extracellular stimuli [31]. They are heterotrimers composed of three subunits, α, β, and γ. In humans, there are 21 Gα subunits encoded by 16 genes, 6 Gβ subunits encoded by 5 genes, and 12 Gγ subunits [31,32]. The β and γ subunits are tightly associated and can be considered one functional unit. G-proteins are identified by their Gα subunit. Based on the sequence and functional similarities, Gα proteins are grouped into four families: Gαs, Gαi, Gαq, and Gα12 [33].

Gαs family includes two members: Gαs (s stands for stimulation) and Gαolf (olf stands for olfaction). Gαs shows a ubiquitous expression, whereas Gαolf is specifically expressed in the olfactory sensory neurons.

Gαi family is the largest family, including Gαi1, Gαi2, Gαi3, Gαo, Gαt, Gαg, and Gαz. Gαi (inhibitory) proteins are ubiquitously expressed, while Gαo (o stands for other) is highly expressed in neurons and has two spliced variants, GαoA and GαoB. Gαt (t stands for transducin) has two isoforms, Gαt1, expressed in the rod cells, and Gαt2, expressed in the cone cells. Gαg (g stands for gustducin) is found in taste receptor cells. Gαz is expressed in neuronal tissues and in platelets.

In humans, the Gαq family consists of Gαq, Gα11, Gα14, and Gα16. Gαq and Gα11 are ubiquitously expressed, while Gα14 and Gα15/16 expression is more restricted (Gα14 in kidney, lung, and liver, Gα15/16 in hematopoietic cells).

Gα12 family proteins, including Gα12 and Gα13, are expressed in most types of cells.

The structure of the Gα subunit reveals a conserved protein fold that is composed of a GTPase domain and a α-helical domain [34]. Between these two domains lies a deep cleft within which guanosine diphosphate (GDP) or (GTP) is tightly bound. The nucleotide-binding pocket of G-proteins is highly conserved. The Ras-like GTPase domain, conserved in all members of the G protein superfamily, hydrolyzes GTP and provides the binding surfaces for the Gβγ dimer, GPCRs and effector proteins. The α-helical domain is unique to Gα proteins and is composed of a six α-helix bundle that forms a lid over the nucleotide-binding pocket.

All Gα subunits, except Gαt, are posttranslationally modified with the fatty acid palmitate at the N-terminus. Members of the Gαi family are also myristoylated at the N-terminus. These modifications regulate membrane localization and protein–protein interactions [35,36].

The crystal structure of a Gαβγ heterotrimer shows that GTPase domain and a α-helical domain of Gα interact with different regions of Gβ.

Biophysical studies of rhodopsin-family GPCRs have shown that receptor activation results in a conformational change involving transmembrane helix VI, which opens a pocket for G protein binding. Structure, conformation, and specificity of the G protein binding site can be regulated by the identity of the bound ligand. Guanine nucleotide loading and unloading take place in the Ras-like domain of G-proteins, which serves as a platform for both guanine nucleotide-binding as well as in GTP hydrolysis [37].

G-proteins biochemically function in a GTPase cycle (Fig. 10.2), with their biological activity determined by the bound nucleotide. When not stimulated by a receptor, inactive Gα is bound to

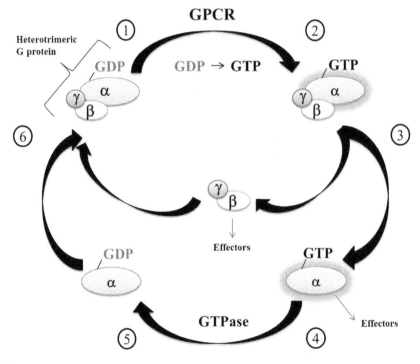

FIGURE 10.2

Heterotrimeric G-proteins cycle. (1) Under basal conditions, G-proteins exist as heterotrimers composed of α, β, and γ subunits and GDP is bound to the α subunit. (2) Agonist-bound GPCRs act as GEFs, leading to the dissociation of GDP from the α subunit. Nucleotide-free Gα then binds GTP. (3) GTP binding induces a conformational change in Gα that induces the dissociation from Gβγ. (4) Free Gα, bound to GTP, and free βγ subunit are able to regulate downstream effector proteins. (5) The intrinsic GTPase activity of Gα causes the hydrolysis of GTP to GDP. (6) This leads to the reassociation of the α and βγ subunits, and signal termination. *GDP,* Guanosine diphosphate; *GEFs,* guanine nucleotide exchange factors; *GPCR,* G protein-coupled receptors.

GDP and to Gβγ to form the inactive G protein trimer. Upon agonist binding, GPCRs act as guanine nucleotide-exchange factor to increase the exchange of GDP bound on the Gα subunit with GTP. This leads to the dissociation of GTP-bound Gα subunit from Gβγ dimer resulting in two functional subunits (Gα and Gβγ). Both Gα and Gβγ subunits are able of initiating signals by interacting with downstream effector proteins. The signal is terminated by the intrinsic GTPase activity of Gα, which hydrolyzes the bound GTP to GDP. GDP-bound Gα then reassembles with Gβγ and becomes inactive until the cycle is reinitiated by an activated heptahelical transmembrane receptor.

10.1.2.2 Gsα and cAMP/PKA pathway activation

The GTP-bound Gsα is able to stimulate different effectors giving rise to various intracellular responses depending on the cellular context.

The most extensively investigated effector stimulated by Gsα is membrane-bound AC, which catalyzes the conversion of ATP into the second messenger cAMP, which, in turn, activates the cAMP-dependent protein kinase (protein kinase A, PKA) (Fig. 10.3).

In pituitary cells, Gsα plays a central role in mediating cAMP pathway activation after stimulation by hormones, such as GH and corticotroph-releasing hormone.

AC protein family includes 10 isoforms, coded by 10 different genes and classified in 5 families [38]. Mouse pituitary expresses mRNAs for AC isoform II, III, VI, and VII [39], depending on the cell type. In particular, GH-producing cells mainly contain AC types I, III (calcium-calmodulin sensitive), and VI, whereas PRL-secreting cells express AC II, IV, VI, and VIII [40,41]. On the contrary, AC IX, but not II and VI, was detected in corticotropes [42].

Although each GTP-bound Gsα can activate only one AC enzyme, amplification of the signal occurs because one receptor can activate multiple copies of Gsα while that receptor remains bound to its activating agonist.

cAMP is an ubiquitous second messenger implicated in the regulation of a wide variety of cell functions, including cell proliferation, that is differently affected depending on the cell type [43]. Indeed, cAMP may either not influence, induce cell arrest, or even suppress mitogenic action of growth factors in some cell lines or conversely promote cell growth in others [44]. The observation activating mutations of Gsα are found in about 30%−40% of GH-secreting pituitary adenomas [45−47] indicates that somatotrophs recognize cAMP as a growth factor.

The best known cAMP downstream effector is PKA. This serine/threonine kinase is a tetrameric enzyme composed of two regulatory (R) and two catalytic (C) subunits (Fig. 10.3).

The R subunit exists in two isoforms, R1 and R2, which give rise to two PKA isozymes (PKA 1 and 2). Moreover, R1 and R2 exist in two isoforms A and B, and the four different R isoforms, R1A, R1B, R2A, and R2B, differ in tissue distribution, subcellular localization, and biological properties. The observed changes in the proportion of R1 and R2 during embryonic development, differentiation processes, and neoplastic transformation indicate distinct roles for these isoenzymes in growth control. In particular, the ratio between R1 and R2 has been found unbalanced in favor of R2 in GH-secreting pituitary tumors, and it has been demonstrated that a low expression of R1A protein may favor cAMP-dependent proliferation of tumoral somatotrophs [48].

Binding of cAMP to high-affinity cAMP-binding sites of each R subunit induces dissociation of C subunits that become free to phosphorylate their substrates [49]. The nuclear transcription factor

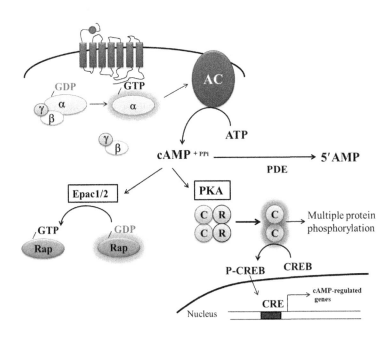

FIGURE 10.3

Schematic representation of cAMP pathway activation by Gsα. Upon membrane GPCR activation, GTP-bound, free Gsα subunit stimulates AC. It catalyzes cAMP production from ATP. cAMP is degraded to AMP by PDEs. The cAMP effector PKA is a tetrameric serine/threonine kinase consisting of two regulatory (R) and two catalytic (C) subunits. cAMP binds both high-affinity cAMP-binding sites of each R subunit and leads to the release of active C subunits that phosphorylate their substrates, including the nuclear transcription factor CREB. P-CREB binding to CRE induces the transcription of several cAMP-responsive genes. Epac1/2 are exchange proteins directly activated by cAMP able to mediate several cAMP effects by acting as guanine nucleotide exchange factors for Rap1 and 2. *AC*, Adenylyl cyclase; *CRE*, cAMP-responsive element; *P-CREB*, phosphorylated CREB; *PDEs*, phosphodiesterases.

cAMP-response element binding protein (CREB) is a target of PKA. Phosphorylated CREB binding to cAMP-responsive element induces the transcription of several cAMP-responsive genes (Fig. 10.3).

Although all the effects exerted by cAMP were initially attributed to the activation of PKA, two exchange proteins directly activated by cAMP (Epac1/2) have been later identified as cAMP targets able to mediate several cAMP effects by acting as guanine nucleotide-exchange factors for Rap1 and 2 [50–52] (Fig. 10.3). The two isoforms of Epac are coded by two distinct genes, RAPGEF3 and RAPGEF4 in mammals. Epac 1 is ubiquitous, whereas Epac 2 expression is restricted to a few tissues, such as the brain, pituitary, adrenal gland, and pancreas. Both PKA and Epac mediate cAMP effects on pituitary cells proliferation [53].

cAMP is degraded to AMP by cyclic nucleotide phosphodiesterases (PDEs) that play a major role in the regulation of intracellular cAMP levels (Fig. 10.3). There are 11 subfamilies of PDEs with different structures, enzymatic properties, sensitivity to inhibitors, and expression. Each subfamily includes distinct genes that, in turn, generate several transcripts by alternative splicing and/or the use of multiple promoters [54]. PDEs are activated by cAMP both by phosphorylation and gene expression induction.

Other effectors of Gsα include Src kinase [55] and calcium channels [56].

10.1.2.3 Gnas knockout models

The physiological functions of G-proteins have been investigated in gene-knockout mice.

Gnas knockout mouse models have demonstrated that both Gsα and XLαs have important roles in the regulation of energy metabolism. It has been observed that mice with heterozygous disruption of *Gnas* exon 2 on the maternal (m − / +) or paternal (+/p −) allele present distinct early phenotypes [57] similar to mice with paternal uniparental disomy/maternal deletion and maternal uniparental disomy/paternal deletion, respectively, of the distal chromosome 2 region, including *Gnas*. m − / + and + /p − mice have opposite metabolic phenotypes: m − / + mice are obese, hypometabolic, and hypoactive, whereas + /p − mice are very lean, hypermetabolic, and hypoactive. Interestingly, both groups of mice have increased whole-body insulin sensitivity, although to a much greater extent in + /p − mice [58]. XLαs knockout mice were shown to have a phenotype similar to that of + /p − exon 2 mice [59], suggesting that Gsα and XLαs have opposite effects on body metabolism, and that the + /p − exon 2 phenotype is caused by XLαs deficiency. It can be hypothesized that XLαs normally acts as a negative regulator of sympathetic nerve activity, and, therefore, XLαs deficiency in + /p − mice leads to a hyperadrenergic state [60].

Gsα signaling relevance in the regulation of energy expenditure and brown adipose tissue activation has been investigated in mice with a Gnas gene deletion disrupting Gsα expression on the maternal allele, but not the paternal allele, in the dorsomedial nucleus of the hypothalamus. These mice developed obesity and reduced energy expenditure without hyperphagia. Although maternal Gnas deletion impaired activation of brown adipose tissue in mice, their responses to cold environment remained intact [61]. Interestingly, complete Gsα deficiency in the dorsomedial nucleus of the hypothalamus impaired leptin signaling [62].

A mouse model with heterozygous germline inactivation of Gsα demonstrated that Gsα is required postnatally to maintain trabecular bone quality by regulating osteoclast differentiation and bone resorption activity [63].

Finally, conditional epidermal-specific Gnas knockout mice demonstrated that Gnas deletion in the skin is sufficient to induce basal-cell carcinoma-like lesions, characterized by thickening of the epidermis and hair loss, suggesting an important tumor-suppressive function of Gsα, limiting the proliferation of epithelial stem cells and maintaining proper hair follicle homeostasis [64].

10.1.3 GNAS-activating mutations

GNAS-activating mutations are missense mutations in exons 8 and 9 that lead to amino acid substitution of either residue Arg201 or Gln227 of Gsα.

Arg201 can be replaced with Cys, His or Ser, and Gln227 with Arg or Leu. These two residues are crucial for GTPase activity of Gsα, and their substitution causes the loss of catalytic activity and prevents the hydrolysis of GTP. Consequently, Gsα is constitutively activated and the AC system is maintained in a permanently turned-on state. In those cells in which cAMP represents a mitogenic signal, such as pituitary somatotrophs, Gsα can be considered a product of a proto-oncogene, converted into an oncogene (called "gsp").

It is of interest that residue Arg201 is the target of cholera toxin, the exotoxin secreted by Vibrio cholera, that, by ADP-ribosylation of this residue, abolished the GTPase activity of Gsα with a consequent constitutive activation. The resulting increase of cAMP in intestinal cells leads to severe secretory diarrhea.

Germline *GNAS*-activating mutations are considered lethal to the embryo [65,66].

When these mutations occur as an early postzygotic event, they cause the MAS [67] (see paragraph 10.2).

Somatic *gsp* mutations were originally found in a subgroup of GH-secreting pituitary tumors [45,47,68,69]. The phenotype resulting from the expression of *gsp* oncogene, observed in about 30%—40% of GH-secreting pituitary tumors, will be discussed subsequently (see paragraph 10.3.8).

These mutations have also been described, although in a small percentage of cases, in other tumors, such as thyroid adenomas, ovarian and testicular Leydig cell tumors, and primary macro-nodular adrenocortical hyperplasia, as well as in rare cases of corticotropinoma, cortisol, and aldosterone-secreting adrenocortical adenoma.

10.2 Somatic activating mutations in *GNAS*: the McCune—Albright syndrome

When activating mutations within the *GNAS* gene, in particular at Arg201, occur early in embryogenesis, they cause MAS (MIM# 174800). This sporadic disease affects the bones [polyostotic fibrous dysplasia (FD)], skin (café-au-lait spots), and several endocrine tissues, such as gonads, pituitary, thyroid, and adrenal cortex, all these glands being sensitive to trophic agents acting through cAMP-dependent pathway [67,70].

10.2.1 Etiology

The cause of MAS is a spontaneous postzygotic missense mutation at Arg201 or Gln227 of the *GNAS* gene during embryogenesis (for description of *GNAS* mutations, see Section 10.1.3). Mutations at Arg201 have been identified in more than 95% of MAS cases, the most frequent variants being Arg201His and Arg201Cys [66]. Very infrequently, arginine is replaced by serine, glycine, or leucine. Gln227 mutations have been rarely reported [71].

These gain-of-function activating mutations lead to constitutive activity of Gsα, and consequently, of the cAMP pathway. Increased Gsα signaling is the underlying cause of tissue-specific alterations in MAS. In particular, in bone marrow stromal cells, Gsα signaling, and cAMP pathway activation promote the osteogenic commitment but inhibit their further differentiation into osteoblasts [72]. This results in the formation of fibrous dysplastic lesions consisting of fibrous cells that express early osteoblastic markers, such as alkaline phosphatase [73,74]. The increased Gsα signaling is also the cause of skin lesions, due to the role of cAMP pathway in mediating the action of α-MSH in the stimulation of melanin production [75]. In addition, Gsα promotes FGF23 transcription in osteoblast and osteocytes, suggesting that constitutive cAMP pathway activation in dysplastic tissues increases FGF23 leading to phosphate wasting [76,77].

In patients, *GNAS* mutations are distributed in a mosaic pattern. It is postulated that the mutation is acquired during early embryonic development [78], and their signs and symptoms depend on the time frame the mutation occurs. To date, no patient with MAS inherited the disease from a parent or transmitted it to the offspring. Germline gain-of-function mutations are presumed to be incompatible with life [78], although in transgenic mice, germline transmission of a mutant Gsα has been reported [79].

The allelic origin of the postzygotically acquired Gsα mutation has an impact on the involvement of affected tissues. This is particularly true for GH-secreting adenomas, reflecting the monoallelic maternal expression of Gsα in the pituitary [80]. Accordingly, GH hypersecretion/acromegaly has been found only in those patients in whom the mutation was located on the maternal allele [26,27,81]. In contrast, both maternal and paternal mutations can be present in affected thyroid or ovaries despite the predominant maternal expression of Gsα in those tissues, as well.

10.2.2 Epidemiology

The estimated prevalence of MAS is between 1 in 100,000 and 1 in 1,000,000 [82], affecting both males and females across ethnic groups.

10.2.3 Diagnosis

MAS is largely a clinical diagnosis. Genetic testing for *GNAS* mutation is widely available; however, due to the mosaic nature of the disease, a negative genetic test result (e.g., in blood) does not exclude the presence of the mutation in other tissues. Moreover, a positive result contributes little to the clinical management. However, newer techniques such as digital PCR may improve the sensitivity of genetic testing in individuals who have clinical signs of MAS [83–85]. The NIH Genetic Testing Registry (GTR) displays genetic tests that are currently available for the *GNAS* gene and MAS.

MAS is characterized by the classic triad of polyostotic FD, café-au-lait skin pigmentation with highly irregular borders, and peripheral precocious puberty [86,87]. Moreover, when mutated cells are present in thyroid, adrenal, and/or pituitary tissues, clinical presentation may include hyperthyroidism, Cushing syndrome, and pituitary gigantism/acromegaly. Renal phosphate wasting with or without rickets/osteomalacia and rarely other organ systems may be involved.

The diagnosis is made when at least two of the following clinical features are present in an individual: café au-lait macules, FD, and autonomous endocrine hyperfunction [88]. To identify and characterize fibrodysplastic lesions, radiographs or computed tomography are used. The diagnosis of FD can be made by radiographs showing an expansive lesion with a "ground glass" appearance; however, bone scans help define and monitor the extent of the disease [82]. In early childhood stress fractures may occur, leading to permanent deformity if not stabilized. CT scan of the skull helps in the diagnosis of craniofacial bone lesions and pituitary lesions. Regular vision and hearing tests can monitor lesions affecting those systems. Baseline bone scan with 99Tc-methyl diphosphate is indicated in endocrinopathies that affect bone development [89].

Precocious puberty in MAS syndrome is frequently the presenting symptom in girls, affecting nearly 80% of patients [80]. Evaluation includes serum estradiol or testosterone levels together with gonadotropins. Recurrent ovarian cysts result in episodic estrogen production and intermittent vaginal bleeding, which may ultimately lead to bone age advancement and reduced adult height. In boys, testicular involvement is common, whereas precocious puberty develops only rarely [90]. Testicular ultrasound is used to assess for hormonally active tumors. Two-thirds of patients present thyroid abnormalities, and half of them are associated with frank hyperthyroidism [91]. Evaluation of hyperthyroidism includes measurement of thyroid-stimulating hormone (TSH) and free T4 and T3 together with thyroid ultrasound. GH excess affects 15%–20% of patients [80] and its evaluation includes IGF-1 measurement, oral glucose tolerance test (OGTT) with serum GH measurements. Hypercortisolism is the rarest endocrinopathy associated with MAS and occurs exclusively during the neonatal period due to cortisol overproduction from the fetal adrenal gland [92]. It may undergo spontaneous resolution after the involution of the fetal adrenal gland. Monitoring levels of serum phosphate and renal reabsorption of phosphate is a recommendation. Periodic ultrasound is recommended with an appropriate follow-up of any abnormal findings.

10.2.4 Clinical characteristics

At least two of the primitive germ layers of the embryo (ectoderm, mesoderm, and endoderm) are affected. Affected tissues commonly include skin, bones, and endocrine tissues. However, as Gsα signaling is ubiquitous, additional sites may be affected.

The severity of the disease ranges from mild to severe. Manifestations vary depending on the tissue involved, the role of Gsα in the affected tissue, and the timing of the mutation during embryogenesis. Mutations occurring early in development lead to widespread disease, while those occurring later in development lead to limited disease.

1. *Dermatologic features*: Skin findings are the light brown café-au-lait macules, which arise from the ectoderm [88]. Café-au-lait macules are common and are typically the first noticeable manifestation of the disease, apparent at or shortly after birth. There is no correlation between the size of the skin lesions and the extent of disease or areas of bone disease. These lesions characteristically respect the midline of the body and have irregular borders resembling the "coast of Maine," in contrast to the smooth-bordered "coast of California" macules seen in neurofibromatosis. Common locations are the posterior neck and the base of the spine, trunk, and face. Attention should be paid as café-au-lait macules may be found in 10% of the healthy population.
2. *Skeletal features*: The skeletal aspect of MAS, arising from the mesoderm, is bone FD [88]. As with skin, FD demonstrates a mosaic pattern: it may involve any site and combination of craniofacial, axial, and/or appendicular skeleton. The most commonly involved bones are the skull base and proximal femurs. FD can manifest along a wide spectrum: from an isolated, asymptomatic monostotic lesion to severe polyostotic disease affecting more than one bone and leading to loss of vision, hearing, and/or mobility. While there is generally a central-to-peripheral gradient, any combination of involved bones is possible. Recurrent fractures and progressive deformity can lead to disability. FD involving the vertebrae is common and may lead to scoliosis, which in rare instances may be severe, progressive, and even lethal [93].

Individual bone lesions typically manifest during the first few years of life and expand during childhood. The vast majority of clinically significant bone lesions are detectable by age 10 years, with few new and almost no clinically significant bone lesions appearing after age 15 years [94]. In adulthood, FD lesions typically become less active, likely related to apoptosis of pathogenic variant-bearing cells [95].

3. *Endocrine features*: Endocrinopathies typically develop during infancy or early childhood and persist into adulthood. Endocrinopathies can include precocious puberty, hyperthyroidism, GH excess, phosphate wasting, and neonatal hypercortisolism. Ovarian cysts lead to estrogen production, resulting in breast development, growth acceleration, and vaginal bleeding. Ovarian cysts typically continue into adulthood, leading to irregular menses. Although less common in males, autonomous testosterone production may result in testicular enlargement, growth acceleration, pubic and axillary hair, acne, and aggressive and/or inappropriate sexual behavior [90].

Hyperthyroidism results from both increased hormone production and increased conversion of thyroxine (T4) to triiodothyronine (T3) [91]. Production of excessive thyroid hormones leads to a hypermetabolic state, advanced skeletal maturity, increased bone turnover, and fractures. Other thyroid abnormalities include goiter, cysts, and nodules. Malignant transformation of affected thyroid tissue has been rarely reported [96].

Approximately 15%–20% of MAS patients harbor *GNAS* pathogenic variants in the anterior pituitary that can lead to autonomous GH production, often accompanied by hyperprolactinemia [97]. High levels of GH may result in linear growth acceleration in children (gigantism is not treated), worsening craniofacial FD, and features of acromegaly.

Renal phosphate wasting with or without hypophosphatemia is an additional manifestation associated with FD [88], due to excessive synthesis and secretion of phosphaturic hormone fibroblast growth factor-23 (FGF23) from the dysplastic tissue.

Neonatal Cushing syndrome symptoms typically develop in the neonatal period leading to critical illness and can result in death. Spontaneous regression has been reported in approximately half of survivors, presumably related to fetal adrenal involution.

10.2.5 **Differential diagnosis**

Neurofibromatosis type 1 (NF1) and MAS have several overlapping features. Both can present with café-au-lait macules and skeletal abnormalities. The café-au-lait macules of neurofibromatosis (six or more café au-lait macules) typically have smooth borders (coast of California), as opposed to irregular appearance of macules in MAS (coast of Maine). Skeletal features of NF1 include kyphoscoliosis, sphenoid dysplasia, cortical thinning of long bones, and bowing and dysplasia, particularly of the tibia, which may result in pseudoarthroses. Distinct features of NF1 include tumors of the nervous system, which are lacking in MAS, such as neurofibromas and optic gliomas, pigmented iris hamartomas, and axillary freckling. NF1 is caused by heterozygous pathogenic variants in *NF1* and is inherited in an autosomal dominant manner.

FD is common as an isolated finding of single or multiple lesions. It arises from postzygotic somatic activating mutations in *GNAS*, as MAS, but in the absence of endocrinopathy and café-au-lait macules, the diagnosis of isolated FD is more likely [98].

Precocious puberty may be central or peripheral and in females is often idiopathic. Hormonally active tumors and congenital adrenal hyperplasia are other differential diagnoses in patients presenting with precocious puberty. Patients presenting with precocious puberty should undergo a skeletal survey to rule out MAS [99].

Cutaneous-skeletal hypophosphatemia syndrome is a mosaic disorder resulting from somatic activating pathogenic variants in HRAS and NRAS [100]. Affected individuals develop cutaneous lesions (epidermal and large congenital melanocytic nevi) following a mosaic distribution; a mosaic skeletal dysplasia; overproduction of FGF23 resulting in rickets/osteomalacia; and variable other associated anomalies of the eye, brain, and vascular system [101].

10.2.6 **Management**

Treatment must be individualized based on each patient's clinical presentation. The first international consensus management guidelines have been published very recently, and we advise the reader to refer to them for an extensive update [102].

There are no available specific medical therapies for FD, and management is palliative, focused on optimizing function and minimizing morbidity related to fractures and deformity. In particular, orthopedic surgery is sometimes needed in order to repair fractures and prevent or correct deformities. Treatment of scoliosis is of particular importance, as it may be rapidly progressive, and surgical fusion has been shown to be effective at stabilizing the spine [103,104]. Physical therapy to optimize function and attenuate loss of mobility is appropriate. Intravenous bisphosphonates such as zoledronic acid and pamidronate seem to be effective for the treatment of FD-related pain, but they did not show long-term effects on the course of the disease. Dosing should be based on symptoms. Although Denosumab, a human monoclonal antibody to RANKL, demonstrated significant reduction in pain, bone turnover markers, and tumor growth rate, it associated with clinically significant disturbances of mineral metabolism both during treatment and after discontinuation [105–107].

Treatment of precocious puberty is important to prevent bone age advancement and compromise of adult height. The aromatase inhibitor Letrozole [108,109] and/or Tamoxifen may be effective for the treatment of precocious puberty in girls.

Medications and/or surgery may be used to treat hyperthyroidism, GH excess, and hypercortisolism. In particular, the first-line treatment for medical management of hyperthyroidism is Methimazole [110], whereas Propylthiouracil is no longer recommended due to the risk of hepatotoxicity in children [111]. Thyroidectomy is the preferred definitive treatment in most of the affected individuals. Radioablation is avoided due to potential preferential uptake by tissues bearing a somatic activating *GNAS* mutation, which may lead to an increased risk of malignancy in the remaining unaffected gland. Additionally, *GNAS* pathogenic variants are associated with a slight increased risk of malignant transformation in both thyroid and nonthyroidal tissues, and the risk is potentially enhanced by radiation exposure. Current therapies for IGF-1 excess are based on the use of somatostatin (SS) analogs (SSA) and, when an insufficient response is observed, the GH-receptor antagonist pegvisomant [97,112] represents the current therapies for IGF-1 excess. Surgery may be technically difficult or precluded due to craniofacial FD. Moreover, the somatotropes may be hyperplastic overall leading to incomplete surgical removal of the affected tissue.

Hyperprolactinemia is generally responsive to treatment with dopamine agonists (DA) (cabergoline and bromocriptine).

As for FGF23-mediated phosphate wasting, treatment of frank hypophosphatemia includes oral phosphorus and calcitriol. Important therapeutic endpoints include growth velocity and radiographic evidence of epiphyseal healing.

Treatment guidelines for hypercortisolism are difficult to establish given the rarity of neonatal Cushing syndrome. Definitive treatment includes surgical removal of the affected adrenal glands. For medical treatment, metyrapone is frequently effective and is preferred over ketoconazole in children with liver abnormalities.

10.2.7 Molecular genetics and genetic counseling

MAS is not inherited. In all affected individuals, it is caused by a de novo *GNAS* mutation that occurs sporadically after conception at an early stage of embryonic development. Verified vertical transmission has been never observed, likely due to embryonic lethality. No parent of a child with MAS has been demonstrated to have clinical manifestations of the disorder. Given the somatic mutational mechanism of MAS, the risk of recurrence for this disorder is not considered to be increased in respect of the general population. For these reasons, prenatal testing for MAS is not indicated.

10.3 GH-producing tumors and acromegaly

GH-secreting pituitary tumors origin within the somatotroph cells of the anterior pituitary gland. GH released by these tumors triggers the production of insulin-like growth factor 1 (IGF-1), leading to a range of signs and symptoms known as acromegaly (or gigantism in children). Treatment options for these patients include surgery, radiation therapy, and medication mainly with SSA.

10.3.1 Etiology

GH-secreting tumors account for 20% of all pituitary tumors and may present with variants, including pure somatotroph or plurihormonal tumors [113]. The latter family of neoplasms expresses multiple hormones, more frequently GH and prolactin, and rarely, GH and thyroid-stimulating hormone or subunit of glycoprotein hormones. GH-secreting pituitary tumors are prevalently diagnosed as macroadenomas with a diameter size >10 mm (75%−85%). Invasion of the cavernous sinus is a clinical issue described in 20%−50% of cases [114]. The increased secretion of GH further stimulates the liver to produce IGF-1, which is involved in bone and tissue growth, metabolism, and other processes. Chronic exposure to GH and IGF-1 hypersecretion causes acromegaly, a rare disorder of disproportionate skeletal, tissue, and organ growth. Gigantism (excessive height) occurs if the GH excess begins before puberty, prior to the closure of epiphyseal plates [115]. More than 95% of patients with acromegaly or gigantism harbor a GH-secreting pituitary tumor. In less than 5% of cases, acromegaly is related to a hypothalamic tumor that produces GH-releasing hormone, again leading to excessive GH. Peripheral GH-secreting tumors are extremely infrequent (Isidro, 2005)[116].

As demonstrated by X-chromosome inactivation analysis, GH-secreting pituitary tumors are monoclonal tumors occurring sporadically in the vast majority of the patients [117]. In approximately 5% of patients, acromegaly can present as part of familiar isolated pituitary adenomas in aryl hydrocarbon receptor-interacting protein (AIP) or GPR101 (G protein-coupled receptor 101) mutation positive and negative cases [118][119] or be associated with familial syndromes occurring concomitantly with other endocrine tumors, such as in multiple endocrine neoplasia type 1 or 4 (MEN1, MEN4) syndromes [120], Carney complex syndrome [121], SDHx-related pituitary tumors, and MAS [122].

10.3.2 Epidemiology

Acromegaly occurs with an estimated annual incidence of 5 cases per 1 million individuals and a prevalence ranging from 2.8 to 13.7 cases (Holdaway, 1999). This endocrine disorder is an insidious disease, and patients are usually diagnosed when they present with an acromegaly-associated comorbidity with an average age ranging from 40 to 50 years and a mean delay in diagnosis that has reduced over the years but is still estimated to be around 3−6 years [123][124,125]. The delay in diagnosis and comorbidities represents the main factors influencing prognosis [115]. Younger patients show more rapidly growing pituitary tumors often associated with more aggressive disease. Both genders are equally affected and the incidence of acromegaly does not vary with ethnicity.

10.3.3 Clinical characteristics

Hypersecretion of GH leads to excessive production of IGF-1. IGF-1 mediates most of the phenotypic features and metabolic effects of GH, but GH excess also has direct detrimental consequences. Acromegaly usually presents with gradual and progressive acral overgrowth (hands and feet), facial changes (such as mandibular prognathism, nose enlargement, and forehead prominence), and soft tissue swelling (of tongue, heart, kidney, colon, periarticular, and cartilaginous thickening) [115]. GH is a potent insulin antagonist, and acromegaly results in impaired glucose tolerance in many patients with diabetes mellitus observed in up to a third. Lipid abnormalities, such as increased serum triglycerides, may be an accompanying feature of insulin resistance [126]. Cardiovascular complications are common and account for as much as 60% of the mortality in acromegaly. Further systemic complications include hypertension, arthritis, osteoporosis, carpal tunnel syndrome, and sleep apnea [127]. Visual impairment is rarely encountered and represents a symptom of acromegaly in the setting of chiasmal compression with macroadenomas.

Due to cardiovascular, metabolic, and respiratory comorbidities, acromegaly is associated with a 2-fold increase in mortality [128]. Moreover, it has become increasingly apparent that high levels of IGF-1 in acromegaly are related to an increased risk of developing neoplasia, particularly colon cancer. However, mortality returns close to that expected in the general population after appropriate treatment and biochemical normalization of GH and IGF-1 levels.

10.3.4 **Diagnosis**

The diagnosis of acromegaly is based on a combination of clinical examination and biochemical assessment. Biochemical screening is the first step to demonstrate a dysregulated and enhanced GH secretion as well as elevated IGF-1 levels. The Endocrine Society guidelines and experts' consensus recommend using age- and sex-adjusted IGF-1 levels in combination with GH nadir during an OGTT to diagnose or rule out acromegaly. In particular, GH nadir ≥ 0.4 µg/L during OGTT is the currently used cutoff to diagnose acromegaly (using ultrasensitive GH assays). IGF-1 levels are relatively stable, correlate with clinical features of acromegaly, and exhibit a log-linear relationship with elevated GH levels. Similar to IGF-1 assays, the GH assay method can impact the absolute GH concentration reported by a laboratory [129]. Magnetic resonance imaging is considered to be the most effective imaging technique to confirm the presence of a GH-secreting pituitary tumor and evaluate tumor invasiveness, proximity to the optic chiasm, and compression of surrounding structures by the tumor mass. In addition, assessment of the integrity of the other pituitary hormones needs to be performed by a combination of basal and dynamic tests.

In the rare patient in whom a nonpituitary etiology is suspected, measurement of serum growth hormone releasing hormone (GHRH) may be performed in order to identify ectopic sources of tumor. Typically, elevated GRHR levels occur in ectopic GHRH syndromes such as neuroendocrine tumors.

10.3.5 **Management**

Treatment for acromegaly is aimed at restoring normal rates of GH and IGF-1 secretion that usually reflects adequate management of disease-associated symptoms and comorbidities. Goals of treatment are also removal of the pituitary tumor, resolution of mass effects, and prevention of recurrence. Data from large observational studies show improved long-term outcomes and reduced mortality in patients who achieve GH <0.4 µg/L [129]. Acromegaly treatment is often multimodal, and three modalities of treatment are currently available: transsphenoidal surgery, medical therapy and pituitary irradiation.

Surgery is the treatment of choice for most patients. However, about 50% of them do not achieve remission after debulking of tumor. Determinants of surgical remission include tumor size, degree of invasiveness, and the experience of the surgeon in resecting challenging tumors [130]. In particular lower remission rate is seen with macroadenomas and tumors invading the cavernous sinus and parasellar area.

Medical therapy is recommended for patients with persistent disease despite surgery as well as for patients for whom surgery is not a suitable option. The first-generation SSA octreotide and lanreotide represent the mainstay of medical treatment. Other pharmacologic agents available include pasireotide, a second-generation SS receptor ligands, pegvisomant, a genetically engineered recombinant GH-receptor antagonist, and, in selected cases, DA such as cabergoline. The SSA bind cognate receptors (SSTR) in the tumor. By coupling to Gαi protein, SSTR inhibits AC and cAMP accumulation, and some have been found to reduce calcium entry by modulating L-type Ca^{2+} and K^+ channels [131], all these events being involved in the and suppression of GH release. SST2 and SST5 are the receptor subtypes mainly involved in the cytostatic effects of SSA through the inhibition of ERK1/2 and other partially different pathways [132], whereas SST2 is considered the major mediator of proapoptotic signals in the somatotrophs [133]. Biochemical control and tumor reduction rate range between 25% and 70% with the first-generation SSA [129]. Normalized levels of IGF-1 are still achieved in fewer than half of patients treated with pasireotide [129].

Pegvisomant blocks GH action by binding peripheral GH receptors resulting in a suppressed IGF-1 production. Pegvisomant monotherapy administered in patients with proven resistance to long-term high-dose SSA treatment after unsuccessful surgery and/or radiotherapy yields biochemical control rates of 60% [134]. Novel growth hormone receptor (GHR) antagonists in development include GHR antibodies, small orally active molecules, and antisense molecules directed against the GHR.

Cabergoline is selective for the dopamine receptor type 2 (D2) whose efficacy in reducing GH and IGF-I levels has been observed in several studies, both as monotherapy and in combination schemes. The recent guidelines suggest that cabergoline can be attempted as a first-line medical therapy in patients with mildly elevated levels of IGF-1 of <2.5 times the upper limit of normal and may have a role as a combination treatment with SSA in patients partially responders or resistant to SSA.

Radiotherapy, administered as conventional external-beam or as stereotactic radiosurgery with the use of gamma-knife, is reserved for patients with recurred or persisted disease activity after unsuccessful surgery and those who are resistant or intolerant to pharmacological treatment. Radiotherapy has been shown to be effective and safe in reducing both serum GH and IGF-I concentrations with the time to remission estimated to range from 12 to 60 months. Rarely encountered local side effects include visual deficits, cerebral radionecrosis, cerebrovascular damage, and cognitive deficits. In more than 40% of patients a functional decline of other pituitary axes has been described 10 years after irradiation [135].

10.3.6 Acromegaly as part of McCune–Albright syndrome

Acromegaly is encountered in about 20%–30% of patients with MAS raising particular diagnostic and therapeutic issues. So far, more than 100 MAS patients with documented acromegaly have been reported in the literature [97].

The diagnosis may be delayed in childhood, given that growth acceleration from GH excess may be initially interpreted as the consequence of comorbid precocious puberty or secondarily masked by the premature closure of the epiphyses due to precocious puberty. When diagnosed during adulthood, the mean age at diagnose is around 30 years (range, 17–64 years), which is, however, about 10 years earlier than in "classic" acromegaly. The diagnosis is more often suggested by shoe size modifications than by the dysmorphic craniofacial effects of acromegaly, features that may be masked by the skeletal deformities of MAS. Acromegaly may be also occasionally assessed during follow-up of MAS, or during routine evaluation of amenorrhea–galactorrhea syndrome or sexual problems. Patients may report visual defects and/or headaches, and the presence of a pituitary tumor may be detected during imaging studies of craniofacial FD. As in classic acromegaly, measurements of IGF-1 and GH levels (randomly or after OGTT) are generally performed to confirm the diagnosis of MAS-associated acromegaly. Life expectancy in these patients may be reduced. MAS is associated with an increased risk of thyroid carcinoma, hepatic, and pancreatic lesions and enhanced the risk of developing sarcoma. The incidence of cancer may be as well increased in patients with MAS and acromegaly [97]. However, the limited number of patients makes it difficult to ascertain whether GH/IGF-1 excess itself is associated with an increased risk of neoplasm development.

Management of morbidities associated with acromegaly in MAS patients is similar to those of classical acromegalic patients. However, due to the thickness of the cranial dysplasia at the skull base (particularly when it involves the sphenoid bone) and to the risk of hemorrhage given the high vascularity of FD, pituitary surgery is technically difficult in patients with MAS; thus medical treatment is preferred and widely used. SSA improves GH/IGF-1 levels in most patients, although hormone normalization achievement fails in the majority. Administered alone, DA has been shown to provide the control of acromegaly in a few patients, whereas SSA alone or combined with DA controlled GH/IGF-1 excess in about 30% of cases. Overall, the pharmacological response to SSA does not vary from that observed in patients with sporadic acromegaly treated with SSA as the primary treatment. The efficacy of medical treatment for MAS-associated acromegaly has greatly improved with the introduction of pegvisomant. Indeed, if IGF-1 levels decrease but do not normalize with SSA, combined therapy with pegvisomant and SSA may be proposed rather than shifting from SSA to pegvisomant, as in sporadic acromegaly. Pituitary therapeutic irradiation has been used to treat acromegaly also in patients with MAS; however, the disappointing results obtained so far may be linked to the limited duration of follow-up in some of these patients [97].

10.3.7 Somatic activating mutations in GNAS and its epigenetic imprinting regulation in GH-secreting tumors

The monoclonal origin of most pituitary tumors indicates that these lesions derive from the replication of a single cell that acquired a specific growth advantage. Despite intensive investigations, still little is known about the genetic causes of GH-secreting pituitary tumors [136]. The only mutational change associated with pituitary tumors identified to date in a subset of sporadic GH-secreting tumors (ranging from 30% to 40%) occurs in the gene encoding the α subunit of the stimulatory G-protein (*GNAS*) [45] [137][138][139][140]. This finding heralded the identification of the same genetic defects in MAS: amino acid substitutions in exons 8 and 9, replacing either Arg201 with Cys, His, or Ser; or Gln227 with Arg or Leu, as described previously. MAS is caused by *GNAS* mutations occurring during early embryonic development. When these mutational events are acquired later during life, they may cause focal diseases, such as acromegaly. *GNAS* mutations are less commonly found in other endocrine tumors such as toxic thyroid adenomas, hyperfunctioning adrenal tumors, and Leydigiomas. By preventing hydrolysis of GTP (a step that is required for the reassembly of the G-protein β heterotrimer and signals the end of the G protein activation, as illustrated in Figs. 10.2 and 10.3), activating *GNAS* mutations lead to a permanently turned-on state of the AC system, resulting in elevated cAMP levels and excess of GH synthesis and secretion.

As already mentioned in this chapter, the GNAS locus is subjected to genomic imprinting, the epigenetic phenomenon occurring either during embryogenesis or postnatal period by which one of the two allelic variants (maternal or paternal) is characterized by a partial or total loss of expression [23]. Due to highly complex imprinted expression pattern, the Gsα transcript mainly derives from the maternal allele in almost all endocrine tissues [28,29]. Accordingly, *GNAS* mutations in sporadic pituitary tumors have been localized to the maternal allele, and the specific maternal origin of *gsp* oncogene was confirmed in GH-secreting tumors associated with the MAS [26,27].

Interestingly, the majority of tumors negative for *gsp* mutations present a partial loss of genomic imprinting and a biallelic expression of Gs transcripts, suggesting that loss of imprinting may represent a secondary feature in pituitary tumorigenesis [141]. Indeed, somatotroph tumors negative for *gsp* mutations and expressing high mRNA and protein levels of Gsα show a clinical phenotype similar to those of *gsp*-positive tumors. Such observations suggest that besides the mutational status of *GNAS*, Gsα expression levels may also play a role in determining tumoral features of somatotroph tumors.

10.3.8 In vivo phenotype of *gsp* mutations in GH-secreting pituitary tumors

Consistent with the well-recognized mitogenic role played by cAMP signaling pathway in pituitary somatotrophs, functional studies on cell lines transfected with mutant Gsα allowed to identify Gsα as a product of a proto-oncogene converted into an oncogene (*gsp*) that confers a proliferative phenotype. At variance with the in vitro observations, *gsp* mutations are generally associated with a benign phenotype in vivo. To date, only one study reported the presence of *gsp* mutation in one patient harboring a lethal prolactinoma: in this case, this change represented the second hit for the transition from prolactinoma to acromegaly [142]. The reduced oncogenic potential of *GNAS* mutations could be explained by the existence of intracellular regulatory molecular mechanisms being able to counteract in vivo the activation of the cAMP pathway. In particular, the expression of *gsp* oncogene correlates with the activation state of PDE enzymes that, in turn, induce cAMP degradation. cAMP-specific PDE4 expression was found to be about significantly higher in gsp-positive GH-secreting tumors than in wild type (WT) tissues [143][144]. Upregulation of inducible early cAMP-repressor (ICER) gene transcription has also been seen in tumors with *gsp* mutations. ICER inhibits the transcription of cAMP-responsive genes by competing with the binding of CREB to CRE [145]. Moreover, the presence of *gsp* mutations negatively affects Gsα protein stability, as demonstrated by Ballaré and colleagues [146]. Furthermore, it is worth mentioning that proliferation of GH-secreting pituitary cells may be inhibited independently of cAMP levels regulation, as the case of SST5, an SSTR able to exert antiproliferative effects in somatotrophs by coupling with the inhibitory G protein GoA [147] [44].

Patients carrying *gsp* mutations show similarity in the clinical and biochemical features to those without the mutations. Screening studies carried out on large series of GH-secreting tumors first indicate no difference in gender, clinical behavior, GH and IGF-1 levels, duration of the disease, cure rate, and clinical outcome in patients positive or negative for *gsp* mutations [148][149], even though more recent published data indicate a higher *GNAS* mutation frequency in tumors from older patients [150]. Moreover, *gsp*-positive tumors frequently present as very small size tumors with a well differentiated and densely granulated aspect. Moreover, serum GH levels of patients with *gsp* mutations are not further increased after GHRH stimulation. These tumors are characterized by abnormally high cAMP levels and respond to cAMP-independent stimuli, consistent with constitutive Gs and AC activation. Finally, *gsp*-positive tumors show increased sensitivity to inhibitory agents, such as SSA, that act by inhibiting cAMP formation, a feature so far unexplained as no increases in SSTRs expression levels have been found in these tumors [151,152][153].

10.3.9 Other cAMP pathway alterations in GH-secreting pituitary tumors

Several lines of evidence support the role of the cAMP signaling pathway GHRH-mediated in promoting somatotroph tumorigenesis. Alterations in other genes involved in cAMP signaling have been identified in sporadic GH-secreting pituitary tumors. As described in Fig. 10.3, cyclic AMP acts mainly through PKA, a heterotetrameric enzyme that is ubiquitously expressed in eukaryotic cells. PKA-dependent serine-threonine phosphorylation of cAMP target molecules is critically restrained by adequate levels and normal function of regulatory subunit type Ia of the PKA (PRKAR1A). Specifically, PKA regulatory subunits bind the PKA catalytic subunits in the absence of cAMP molecules. Interestingly, a reduced PRKAR1A expression due to increased proteasomal degradation has been identified in sporadic GH-secreting pituitary tumors [154][155]. The newly identified mutations affecting the gene encoding the catalytic subunit α of the PKA (PRKACA) resulting in an increased PKA activity [156] have been detected in a large proportion of cortisol-secreting adrenocortical adenomas [157][158][159], but not in the large cohort of sporadic GH-secreting pituitary tumors studied [160].

Furthermore, recent whole-genome studies in GH-secreting tumors highlighted a number of altered genes associated with the Ca^{2+} signaling [140][161]. This is in line with the notion that GHRH activates not only Gsα and cAMP-dependent signaling pathways but also Gα-I, Gβ, and Gγ, leading to the release of intracellular Ca^{2+}, which further promotes GH secretion. These data support the idea that dysregulation of the calcium signaling might be involved in pituitary tumorigenesis.

Genetic variants of PDE11A4 determining a reduced cAMP degradation rate have been described to marginally contribute to the development of pituitary somatotroph tumors [162]. Finally, a recurrent somatic mutation in the GPR101 gene, which encodes an orphan GPCR, has been recently found in some adults with acromegaly (4% of cases) [163].

References

[1] Blatt C, Eversole-Cire P, Cohn VH, et al. Chromosomal localization of genes encoding guanine nucleotide-binding protein subunits in mouse and human. Proc Natl Acad Sci USA 1988;85:7642−6.

[2] Gejman PV, Weinstein LS, Martinez M, et al. Genetic mapping of the Gs-α subunit gene (GNAS) to the distal long arm of chromosome 20 using a polymorphism detected by denaturing gradient gel electrophoresis. Genomics 1991;9:782−3.

[3] Rao VV, Schnittger S, Hansmann I. G protein Gs alpha (GNAS 1), the probable candidate gene for Albright hereditary osteodystrophy, is assigned to human chromosome 20q12-q13.2. Genomics 1991;10 (1):257−61.

[4] Levine MA, Modi WS, O'Brien SJ. Mapping of the gene encoding the alpha subunit of the stimulatory G protein of adenylyl cyclase (GNAS) to 20q13.2 → q13.3 in human by in situ hybridization. Genomics 1991;11(2):478−9.

[5] Peters J, Beechey CV, Ball ST, et al. Mapping studies of the distal imprinting region of mouse chromosome 2. Genet Res 1994;63(3):169−74.

[6] Kozasa T, Itoh H, Tsukamoto T, et al. Isolation and characterization of the human Gsα gene. Proc Natl Acad Sci USA 1988;85:2081−5.

[7] Kehlenbach RH, Matthey J, Huttner WB. XL alpha s is a new type of G protein. Nature 1994;372:804−9.

[8] Ischia R, Lovisetti-Scamihorn P, Hogue-Angeletti R, et al. Molecular cloning and characterization of NESP55, a novel chromogranin-like precursor of a peptide with 5-HT1B receptor antagonist activity. J Biol Chem 1997;272:11657—62.

[9] Ishikawa Y, Bianchi C, Nadal-Ginard B, et al. Alternative promoter and 5′ exon generate a novel $G_s\alpha$ mRNA. J Biol Chem 1990;265:8458—62.

[10] Swaroop A, Agarwal N, Gruen JR, et al. Differential expression of novel Gs alpha signal transduction protein cDNA species. Nucleic Acids Res 1991;19:4725—9.

[11] Bastepe M, Jüppner H. GNAS locus and pseudohypoparathyroidism. Horm Res 2005;63(2):65—74.

[12] Hayward B, Kamiya M, Strain L, et al. The human GNAS gene is imprinted and encodes distinct paternally and biallelically expressed G proteins. Proc Natl Acad Sci USA 1998;95:10038—43.

[13] Peters J, Wroe SF, Wells CA, et al. A cluster of oppositely imprinted transcripts at the Gnas locus in the distal imprinting region of mouse chromosome 2. Proc Natl Acad Sci USA 1999;96:3830—5.

[14] Pasolli HA, Klemke M, Kehlenbach R, et al. Characterization of the extra-large G protein alpha-subunit XLalphas. I. Tissue distribution and subcellular localization. J Biol Chem 2000;275(43):33622—32.

[15] Klemke M, Pasolli HA, Kehlenbach RH, et al. Characterization of the extra-large G protein alpha-subunit XLalphas. II. Signal transduction properties. J Biol Chem 2000;275(43):33633—40.

[16] Bastepe M, Gunes Y, Perez-Villamil B, et al. Receptor-mediated adenylyl cyclase activation through XLalpha(s), the extra-large variant of the stimulatory G protein alpha-subunit. Mol Endocrinol 2002; 16(8):1912—19.

[17] Kaya AI, Ugur O, Oner SS, et al. Coupling of beta2-adrenoceptors to XLalphas and Galphas: a new insight into ligand-induced G protein activation. J Pharmacol Exp Ther 2009;329(1):350—9.

[18] Lovisetti-Scamihorn P, Fischer-Colbrie R, Leitner B, et al. Relative amounts and molecular forms of NESP55 in various bovine tissues. Brain Res 1999;829(1—2):99—106.

[19] Li T, Vu TH, Zeng ZL, et al. Tissue-specific expression of antisense and sense transcripts at the imprinted Gnas locus. Genomics 2000;69(3):295—304.

[20] Hayward B, Bonthron D. An imprinted antisense transcript at the human GNAS locus. Hum Mol Genet 2000;9:835—41.

[21] Wroe SF, Kelsey G, Skinner JA, et al. An imprinted transcript, antisense to Nesp, adds complexity to the cluster of imprinted genes at the mouse Gnas locus. Proc Natl Acad Sci USA 2000;97:3342—6.

[22] Puzhko S, Goodyer C, Kerachian M, et al. Parathyroid hormone signaling via $G\alpha$s is selectively inhibited by an NH2-terminally truncated $G\alpha$s: implications for pseudohypoparathyroidism. J Bone Min Res 2011;26:2473—85.

[23] Bartolomei MS, Tilgham SM. Genomic imprinting in mammals. Annu Rev Genet 1997;31:493—525.

[24] Liu J, Yu S, Litman D, et al. Identification of a methylation imprint mark within the mouse Gnas locus. Mol Cell Biol 2000;20(16):5808—17.

[25] Germain-Lee EL, Ding CL, Deng Z, et al. Paternal imprinting of Galpha(s) in the human thyroid as the basis of TSH resistance in pseudohypoparathyroidism type 1a. Biochem Biophys Res Commun 2002;296(1):67—72.

[26] Mantovani G, Bondioni S, Lania AG, et al. Parental origin of Gsalpha mutations in the McCune-Albright syndrome and in isolated endocrine tumors. J Clin Endocrinol Metab 2004;89:3007—9.

[27] Mantovani G, Bondioni S, Locatelli M, et al. Biallelic expression of the Gsalpha gene in human bone and adipose tissue. J Clin Endocrinol Metab 2004;89:6316—19.

[28] Hayward BE, Barlier A, Korbonits M, et al. Imprinting of the G(s)alpha gene GNAS in the pathogenesis of acromegaly. J Clin Invest 2001;107(6):R31—6.

[29] Mantovani G, Ballare E, Giammona E, et al. The $G_s\alpha$ gene: predominant maternal origin of transcription in human thyroid gland and gonads. J Clin Endocrinol Metab 2002;87(10):4736—40.

[30] Mantovani G, Spada A, Elli FM. Pseudohypoparathyroidism and Gsα-cAMP-linked disorders: current view and open issues. Nat Rev Endocrinol 2016;12(6):347−56.

[31] Oldham WM, Hamm HE. Heterotrimeric G protein activation by G-protein-coupled receptors. Nat Rev Mol Cell Biol 2008;9(1):60−71.

[32] Downes GB, Gautam N. The G protein subunit gene families. Genomics 1999;62(3):544−52.

[33] Simon MI, Strathmann MP, Gautam N. Diversity of G proteins in signal transduction. Science 1991; 252(5007):802−8.

[34] Sprang SR, Chen Z, Du X. Structural basis of effector regulation and signal termination in heterotrimeric Galpha proteins. Adv Protein Chem 2007;74:1−65.

[35] Smotrys JE, Linder ME. Palmitoylation of intracellular signaling proteins: regulation and function. Annu Rev Biochem 2004;73:559−87.

[36] Chen CA, Manning DR. Regulation of G proteins by covalent modification. Oncogene 2001; 20(13):1643−52.

[37] Sprang SR. G protein mechanisms: insights from structural analysis. Annu Rev Biochem 1997;66:639−78.

[38] Sunahara RK, Taussig R. Isoforms of mammalian adenylyl cyclase: multiplicities of signaling. Mol Interv 2002;2(3):168−84.

[39] Pronko SP, Saba LM, Hoffman PL, et al. Type 7 adenylyl cyclase-mediated hypothalamic-pituitary-adrenal axis responsiveness: influence of ethanol and sex. J Pharmacol Exp Ther 2010;334(1):44−52.

[40] Paulssen RH, Johansen PW, Gordeladze JO, et al. Cell-specific expression and function of adenylyl cyclases in rat pituitary tumour cell lines. Eur J Biochem 1994;222(1):97−103.

[41] Wachten S, Masada N, Ayling LJ, et al. Distinct pools of cAMP centre on different isoforms of adenylyl cyclase in pituitary-derived GH3B6 cells. J Cell Sci 2010;123(Pt 1):95−106.

[42] Antoni FA, Sosunov AA, Haunso A, et al. Short-term plasticity of cyclic adenosine 3′,5′-monophosphate signaling in anterior pituitary corticotrope cells: the role of adenylyl cyclase isotypes. Mol Endocrinol 2003;17(4):692−703.

[43] Stork PJ, Schmitt JM. Crosstalk between cAMP and MAP kinase signaling in the regulation of cell proliferation. Trends Cell Biol 2002;12(6):258−66.

[44] Peverelli E, Mantovani G, Lania AG, Spada A. cAMP in the pituitary: an old messenger for multiple signals. J Mol Endocrinol 2013;52(1):R67−77.

[45] Landis C, Masters SB, Spada A, et al. GTPase inhibiting mutations activate the alpha chain of Gs and stimulate adenylyl cyclase in human pituitary tumours. Nature 1989;340(6236):692−6.

[46] Lania A, Mantovani G, Spada A. G protein mutations in endocrine diseases. Eur J Endocrinol 2001;145 (5):543−59.

[47] Vallar L, Spada A, Giannattasio G. Altered Gs and adenylate cyclase activity in human GH-secreting pituitary adenomas. Nature 1987;330(6148):566−8.

[48] Lania AG, Mantovani G, Ferrero S, et al. Proliferation of transformed somatotroph cells related to low or absent expression of protein kinase a regulatory subunit 1A protein. Cancer Res 2004;64(24):9193−8.

[49] Skalhegg BS, Tasken K. Specificity in the cAMP/PKA signaling pathway. Differential expression, regulation, and subcellular localization of subunits of PKA. Front Biosci 2000;5:D678−93.

[50] Bos JL. Epac: a new cAMP target and new avenues in cAMP research. Nat Rev Mol Cell Biol 2003;4 (9):733−8.

[51] de Rooij J, Zwartkruis FJ, Verheijen MH, et al. Epac is a Rap1 guanine-nucleotide-exchange factor directly activated by cyclic AMP. Nature 1998;396(6710):474−7.

[52] Kawasaki H, Springett GM, Mochizuki N, et al. A family of cAMP-binding proteins that directly activate Rap1. Science 1998;282(5397):2275−9.

[53] Vitali E, Peverelli E, Giardino E, et al. Cyclic adenosine 3′-5′-monophosphate (cAMP) exerts prolifer-ative and anti-proliferative effects in pituitary cells of different types by activating both cAMP-dependent protein kinase A (PKA) and exchange proteins directly activated by cAMP (Epac). Mol Cell Endocrinol 2014;383(1−2):193−202.

[54] Bender AT, Beavo JA. Cyclic nucleotide phosphodiesterases: molecular regulation to clinical use. Pharmacol Rev 2006;58(3):488−520.

[55] Ma Y, Huang J, Ali S, et al. Src tyrosine kinase is a novel direct effector of G proteins. Cell 2000;102:635−46.

[56] Mattera R, Graziano MP, Yatani A, et al. Splice variants of the α subunit of the G protein Gs activate both adenylyl cyclase and calcium channels. Science 1989;243:804−7.

[57] Yu S, Yu D, Lee E, et al. Variable and tissue-specific hormone resistance in heterotrimeric Gs protein α-subunit (Gsα) knockout mice is due to tissue-specific imprinting of the Gsα gene. Proc Natl Acad Sci USA 1998;95:8715−20.

[58] Yu S, Castle A, Chen M, et al. Weinstein LS increased insulin sensitivity in Gsα knockout mice. J Biol Chem 2001;276:19994−8.

[59] Plagge A, Gordon E, Dean W, et al. The imprinted signaling protein XLαs is required for postnatal adaptation to feeding. Nat Genet 2004;36:818−26.

[60] Chen M, Haluzik M, Wolf NJ, et al. Increased insulin sensitivity in paternal Gnas knockout mice is asso-ciated with increased lipid clearance. Endocrinology 2004;145:4094−102.

[61] Chen M, Shrestha YB, Podyma B, et al. Gsα deficiency in the dorsomedial hypothalamus underlies obe-sity associated with Gsα mutations. J Clin Invest 2017;127(2):500−10.

[62] Chen M, Wilson EA, Cui Z, et al. Gsα deficiency in the dorsomedial hypothalamus leads to obesity, hyperphagia, and reduced thermogenesis associated with impaired leptin signaling. Mol Metab 2019;25:142−53.

[63] Ramaswamy G, Fong J, Brewer N, et al. Ablation of Gsα signaling in osteoclast progenitor cells adversely affects skeletal bone maintenance. Bone 2018;109:86−90.

[64] Iglesias-Bartolome R, Torres D, Marone R, et al. Inactivation of a Gα(s)-PKA tumour suppressor path-way in skin stem cells initiates basal-cell carcinogenesis. Nat Cell Biol 2015;17(6):793−803.

[65] Aldred MA, Trembath RC. Activating and inactivating mutations in the human GNAS gene. Hum Mutat 2000;16(3):183−9.

[66] Lumbroso S, Paris F, Sultan C, et al. Activating Gsalpha mutations: analysis of 113 patients with signs of McCune-Albright syndrome−a European Collaborative Study. J Clin Endocrinol Metab 2004;89:2107−13.

[67] Weinstein LS, Shenker A, Gejman PV, et al. Activating mutations of the stimulatory G protein in the McCune−Albright syndrome. N Engl J Med 1991;325(24):1688−95.

[68] Lyons J, Landis CA, Harsh G, et al. Two G protein oncogenes in human endocrine tumors. Science 1990;249(4969):655−9.

[69] Clementi E, Malgaretti N, Meldolesi J, et al. A new constitutively activating mutation of the Gs protein alpha subunit-gsp oncogene is found in human pituitary tumours. Oncogene 1990;5(7):1059−61.

[70] Schwindinger WF, Francomano CA, Levine MA. Identification of a mutation in the gene encoding the alpha subunit of the stimulatory G protein of adenylyl cyclase in McCune-Albright syndrome. Proc Natl Acad Sci USA 1992;89(11):5152−6.

[71] Idowu BD, Al-Adnani M, O'Donnell P, et al. A sensitive mutation-specific screening technique for GNAS mutations in cases of fibrous dysplasia: the first report of a codon 227 mutation in bone. Histopathology 2007;50:691−704.

[72] Wu JY, Aarnisalo P, Bastepe M, et al. Gs α enhances commitment of mesenchymal progenitors to the osteoblast lineage but restrains osteoblast differentiation in mice. J Clin Invest 2011;121:3492−504.

[73] Riminucci M, Fisher LW, Shenker A, et al. Fibrous dysplasia of bone in the McCune-Albright syndrome: abnormalities in bone formation. Am J Pathol 1997;151(6):1587−600.

[74] Marie PJ, de Pollak C, Chanson P, et al. Increased proliferation of osteoblastic cells expressing the activating Gs alpha mutation in monostotic and polyostotic fibrous dysplasia. Am J Pathol 1997; 150(3):1059−69.

[75] Cone RD, Lu D, Koppula S, et al. The melanocortin receptors: agonists, antagonists, and the hormonal control of pigmentation. Recent Prog Horm Res 1996;51:287−317.

[76] Rhee Y, Bivi N, Farrow E, et al. Parathyroid hormone receptor signaling in osteocytes increases the expression of fibroblast growth factor-23 in vitro and in vivo. Bone 2011;49:636−43.

[77] Lavi-Moshayoff V, Wasserman G, Meir T, et al. PTH increases FGF23 gene expression and mediates the high-FGF23 levels of experimental kidney failure: a bone parathyroid feedback loop. Am J Physiol Ren Physiol 2010;299:F882−9.

[78] Happle R. The McCune-Albright syndrome: a lethal gene surviving by mosaicism. Clin Genet 1986;29:321−4.

[79] Saggio I, Remoli C, Spica E, et al. Constitutive expression of Gs α (R201C) in mice produces a heritable, direct replica of human fibrous dysplasia bone pathology and demonstrates its natural history. J Bone Min Res 2014;29:2357−68.

[80] Collins MT, Singer FR, Eugster E. McCune-Albright syndrome and the extraskeletal manifestations of fibrous dysplasia. Orphanet J Rare Dis 2012;7(Suppl. 1):S4.

[81] Mariot V, Wu JY, Aydin C, et al. Potent constitutive cyclic AMP-generating activity of XL α s implicates this imprinted GNAS product in the pathogenesis of McCune-Albright syndrome and fibrous dysplasia of bone. Bone 2011;48:312−20.

[82] Dumitrescu CE, Collins MT. McCune-Albright syndrome. Orphanet J Rare Dis 2008;3:12.

[83] Vasilev V, Daly AF, Thiry A, et al. McCune-Albright syndrome: a detailed pathological and genetic analysis of disease effects in an adult patient. J Clin Endocrinol Metab 2014;99(10):E2029−38.

[84] Rostomyan L, Beckers A. Screening for genetic causes of growth hormone hypersecretion. Growth Horm IGF Res 2016;30-31:52−7.

[85] Elli FM, de Sanctis L, Bergallo M, et al. Improved molecular diagnosis of McCune-Albright syndrome and bone fibrous dysplasia by digital PCR. Front Genet 2019;10:862.

[86] Albright F, Butler AM, Hampton AO, et al. Syndrome characterized by osteitis fibrosa disseminata, areas, of pigmentation, and endocrine dysfunction, with precocious puberty in females: report of 5 cases. N Engl J Med 1937;216:727−46.

[87] McCune DJ. Osteitis fibrosa cystica: the case of a nine-year-old girl who also exhibits precocious puberty, multiple pigmentation of the skin and hyperthyroidism. Am J Dis Child 1936;52:743−4.

[88] Boyce A.M., Florenzano P., de Castro L.F., Collins M.T.: Fibrous dysplasia/McCune−Albright syndrome. In: GeneReviews® [Internet]. Seattle, WA: University of Washington, Seattle; 1993−2020. 2015 Feb 26 [updated 2019 Jun 27].

[89] Benhamou J, Gensburger D, Messiaen C, et al. Prognostic factors from an epidemiologic evaluation of fibrous dysplasia of bone in a modern cohort: the FRANCEDYS study. J Bone Min Res 2016; 31(12):2167−72.

[90] Boyce AM, Chong WH, Shawker TH, et al. Characterization and management of testicular pathology in McCune-Albright syndrome. J Clin Endocrinol Metab 2012;97(9):E1782−90.

[91] Celi FS, Coppotelli G, Chidakel A, et al. The role of type 1 and type 2 5′-deiodinase in the pathophysiology of the 3,5,3′-triiodothyronine toxicosis of McCune-Albright syndrome. J Clin Endocrinol Metab 2008;93(6):2383−9.

[92] Brown RJ, Kelly MH, Collins MT. Cushing syndrome in the McCune-Albright syndrome. J Clin Endocrinol Metab 2010;95(4):1508−15.

[93] Leet AI, Chebli C, Kushner H, et al. Fracture incidence in polyostotic fibrous dysplasia and the McCune-Albright syndrome. J Bone Min Res 2004;19(4):571−7.

[94] Hart ES, Kelly MH, Brillante B, et al. Onset, progression, and plateau of skeletal lesions in fibrous dysplasia and the relationship to functional outcome. J Bone Min Res 2007;22(9):1468−74.

[95] Kuznetsov SA, Cherman N, Riminucci M, et al. Age-dependent demise of GNAS-mutated skeletal stem cells and "normalization" of fibrous dysplasia of bone. J Bone Min Res 2008;23(11):1731−40.

[96] Collins MT, Sarlis NJ, Merino MJ, et al. Thyroid carcinoma in the McCune-Albright syndrome: contributory role of activating Gs alpha mutations. J Clin Endocrinol Metab 2003;88(9):4413−17.

[97] Salenave S, Boyce AM, Collins MT, et al. Acromegaly and McCune-Albright syndrome. J Clin Endocrinol Metab 2014;99(6):1955−69.

[98] Turan S, Bastepe M. GNAS spectrum of disorders. Curr Osteoporos Rep 2015;13(3):146−58.

[99] Bercaw-Pratt JL, Moorjani TP, Santos XM, et al. Diagnosis and management of precocious puberty in atypical presentations of McCune-Albright syndrome: a case series review. J Pediatr Adolesc Gynecol 2012;25(1):e9−e13.

[100] Lim YH, Ovejero D, Sugarman JS, et al. Multilineage somatic activating mutations in HRAS and NRAS cause mosaic cutaneous and skeletal lesions, elevated FGF23 and hypophosphatemia. Hum Mol Genet 2014;23:397−407.

[101] Ovejero D, Lim YH, Boyce AM, et al. Cutaneous skeletal hypophosphatemia syndrome: clinical spectrum, natural history, and treatment. Osteoporos Int 2016;27:3615−26.

[102] Javaid MK, Boyce A, Appelman-Dijkstra N, et al. Best practice management guidelines for fibrous dysplasia/McCune-Albright syndrome: a consensus statement from the FD/MAS international consortium. Orphanet J Rare Dis 2019;14(1):139.

[103] Leet AI, Magur E, Lee JS, et al. Fibrous dysplasia in the spine: prevalence of lesions and association with scoliosis. J Bone Jt Surg Am 2004;86(3):531−7.

[104] Mancini F, Corsi A, De Maio F, et al. Scoliosis and spine involvement in fibrous dysplasia of bone. Eur Spine J 2009;18(2):196−202.

[105] Boyce AM, Chong WH, Yao J, et al. Denosumab treatment for fibrous dysplasia. J Bone Min Res 2012;27(7):1462−70.

[106] Benhamou J, Gensburger D, Chapurlat R. Transient improvement of severe pain from fibrous dysplasia of bone with denosumab treatment. Joint Bone Spine 2014;81(6):549−50.

[107] Ganda K, Seibel MJ. Rapid biochemical response to denosumab in fibrous dysplasia of bone: report of two cases. Osteoporos Int 2014;25(2):777−82.

[108] Feuillan P, Calis K, Hill S, et al. Letrozole treatment of precocious puberty in girls with the McCune-Albright syndrome: a pilot study. J Clin Endocrinol Metab 2007;92(6):2100−6.

[109] Estrada A, Boyce AM, Brillante BA, et al. Long-term outcomes of letrozole treatment for precocious puberty in girls with McCune-Albright syndrome. Eur J Endocrinol 2016;175(5):477−83.

[110] Tessaris D, Corrias A, Matarazzo P, et al. Thyroid abnormalities in children and adolescents with McCune-Albright syndrome. Horm Res Paediatr 2012;78(3):151−7.

[111] Ross DS, Burch HB, Cooper DS, Greenlee MC, Laurberg P, Maia AL, et al. 2016 American Thyroid Association guidelines for diagnosis and management of hyperthyroidism and other causes of thyrotoxicosis. Thyroid 2016;26:1343−421.

[112] Boyce AM, Glover M, Kelly MH, et al. Optic neuropathy in McCune-Albright syndrome: effects of early diagnosis and treatment of growth hormone excess. J Clin Endocrinol Metab 2013;98(1):E126−34.

[113] Heaney AP, Melmed S. Molecular targets in pituitary tumours. Nat Rev Cancer 2004;4(4):285−95.

[114] Anik I, Cabuk B, Gokbel A, et al. Endoscopic transsphenoidal approach for acromegaly: remission rates in 401 patients: 2010 consensus criteria. World Neurosurg 2017;108:278−90.

[115] Melmed S. Acromegaly pathogenesis and treatment. J Clin Invest 2009;119(11):3189−202.

[116] Isidro ML, Iglesias Díaz P, Matías-Guiu X, et al. Acromegaly due to a growth hormone-releasing hormone-secreting intracranial gangliocytoma. J Endocrinol Invest 2005;28(2):162−5.

[117] Asa SL, Ezzat S. The pathogenesis of pituitary tumours. Nat Rev Cancer 2002;2(11):836−49.

[118] Chahal HS, Chapple JP, Frohman LA, et al. Clinical, genetic and molecular characterization of patients with familial isolated pituitary adenomas (FIPA). Trends Endocrinol Metab 2010;21(7):419−27.

[119] Iacovazzo D, Caswell R, Bunce B, et al. Germline or somatic GPR101 duplication leads to X-linked acrogigantism: a clinico-pathological and genetic study. Acta Neuropathol Commun 2016;4(1):56.

[120] Thakker RV. Multiple endocrine neoplasia—syndromes of the twentieth century. J Clin Endocrinol Metab 1998;83:2617−20.

[121] Carney JA, Gordon H, Carpenter PC, et al. The complex of myxomas, spotty pigmentation, and endocrine overactivity. Medicine (Baltimore) 1985;64(4):270−83.

[122] Akintoye SO, Chebli C, Booher S, et al. Characterization of gsp-mediated growth hormone excess in the context of McCune−Albright syndrome. J Clin Endocrinol Metab 2002;87(11):5104−12.

[123] Nachtigall L, Delgado A, Swearingen B, et al. Extensive clinical experience: changing patterns in diagnosis and therapy of acromegaly over two decades. J Clin Endocrinol Metab 2008;93(6):2035−41.

[124] Reid TJ, Post KD, Bruce JN, et al. Features at diagnosis of 324 patients with acromegaly did not change from 1981 to 2006: acromegaly remains under-recognized and under-diagnosed. Clin Endocrinol (Oxf) 2010;72(2):203−8.

[125] Kreitschmann-Andermahr I, Siegel S, Kleist B, et al. Diagnosis and management of acromegaly: the patient's perspective. Pituitary 2016;19(3):268−76.

[126] Rosario PW. Frequency of acromegaly in adults with diabetes or glucose intolerance and estimated prevalence in the general population. Pituitary 2011;14:217−21.

[127] Sesmilo G, Resmini E, Sambo M, et al. Prevalence of acromegaly in patients with symptoms of sleep apnea. PLoS One 2017;12(9)e0183539.

[128] Holdaway IM, Rajasoorya C. Epidemiology of acromegaly. Pituitary 1999;2(1):29−41.

[129] Melmed S, Bronstein MD, Chanson P, et al. A Consensus Statement on acromegaly therapeutic outcomes. Nat Rev Endocrinol 2018;14(9):552−61.

[130] Shimon I, Cohen ZR, Ram Z, et al. Transsphenoidal surgery for acromegaly: endocrinological follow-up of 98 patients. Neurosurgery 2001;48(6):1239−43.

[131] Yang SK, Chen C. Involvement of somatostatin receptor subtypes in membrane ion channel modification by somatostatin in pituitary somatotropes. Clin Exp Pharmacol Physiol 2007;34(12):1221−7.

[132] Florio T. Molecular mechanisms of the antiproliferative activity of somatostatin receptors (SSTRs) in neuroendocrine tumors. Front Biosci 2008;13:822−40.

[133] Ferrante E, Pellegrini C, Bondioni S, et al. Octreotide promotes apoptosis in human somatotroph tumor cells by activating somatostatin receptor type 2. Endocr Relat Cancer 2006;13(3):955−62.

[134] Colao A, Auriemma RS, Lombardi G, et al. Resistance to somatostatin analogs in acromegaly. Endocr Rev 2011;32(2):247−71.

[135] Jenkins PJ, Bates P, Carson MN, et al. Conventional pituitary irradiation is effective in lowering serum growth hormone and insulin-like growth factor-I in patients with acromegaly. J Clin Endocrinol Metab 2006;91(4):1239−45.

[136] Horvath A, Stratakis CA. Clinical and molecular genetics of acromegaly: MEN1, Carney complex, McCune-Albright syndrome, familial acromegaly and genetic defects in sporadic tumors. Rev Endocr Metab Disord 2008;9(1):1−11.

[137] Yoshimoto K, Iwahana H, Fukuda A, et al. Rare mutations of the Gs alpha subunit gene in human endocrine tumors. Mutation detection by polymerase chain reaction primer-introduced restriction analysis. Cancer 1993;72(4):1386−93.

[138] Song ZJ, Reitman ZJ, Ma ZY, et al. The genome-wide mutational landscape of pituitary adenomas. Cell Res 2016;26(11):1255–9.

[139] Taboada GF, Tabet AL, Naves LA, et al. Prevalence of gsp oncogene in somatotropinomas and clinically non-functioning pituitary adenomas: our experience. Pituitary 2009;12(3):165–9.

[140] Ronchi CL, Peverelli E, Herterich S, et al. Landscape of somatic mutations in sporadic GH-secreting pituitary adenomas. Eur J Endocrinol 2016;174(3):363–72.

[141] Picard C, Silvy M, Gerard C, et al. Gs alpha overexpression and loss of Gs alpha imprinting in human somatotroph adenomas: association with tumor size and response to pharmacologic treatment. Int J Cancer 2007;121(6):1245–52.

[142] Lania AG, Ferrero S, Pivonello R, et al. Evolution of an aggressive prolactinoma into a growth hormone secreting pituitary tumor coincident with GNAS gene mutation. J Clin Endocrinol Metab 2010;95(1):13–17.

[143] Lania AG, Persani L, Ballarè E, et al. Constitutively active Gsa is associated with an increased phosphodiesterase activity in human growth hormone secreting adenomas. J Clin Endocrinol Metab 1998;83(5):1624–8.

[144] Persani L, Lania AG, Borgato S, et al. Relevant cAMP-specific phosphodiesterase isoforms in human pituitary: effect of Gsa mutations. J Clin Endocrinol Metab 2001;86(8):3795–800.

[145] Peri A, Conforti B, Baglioni-Peri S, et al. Expression of cyclic adenosine 3,5-monophosphate (cAMP)-responsive element binding protein and inducible-cAMP early repressor genes in growth hormone-secreting pituitary adenomas with or without mutations of the Gs alpha gene. J Clin Endocrinol Metab 2001;86(5):2111–17.

[146] Ballarè E, Mantovani S, Lania A, et al. Activating mutations of the Gs alpha gene are associated with low levels of Gs alpha protein in growth hormone-secreting tumors. J Clin Endocrinol Metab 1998;83(12):4386–90.

[147] Peverelli E, Lania AG, Mantovani G, et al. Characterization of intracellular signaling mediated by human somatostatin receptor 5: role of the DRY motif and the third intracellular loop. Endocrinology 2009a;150(7):3169–76.

[148] Spada A, Arosio M, Bochicchio D, et al. Clinical, biochemical, and morphological correlates in patients bearing growth hormone-secreting pituitary tumors with or without constitutively active adenylyl cyclase. J Clin Endocrinol Metab 1990;71(6):1421–6.

[149] Adams EF, Brockmeier S, Friedmann E, et al. Clinical and biochemical characteristics of acromegalic patients harboring gsp-positive and gsp-negative pituitary tumors. Neurosurg 1993;33(2):198–203.

[150] Gadelha MR, Trivellin G, Hernández-Ramírez LC, et al. Genetics of pituitary adenomas. Front Horm Res 2013;41:111–40.

[151] Barlier A, Gunz G, Zamora AJ, et al. Pronostic and therapeutic consequences of Gs alpha mutations in somatotroph adenomas. J Clin Endocrinol Metab 1998;83(5):1604–10.

[152] Barlier A, Pellegrini-Bouiller I, Gunz G, et al. Impact of gsp oncogene on the expression of genes coding for Gsalpha, Pit-1, Gi2alpha, and somatostatin receptor 2 in human somatotroph adenomas: involvement in octreotide sensitivity. J Clin Endocrinol Metab 1999;84(8):2759–65.

[153] Corbetta S, Ballaré E, Mantovani G, et al. Somatostatin receptor subtype 2 and 5 in human GH-secreting pituitary adenomas: analysis of gene sequence and mRNA expression. Eur J Clin Invest 2001;31(3):208–14.

[154] Sandrini F, Kirschner LS, Bei T, et al. PRKAR1A, one of the Carney complex genes, and its locus (17q22–24) are rarely altered in pituitary tumours outside the Carney complex. J Med Genet 2002;39(12):e78.

[155] Yamasaki H, Mizusawa N, Nagahiro S, et al. GH-secreting pituitary adenomas infrequently contain inactivating mutations of PRKAR1A and LOH of 17q23-24. Clin Endocrinol (Oxf) 2003;58(4):464–70.

[156] Calebiro D, Hannawacker A, Lyga S, et al. PKA catalytic subunit mutations in adrenocortical Cushing's adenoma impair association with the regulatory subunit. Nat Commun 2014;5(5):5680.

[157] Beuschlein F, Fassnacht M, Assie G, et al. Constitutive activation of PKA catalytic subunit in adrenal Cushing's syndrome. N Engl J Med 2014;370(11):1019−28.

[158] Goh G, Scholl UI, Healy JM, et al. Recurrent activating mutation in PRKACA in cortisol-producing adrenal tumors. Nat Genet 2014;46(6):613−17.

[159] Di Dalmazi G, Kisker C, Calebiro D, et al. Novel somatic mutations in the catalytic subunit of the protein kinase A as a cause of adrenal Cushing's syndrome: a European multicentric study. J Clin Endocrinol Metab 2014;99(10):E2093−100.

[160] Larkin SJ, Ferrau F, Karavitaki N, et al. Sequence analysis of the catalytic subunit of PKA in somatotroph adenomas. Eur J Endocrinol 2014;171(6):705−10.

[161] Valimaki N, Demir H, Pitkanen E, et al. Whole-genome sequencing of growth hormone (GH)-secreting pituitary adenomas. J Clin Endocrinol Metab 2015;100(10):3918−27.

[162] Peverelli E, Ermetici F, Filopanti M, et al. Analysis of genetic variants of phosphodiesterase 11A in acromegalic patients, Eur J Endocrinol 2009b;161(5):687−694.

[163] Trivellin G, Daly AF, Faucz FR, et al. Gigantism and acromegaly due to Xq26 microduplications and GPR101 mutation. N Engl J Med 2014;371(25):2363−74.

Surgical management of growth hormone-secreting adenomas

Elizabeth Hogan[1,2] and Prashant Chittiboina[1,3]

[1]*Neuroendocrine Surgery Unit, National Institute of Neurological Diseases and Stroke, Bethesda, MD, United States*
[2]*Department of Neurosurgery, George Washington University, Washington, DC, United States* [3]*Surgical Neurology Branch, National Institute of Neurological Diseases and Stroke, Bethesda, MD, United States*

11.1 Introduction

Growth hormone (GH)-secreting adenomas of the pituitary gland can cause acromegaly in adults and gigantism in children before fusion of bony epiphysis [1]. Gigantism is an extremely rare disorder with an estimated incidence of 8 per million person-years with the total number of reported cases numbering only in the hundreds to date [2]. Although rare, acromegaly is slightly more common with a total prevalence ranging between 2.8 and 13.7 cases per 100,000 people and the annual incidence rates ranging between 0.2 and 1.1 cases/100,000 people [3]. The mortality in this patient population is significantly higher than the general population with the most likely cause of death stemming from cardiac or vascular complications [4,5]. However, with the acceptable reduction in GH levels, this mortality can be decreased [4,6,7]. The first-line therapy for the treatment of GH-secreting adenomas remains surgical for the removal of the tumor mass and normalization of GH and IGF-1 levels while preserving the normal pituitary function [4].

11.2 History of surgical treatment of pituitary pathology

Interest in pituitary pathology was initially started in 1886 when acromegaly was described by Professor Marie [8]. However, it was not until 1893 that Caton and Paul recorded the first attempted resection of a pituitary tumor via a transcranial approach originally described by Sir Victory Horsely [9–11]. Unfortunately, the tumor was unreachable, and the patient died 3 months later. With that experience, Horsely later operated on 10 pituitary tumors by using a combined subfrontal and lateral middle fossa approach [9,10]. The first reported successful surgery for acromegaly was in Vienna in 1907 by Dr. Schloffer [12,13]. He successfully completed this surgery by mobilizing the whole nose on a pedicle and removed bone and mucous membrane to access the pituitary fossa [13,14]. With the initial success of Dr. Schloffer's surgery, other Viennese surgeons that specialized in nasal and sinus surgery began to approach pituitary tumors, such as Hirsch [15]. Other surgeons quickly followed adding multiple refinements to surgery, such as an infranasal or gingival approach in order to reduce infection and improve cosmesis [14]. It was the gingival

Gigantism and Acromegaly. DOI: https://doi.org/10.1016/B978-0-12-814537-1.00012-9

approach that was then used by the first neurosurgeon Dr. Harvey Cushing in 1910 in Baltimore [16]. After that first surgery, he modified this approach in 1912 to a sublabial incision and submucous septal resection as described by Haltsead and Kocher, respectively [16,17]. Cushing performed this modified procedure over 200 times until 1929 when he abandoned this approach in favor of a transcranial route [14,18]. Most neurosurgeons followed Cushing's lead and the transsphenoidal approach fell out of favor [12]. However, during that period, Professor Norman Dott of Edinburgh, who was trained by Cushing, continued to use the transsphenoidal route in Scotland [19]. During his career, he trained Dr. Gerard Guiot of France in the transsphenoidal approach [19,20]. Dr. Guiot went on to perform over a 1000 cases and further enhanced his surgical accuracy by using intraoperative radiofluoroscopy to clearly define the anatomy of the nasal passage while maneuvering his surgical instruments [21]. During his career, he trained Jules Hardy of Montreal in his technique of transsphenoidal surgery with intraoperative radiofluoroscopy.

It was Hardy who then reintroduced the transsphenoidal approach in the late 1970s along with a new concept of microsurgery, which lead to the modern era of pituitary surgery. He was the first person to use an operating microscope for pituitary surgery along with instruments of his own design [14,22,23]. He was also the first person to present the idea of the microadenoma causing endocrine dysfunction without deforming the sella and reported cases where he removed microadenomas with evident benefit. The first of these operations was performed on a patient with acromegaly [14,22]. Therefore the procedure described by Hardy has been the backbone of the leading surgical approach to both macro and micro GH-secreting adenomas by neurosurgeons over the past 50 years [24,25].

11.3 Approach to the sella

The classic transsphenoidal approach that was described by Hardy [26] has undergone further transformations over the past five decades. Areas of the skull base that were initially believed to be only accessible through a transcranial approach, such as the cavernous sinus or suprasellar space, are now being successfully approached transfacially. This advancement can be attributed to the better knowledge of microsurgical anatomy and modern microinstrumentation [9].

Sabit et al. [27] describe a minimally invasive technique to approach the cavernous sinus via a combined transsphenoidal transmaxillary approach that is both extraudal and intranasal. They achieve this through a sublabial incision and maxillotomy. This approach provides adequate medial to lateral reach in the infrasellar and parasellar regions. Additionally, it allows visualization of the complete ipsilateral cavernous sinus and medial aspect of the contralateral sinus. Another example would be the modified transsphenoidal approach described by Mason et al. [28] and Kouri et al. [29], which involves wide bone exposure of the anterior surface of the sella, followed by removal of the posterior portion of the planum sphenoidale. This extension of transsphenoidal approach allowed access to the parasellar and clival regions. Clouldwell et al. [30] describe a refined method of using a asymmetric retractor and alternating the position of the self-retaining retractor in order to enable lateral visualization. This allows for various sections of the skull base to be exposed, and the boney opening can be extended in both the superior/inferior and lateral directions. Their extended approach provides access to tumors that may extend anteriorly in the suprasellar region, laterally into the cavernous sinus, or inferiorly to the clivus [9].

Another modification to Hardy's classical approach is the use of an endoscope instead of a microscope. The first reported use of an endoscope was back in 1978 by Bushe and Halves [31]. Nevertheless, its use in transsphenoidal surgery did not gain popularity until the mid to late 1990s when otolaryngologist switched from the traditional open approach to a completely endoscopic one for sinus surgery. The excellent visualization of cranial sinuses provided by the endoscope prompted neurosurgeons to explore the use of the endoscope during transsphenoidal surgery [32−34]. Since then, several variations of the procedure using the endoscope have been reported. Yaniv and Rappaport [35] document a combined transsphenoidal approach where they used the endoscope for the initial exposure and then transition to the microscope for the tumor resection. Jho et al. [34] document a large series of patients who underwent successful purely endoscopic surgeries for pituitary tumors utilizing only one nostril for the entire case. They would initially hold the scope in the nondominate hand and use the dominant hand for the exposure. Then they would use an endoscope arm to hold the scope, freeing up both hands for tumor resection. They argued that with their approach patients had a faster postoperative recovery. Others have described a purely endoscopic approach using both nostrils and two clinicians, with a four-hand approach. The minimally invasive transsphenoidal approach, whether using a microscope or an endoscope, can be used effectively for the resection of approximately 95% of pituitary tumors. Reserving the craniotomy for larger tumors with significant temporal or anterior cranial fossa extension [36].

11.3.1 Current practices for approach to the sella

In the 21st century, the modern approach for the treatment of GH-secreting pituitary macro and micro adenomas is now surgeon specific, with both a purely endoscopic and a purely microscopic approach being utilized (Fig. 11.1). Recent studies have demonstrated that there is no difference in surgical results when comparing the microscopic and endoscopic techniques in acromegaly when surgery is performed by an experienced surgeon [37−39]. For the microscopic approach, the sella can be approached either via an endonasal incision or through a sublabial incision. The sublabial approach begins with a horizontal sublabial incision executed about 1 cm above the gingivobuccal sulcus, extending between the alae nasalis. The incision is then extended through the periosteum of the premaxilla and elevated to the pyriform aperture and anterior edge of the nasal septum with blunt dissection. Next a subperichondrial space is identified, and a left superior tunnel is opened using blunt dissection until the boney septum is identified. The quadrangular cartilage is then separated from the floor of maxilla and posterior bony septum and pushed laterally connecting the left and right tunnels. The vomer bone is then removed and kept for use at the end of the case. At this point, a Hardy retractor is inserted, and the operating microscope is utilized. Next, the sphenoid sinus ostia are located and used as anatomical landmarks for opening the anterior wall of the sphenoid sinus. The anterior wall is removed exposing the sellar floor. The mucosa is removed from the sella floor, and using a drill and Kerrisons, the sellar floor is removed. The dura is then incised, and the resection of the pituitary tumor is performed either using variously shaped curettes and microforceps or by the pseudocapsular method described below [4,40].

For the endonasal microscopic route, the dissection of a submucosal tunnel is generally performed similar to the most nasal septal surgeries. Then a medial nasal incision is performed, and the mucosa is unilaterally separated from the cartilaginous and osseous nasal septae. This tunnel is then maintained open using a nasal speculum, and the operating microscope is brought in. The

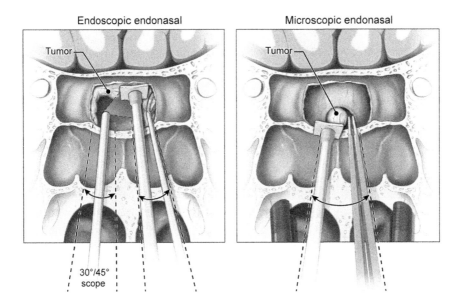

FIGURE 11.1

The drawing on the left illustrates the endoscopic endonasal approach to the sella and pituitary gland. It demonstrates how the use of an angled scope allows for better visualization of the lateral aspect of the sella. The drawing on the right illustrates the microscopic endonasal approach to the sella and pituitary gland. It demonstrates the larger degree of freedom of moment the surgeon possesses using this approach when accessing midline tumors compared to the endoscopic approach.

remainder of the surgery is carried out in similar fashion as the other approaches [4]. In contrast, the endoscopic approach is performed via a direct endonasal route where the incision is made directly in front of the vomer, the osseous nasal septum is fractured, and the operation is carried out in a similar fashion to what is previously described. The endoscopic approach does not utilize a nasal speculum and can use either one nostril or both for access to the sella [4].

11.4 Pituitary adenoma dissection

The pseudocapsule is the layer of compressed normal gland and distinct from the capsule of the pituitary gland. As an adenoma grows, the adjacent normal gland is compressed, and multiple layers of compressed reticulin form a pseudocapsule with higher tensile strength than the adenoma tissue. Dr. Costello first described this pseudocapsule in his 1936 report on pituitary adenomas found incidentally during autopsy [41]. Oldfield then described the pseudocapsular technique for the resection of pituitary microadenomas based on his extensive surgical experience. He described how to create a plane between the pseudocapsule and normal gland and then to carry this plane circumferentially around the tumor, removing the tumor en bloc when possible [42]. The pseudo-capsule technique was later adapted by Oldfield and others to remove pituitary macroadenomas as well [43–45]. Once the sella is adequately exposed and the dura is opened, the tumor resection is performed using a pseudocapsular method when possible (Fig. 11.2). Once the tumor is removed, hemostasis is obtained, and the reconstruction of the sellar floor can be performed using several methods that are determined based on whether or not there is an intraoperative CSF leak. If there is no leak, a small amount of gel foam can be used to fill the sella, and the floor is reconstructed using an autologous bone graft from the previously resected vomer. Once in place, the sella is covered with a form of human fibrin glue or other adhesives. If there is a leak, the sella floor should be reconstructed using an autologous graft of fat or muscle to fill the sella and then bone and fibrin glue to complete the reconstruction. If leaking is still appreciated, a nasal septal flap can be utilized to further reconstruct the sella.

With recent advances in imaging and technology, novel surgical techniques or variants of established techniques have been possible, like the use of neuronavigation or intraoperative MRI [46]. Neuronavigation is being utilized during surgery to more accurately localize anatomic structures. The pointer-based localization is allowing for the three-dimensional reconstruction of the tumor and its adjacent structures based on the preoperative MRI imaging. This allows for a safer more direct approach to the sella even in cases with poor pneumatization of the sphenoid sinus, narrow sella floors, and/or vascular anomalies that would otherwise make the surgical approach and tumor resection more difficult. It is also useful in reoperation cases where the normal anatomy is no longer present [4,46,47]. Intraoperative MRI is becoming increasingly available and allows for intraoperative determination of tumor resection, especially in cases of large tumors with cavernous sinus and suprasellar extension. Multiple studies have demonstrated a higher rate of total tumor resection in patients where intraoperative MRI was utilized [48–51]. Given these findings, its application during transsphenoidal surgery for acromegalic patients with complex adenomas has been strongly advised in order to increase the rate of total tumor resection and decrease patient mortality [52–54].

11.5 Complications

Given the advances in both surgical technique and microsurgical equipment over the past 50 years, complication rates after surgery in experienced hands are extremely low. The mortality rate is less than 0.5%, and serious complications occur in less than 1.5% of cases [55]. The data show that the strongest predictor of postoperative complications is surgeon experience with the rate of complications decreasing in an inverse relationship with increasing annual numbers of pituitary cases [56]. The most common complications after surgery include CSF leak, nasal complications, or endocrine abnormality. In the literature, the incidence of CSF leak after surgery is 0.9%–6% [57–61]. Nasal complications, more commonly epistaxis and less common sinusitis, mucocele, and septum perforation have a rate of 0.7%–1.7% [57,59–61]. The most common endocrine abnormality after surgery is transient versus permanent diabetes insipidus with an incidence of 0.7%–2.3% and 1%–3.5%, respectively [57–61]. The next most common endocrine

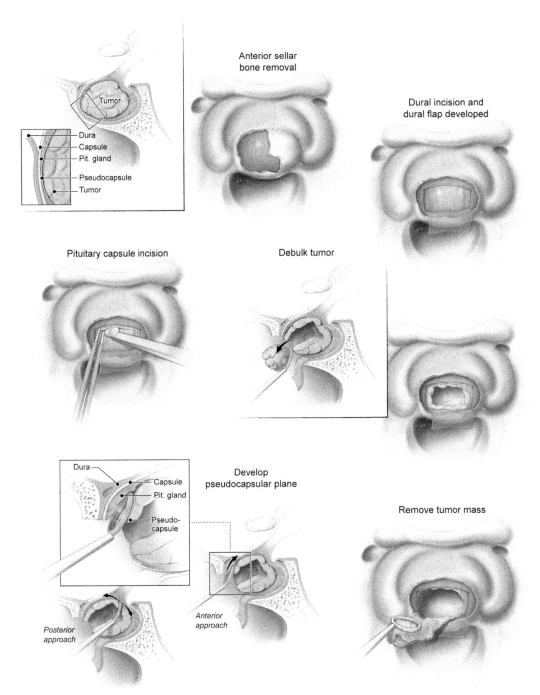

FIGURE 11.2

This drawing depicts the stepwise approach to removing macroadenomas using the tumor's pseudocapsule. This approach ensures that the entire tumor is resected, and that no remanence is left behind. If the pseudocapsule is breeched, small pieces of tumor could still remain and lead to recurrence of acromegaly/gigantism.

abnormality is iatrogenic hypopituitarism with an incidence of less than 3% in the current literature [55]. More rare complications after surgery include meningitis, vascular injury, stroke, ophthalmoplegia, and/or intracerebral hemorrhages [60].

11.6 Endocrine outcomes

The remission rate after surgical intervention in the treatment of GH-secreting adenomas is difficult to elucidate given the ever-changing definition of cure coupled with the absence of clear clinical parameters with which to monitor disease activity. The biochemical markers GH and IGF-1 have been associated with increased mortality in acromegaly and therefore are relied upon as the best indicators of disease burden and remission [62]. Over the past 50 years, the GH cutoff has been progressively lowered as GH assays have become more sensitive and specific. For example, in the 1960s, the postoperative basal GH criteria for remission were 10−20 times less stringent as compared with today [63]. Until the 1990s, GH levels were measured using polyclonal radioimmunoassay, and the treatment target was less than 2.5 µg/L. Then with the switch to the newer monoclonal GH assays, stricter criteria have been created. In the current literature, the most commonly used criteria for remission is called the Cortina criteria and a normal IGF-1 level and nadir GH level after oral glucose tolerance test of less than 1 µg/L defines cure [64−66].

Taking into account the variability between remission criteria when examining the current literature, the surgical cure rate for acromegaly is approximately 70%−80% in experienced surgical hands. The success rate of surgery decreases progressively with increase in tumor size and extent of invasion to as low as 10% in giant tumors >4 cm [67−70]. Additionally, remission rates are also influenced by the preoperative GH levels with basal levels between 40 and 70 µg/L associated with a poor prognosis and lower likelihood of postoperative remission [71,72].

When evaluating the recurrence rates after surgery, no clear consensus exists in the literature secondary to the variability between remission criteria over time. Patients labeled with recurrence in older studies would likely have been categorized as persistent disease using the new stricter criteria [73]. Looking at new literature that uses the current stricter criteria, recurrence rates are between 0% and 8.4% [6,74,75]. Recurrence can develop early or even very late with a latency of greater than 10 years reported [76]. A postoperative prognostic factor that has been associated with a low risk of recurrence in Biermasz et al. [77] is a lack of increase in serum GH levels after injection of thyroid-releasing hormone (TRH). Their study revealed that an increase of more than 1.44 µg/L in serum GH 20 minutes after injection of TRH was predictive of recurrence (75% sensitivity and 100% specificity). Another potential predictor is a higher although still considered normal level of GH (0.19−1.0 µg/L) after an oral glucose suppression test immediately following surgery. It is postulated that this might indicate residual tumor and risk of recurrent disease [77−79].

11.7 Adjunct treatments

Almost one-third of patients with acromegaly will not be cured by initial surgery [80] and will continue to have persistent disease and associated morbidity/mortality. Therefore adjunct therapy is a necessary aspect of the treatment paradigm in order to help work toward normalization of IGF-1 and GH levels. Adjuvant therapy can consist of medication treatment, reoperation, radiation, or any combination of the three and is generally tailored to each individual patient.

11.7.1 Medical treatment

Over the past 30 years, the therapeutic armamentarium for acromegaly has rapidly evolved and currently includes agents that target the GH-secreting adenoma itself (dopamine agonist and somatostatin analog) and that block the effects of GH on its receptor (pegvisomant) [67]. The specific drug class to use for a specific patient can depend on various factors including tumor characteristics on MRI, cost, underlying comorbidities, and degree of active disease. Nonetheless, somatostatin analogs are considered the first-line medical therapy given their long track record of efficacy and lower overall cost [66]. They can improve a patient's biochemical response rate up to 80% of patients and achieve normalization of hormones in about 30% of patients based on the strict Cortina criteria for remission [81–83]. Dose optimization of somatostatin analogs can also increase the rate of response to medical treatment in patients and can lead to clinically significant tumor shrinkage of >20% in about 75% of patients [84–86]. Currently available somatostatin analogs: octreotide-LAR and lanreotide autogel have been reported to be similar in efficacy in both biochemical and tumor control and slightly better than lanreotide-SR [87,88].

Another adjuvant use of somatostatin analogs in the treatment of acromegaly is their use in the preoperative period as an adjuvant to improve surgical outcomes. Theoretically, preoperative use could potentially reduce GH levels and alter tumor characteristics and positively influence surgical outcomes. Additionally, it has been postulated that they can also reduce the risk of surgery by reducing the patient's preoperative comorbidities associated with acromegaly. However, there is no clear evidence to support these claims, and results from published studies are still controversial because of the conflicting results and inconsistent preoperative treatment length [89]. In the literature, there are some studies that show preoperative use of a somatostatin analog, which is associated with a higher likelihood of achieving biochemical cure after surgery as compared to those who did not receive the drug [90–92]. In contrast, there are other studies that show no difference in outcomes between the two cohorts [93,94]. Many clinicians agree that regardless of a potential benefit on the outcome of surgery, preoperative somatostatin analogue use is appropriate as it can potentially reduce the cardiopulmonary anesthetic risk by reducing soft tissue swelling of the tongue and upper airway [67].

Dopamine agonist has been shown to be useful in both pure GH-secreting tumors and mixed prolactin and GH-secreting tumors because GH adenomas have been shown to have a dopamine receptor on their surface. There are limited data on dopamine agonist as a monotherapy, but they have been shown to have an efficacy of about 50% when combined with somatostatin therapy. There are no data on their effect on tumor shrinkage. The most commonly used dopamine agonist is cabergoline and less commonly used is bromocriptine [95,96].

Pegvisomat is a genetically modified GH analogue that acts as a competitive inhibitor at the receptor level to block the action of native GH. In initial studies, it has been shown to be a very effective monotherapy at controlling acromegalic symptoms and normalizing IGF-1 levels in 90%–95% of patients. However, given minimal effect on tumor shrinkage and high cost,

Pegvisomant is generally reserved as a second-line therapy in patients who are suboptimally controlled with somatostatin analogs [97,98].

11.7.2 **Reoperation**

Although surgery is the first line of treatment for patients with acromegaly, repeat surgery is generally reserved for debulking purposes to increase the likelihood of remission with adjuvant medical therapies or to relief mass effect on the parasellar structures. There is minimal evidence of the effectiveness of repeat surgery in the literature. In studies using the Cortina remission criteria, the mean remission rate following a reoperation was 37% (range: 8%−59%) [71,99]. Reoperation for debulking is often offered to increase the effectiveness of adjuvant medical therapy. In a recent study of medically refractory patients with preoperative GH levels greater than 10 µg/L who underwent reoperations were found to have reduction in GH levels below 10, 5, and 2.5 µg/L in 50%, 35%, and 21% of patients, respectively [66,100].

11.7.3 **Radiation therapy**

Radiotherapy has been used for decades in the treatment of GH-secreting adenomas, but currently most experts would argue that it is considered a last resort for patients with persistent postoperative and medically refractory acromegaly given its unfavorable side effect profile and delayed time to achieve biochemical effects [66,101]. However, it has been shown to be an effective treatment and is cost effective, and advances in technology and techniques have improved the side effect profile, and therefore it should still remain as a treatment tool for acromegaly.

The method of radiation therapy with the longest experience in treating acromegaly is conventional radiotherapy (CT). CT is generally administered by a linear accelerator with a total dose of 40−45 Gy, fractionated in at least 20 sessions. A single field, two opposing fields, or a three-field technique can be used to focus a single beam of high-energy radiation onto the treatment area [102]. Treatment of acromegaly patients with CT leads to eventual normalization of GH and IGF-1 in 60%−80% and long-term tumor control in 80%−100% of patients [103−107]. However, the benefits of CT are overshadowed by the risk of adverse effects and the very slow onset of effect, generally 5−15 years after treatment [103,108]. The adverse effects associated with CT are hypopituitarism in 30%−80% of patients, radiation-induced optic neuropathy in 0%−5%, secondary brain tumors in up to 2% at 10−20 years postradiation, cerebral vascular accidents in 4% and 21% at 5 and 20 years, respectively, cranial nerve deficits in 0%−3%, and overall psycho-cognitive impairment associated with a dose and field-size related impairment on verbal memory and executive function when compared to nonirradiated patients [105−107,109−112]. All of these adverse effects lead to an increased mortality 1.6−2.2 times higher than the general population [102,113−115].

The other form of radiation used in the treatment of GH-secreting adenomas is stereotactic radiotherapy. This form of radiation has two modalities: (1) fractionated stereotactic radiotherapy, in which the total dose is delivered in 20−25 fractions of 1−2 Gy each [116]; (2) stereotactic radiosurgery can be delivered by gamma knife, Cyberknife, and linear accelerator-based platforms. Gamma knife usually consists of an array of 192−201 cobalt-60 sources arranged in a hemisphere and focused with a collimator helmet onto an isocenter and radiation is delivered as a single

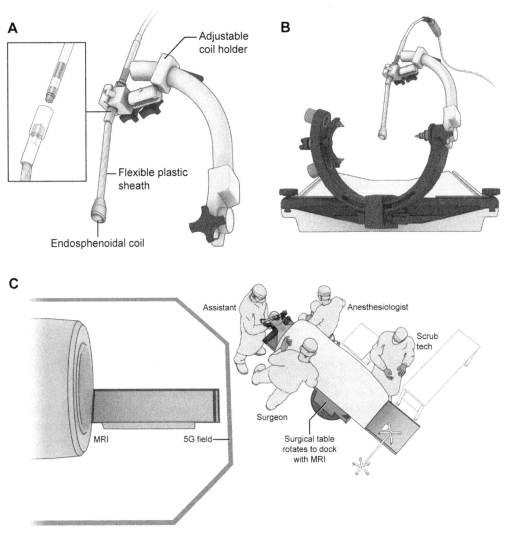

FIGURE 11.3

A Shows a drawing of the first-in-human MRI coil that is placed intraoperatively next to the sella in order to obtain better resolution images. B Demonstrates how the coil attaches to the MRI compatible Mayfield head clamp during surgery. C Depicts how the patient, surgeon, anesthesia and surgical staff are positioned in the MRI surgical suite during the actual surgery.

fraction. Cyberknife combines a mobile linear accelerator attached to a robotic arm and delivers a single fraction dose of radiation via an image-guided robotic system. Linear accelerator utilizes X-rays to deliver multiple arcs or beams of radiation formed from colliding accelerated electrons with a target metal and is usually delivered as a single fraction [117–119]. Radiosurgical methods are able to achieve up to 100% tumor control in patients but cannot be used in tumors that are within 3 mm of the optic apparatus. Additionally, biochemical control is achieved in 30%−60% of patients at 5-year follow-up [116,120,121]. Therefore when compared to CT, stereotactic radiotherapy has a faster interval from therapy to biochemical effects. Radiosurgery has a similar side effect profile as compared to CT except for a lower incidence of hypopituitarism with reported rates around 47% [122].

11.8 New technologies

At the National Institutes of Health Clinical Center in Bethesda, we are currently working on techniques to improve intraoperative adenoma visualization. Intraoperative MRI potentially allows significant improvements in visualization during surgery with visualization proportional to magnetic field strength of the intraoperative MRI magnet. However, increasing the magnetic field strength over 3 T is expensive and is currently impractical for intraoperative MRI studies. Therefore to improve on the imaging resolution of the pituitary gland and the parasellar structures, we developed an intraoperative coil that can be placed within the sphenoid sinus [the endosphenoidal coil (ESC)] during sublabial transsphenoidal surgery [123]. We are currently in the early stages of the first-in-human intraoperative MRI coil for high resolution of imaging during surgery (Fig. 11.3).

11.9 Conclusions

Growth hormone excess from pituitary adenomas leads to rare but highly morbid conditions including acromegaly and gigantism. The diagnosis of GH excess is made by biochemical testing, and the source of GH is localized with preoperative MRI imaging. Currently, surgical tumor extirpation remains the treatment of first choice, with high rates of durable endocrinological remission and low rates of adverse events. In patients with recurrence of tumor or endocrinopathy, or those in poor general health, long-term targeted therapy with somatostatin analogues may be advised. Reoperation or radiation therapy remains as options for medically intractable growth hormone excess. Current advances in surgical techniques including endoscopy and intraoperative MRI imaging make the initial and repeat surgeries safe and effective in the management of acromegaly/gigantism.

References

[1] Wass JA, Laws Jr. ER, Randall RV, Sheline GE. The treatment of acromegaly Clin Endocrinol Metab [Internet] 1986;15(3):683−7071986/08/01. Available from: https://www.ncbi.nlm.nih.gov/pubmed/2876792.

[2] Eugster E. Gigantism. In: Feingold KR, Anawalt B, Boyce A, et al., editors. Endotext [internet] [Internet]. South Dartmouth (MA): MDtext.com, Inc; 2018. Available from: https://www.ncbi.nlm.nih.gov/books/NBK279155/.

[3] Lavrentaki A, Paluzzi A, Wass JA, Karavitaki N. Epidemiology of acromegaly: review of population studies Pituit [Internet] 2017;20(1):4−92016/10/16. Available from: https://www.ncbi.nlm.nih.gov/pubmed/27743174.

[4] Buchfelder M, Feulner J. Neurosurgical treatment of acromegaly. Prog Mol Biol Transl Sci 2016;138:115−39.

[5] Orme SM, McNally RJ, Cartwright RA, Belchetz PE. Mortality and cancer incidence in acromegaly: a retrospective cohort study. United Kingdom Acromegaly Study Group J Clin Endocrinol Metab [Internet] 1998;83(8):2730−41998/08/26. Available from: https://www.ncbi.nlm.nih.gov/pubmed/9709939.

[6] Beauregard C, Truong U, Hardy J, Serri O. Long-term outcome and mortality after transsphenoidal adenomectomy for acromegaly Clin Endocrinol [Internet] 2003;58(1):86−912003/01/10. Available from: https://www.ncbi.nlm.nih.gov/pubmed/12519417.

[7] Swearingen B, Barker 2nd FG, Katznelson L, Biller BM, Grinspoon S, Klibanski A, et al. Long-term mortality after transsphenoidal surgery and adjunctive therapy for acromegaly J Clin Endocrinol Metab [Internet] 1998;83(10):3419−261998/10/13. Available from: https://www.ncbi.nlm.nih.gov/pubmed/9768641.

[8] Marie P. Sur deux cas d'acromégalie; hypertrophie singulière non congénitale des extrémités supérieures, inférieures et céphalique. Rev Med Liege 1886;6:297−333.

[9] Liu JK, Das K, Weiss MH, Laws ER, Couldwell WT. The history and evolution of transsphenoidal surgery. J Neurosurg 2001;95(6):1083−96. Available from: https://doi.org/10.3171/jns.2001.95.6.1083.

[10] Horsley V. On the technique of operations on the central nervous system. BMJ 1906;2:411−23.

[11] FT CRP. Notes of a case of acromegaly treated by operation. BMJ 1893;2:1421−14231.

[12] Laws ER. Surgery for acromegaly: evolution of the techniques and outcomes. Rev Endocr Metab Disord 2008;9(1):67−70.

[13] Schloffer H. Erfolgreiche Operation eines Hypophysentumors auf nasalem Wege. Wien Klin Wchnschr 1907;20:621−4.

[14] Welbourn RB. The evolution of transsphenoidal pituitary microsurgery. Surgery 1986;100(6):1185−90.

[15] Hirsch O. Pituitary tumors; a borderland between cranial and trans-sphenoidal surgery N Engl J Med [Internet] 1956;254(20):937−91956/05/17. Available from: https://www.ncbi.nlm.nih.gov/pubmed/13322189.

[16] Cushing H. The pituitary body and its disorders: clinical states produced by disorders of the hypophysis cerebri. JB Lippincott; 1912. p. 296−305.

[17] Cushing H. The Weir Mitchell lecture. Surgical experiences with pituitary disorders. JAMA 1914;63:1515−25.

[18] Rosegay H. Cushing's legacy to transsphenoidal surgery J Neurosurg [Internet] 1981;54(4):448−541981/04/01. Available from: https://www.ncbi.nlm.nih.gov/pubmed/7009800.

[19] Guiot G, Thibaut B. [Excision of pituitary adenomas by trans-sphenoidal route] Neurochir [Internet] 1959;1(2):133−501959/02/01. Available from: https://www.ncbi.nlm.nih.gov/pubmed/13632863.

[20] Hardy J, Vezina JL. Transsphenoidal neurosurgery of intracranial neoplasm Adv Neurol [Internet] 1976;15:261−731976/01/01. Available from: https://www.ncbi.nlm.nih.gov/pubmed/945663.

[21] Hardy J, Wigser SM. Trans-sphenoidal surgery of pituitary fossa tumors with televised radiofluoroscopic control. J Neurosurg [Internet] 1965;23(6):612−19. Available from: http://www.ncbi.nlm.nih.gov/pubmed/5861144.

[22] Hardy J. Transsphenoidal microsurgery for selective removal of functional pituitary microadenomas. Invited commentary World J Surg [Internet] 1977;1(1):841977/01/01. Available from: https://www.ncbi.nlm.nih.gov/pubmed/868065.

[23] Hardy J. The transsphenoidal surgical approach to the pituitary Hosp Pr [Internet] 1979;14(6):81−91979/06/01. Available from: https://www.ncbi.nlm.nih.gov/pubmed/571837.

[24] Couldwell WT, Simard MF, Weiss MH, Schmidek HH, Sweet WH. Surgical management of growth hormone-secreting and prolactin-secreting pituitary adenomas. WB Saunders; 1995. p. 305—13.

[25] Laws ER, Piepgras DG, Randall RV. Neurosurgical management of acromegaly. Results in 82 patients treated between 1972 and 1977. J Neurosurg 1979;50:454—61.

[26] Hardy J. Transsphenoidal hypophysectomy. J Neurosurg [Internet] 1971;34(4):582—94. Available from: http://www.ncbi.nlm.nih.gov/pubmed/5554367.

[27] Sabit I, Schaefer SD, Couldwell WT. Extradural extranasal combined transmaxillary transsphenoidal approach to the cavernous sinus: a minimally invasive microsurgical model. Laryngoscope 2000;110(2 Pt 1):286—91. Available from: https://doi.org/10.1097/00005537-200002010-00019.

[28] Mason RB, Nieman LK, Doppman JL. Selective excision of adenomas originating in or extending into the pituitary stalk with preservation of pituitary function. J Neurosurg 1997;87:343—51.

[29] Kouri JG, Chen MY, Watson JC. Resection of suprasellar tumors by using a modified transsphenoidal approach Report of four cases J Neurosurg 2000;92:1028—35.

[30] Couldwell WT, Weiss MH, Apuzzo MLJ. The transnasal transsphenoidal approach. Williams & Wilkins; 1998. p. 553—74.

[31] Bushe KA, Halves E. Modifizierte Technik bei transnasaler Operation der Hypophysengeschwulste. Acta Neurochir 1978;41:163—75.

[32] Alfieri A. Endoscopic endonasal transsphenoidal approach to the sellar region: technical evolution of the methodology and refinement of a dedicated instrumentation. J Neurosurg Sci 1999;43:85—92.

[33] Carrau RL, Jho HD, Ko Y. Transnasal-transsphenoidal endoscopic surgery of the pituitary gland. Laryngoscope 1996;106:914—18.

[34] Jho HD, Carrau RL, Ko Y. Endoscopic pituitary surgery: an early experience. Surg Neurol 1997;47:213—23.

[35] Yaniv E, Rappaport ZH. Endoscopic transseptal transsphenoidal surgery for pituitary tumors. Neurosurgery 1997;40:944—6.

[36] Jane Jr. JA, Catalino MP, Laws Jr. ER. Surgical treatment of pituitary adenomas. In: Feingold KR, Anawalt B, Boyce A, Chrousos G, Dungan K, Grossman A, et al., (eds) Endotext [Internet]. South Dartmouth (MA); 2000. Available from: https://www.ncbi.nlm.nih.gov/pubmed/25905217.

[37] Chen C-J, Ironside N, Pomeraniec IJ, Chivukula S, Buell TJ, Ding D, et al. Microsurgical versus endoscopic transsphenoidal resection for acromegaly: a systematic review of outcomes and complications. Acta Neurochir (Wien) [Internet] 2017;159(11):2193—207. Available from: http://link.springer.com/10.1007/s00701-017-3318-6.

[38] Komotar RJ, Starke RM, Raper DM, Anand VK, Schwartz TH. Endoscopic endonasal compared with microscopic transsphenoidal and open transcranial resection of giant pituitary adenomas Pituit [Internet] 2012;15(2):150—92011/11/01. Available from: https://www.ncbi.nlm.nih.gov/pubmed/22038033.

[39] Starke RM, Raper DM, Payne SC, Vance ML, Oldfield EH, Jane Jr. JA. Endoscopic vs microsurgical transsphenoidal surgery for acromegaly: outcomes in a concurrent series of patients using modern criteria for remission J Clin Endocrinol Metab [Internet] 2013;98(8):3190—82013/06/06. Available from: https://www.ncbi.nlm.nih.gov/pubmed/23737543.

[40] Er U, Gurses L, Saka C, Belen D, Yigitkanli K, Simsek S, et al. Sublabial transseptal approach to pituitary adenomas with special emphasis on rhinological complications Turk Neurosurg [Internet] 2008;18 (4):425—302008/12/25. Available from: https://www.ncbi.nlm.nih.gov/pubmed/19107694.

[41] Costello RT. Subclinical adenoma of the pituitary gland Am J Pathol [Internet] 1936;12(2):205—161. 1936/03/01. Available from: https://www.ncbi.nlm.nih.gov/pubmed/19970261.

[42] Oldfield EH. Cushing's disease: lessons learned from 1500 cases Neurosurg [Internet] 2017;64 (CN_Suppl_1):27—362017/09/14. Available from: https://www.ncbi.nlm.nih.gov/pubmed/28889067.

[43] Taylor DG, Jane JA, Oldfield EH. Resection of pituitary macroadenomas via the pseudocapsule along the posterior tumor margin: a cohort study and technical note. J Neurosurg 2018;128(2):422—8. Available from: https://doi.org/10.3171/2017.7.JNS171658.

[44] Chamoun R, Takashima M, Yoshor D. Endoscopic extracapsular dissection for resection of pituitary macroadenomas: technical note. J Neurol Surg Part A Cent Eur Neurosurg 2014.

[45] Prevedello DM, Ebner FH, De Lara D, Filho LD, Otto BA, Carrau RL. Extracapsular dissection technique with the Cotton Swab for pituitary adenomas through an endoscopic endonasal approach - how i do it. Acta Neurochir (Wien) 2013.

[46] Buchfelder M, Schlaffer S-M. Novel techniques in the surgical treatment of acromegaly: applications and efficacy. Neuroendocrinology [Internet] 2016;103(1):32—41. Available from: http://noveltechethics. ca/what-we-do/governance-of-research#Ethics.

[47] Buchfelder M, Schlaffer SM. The surgical treatment of acromegaly. Pituitary 2017;20(1):76—83.

[48] Buchfelder M, Schlaffer S-M. Intraoperative magnetic resonance imaging during surgery for pituitary adenomas: pros and cons. Endocr [Internet] 2012;42(3):483—95. Available from: http://www.ncbi.nlm. nih.gov/pubmed/22833429.

[49] Netuka D, Masopust V, Belšán T, Kramář F, Beneš V. One year experience with 3.0 T intraoperative MRI in pituitary surgery. Acta Neurochir Suppl [Internet] 2011;109(10):157—9. Available from: http:// www.ncbi.nlm.nih.gov/pubmed/20960336.

[50] Schwartz TH, Stieg PE, Anand VK. Endoscopic transsphenoidal pituitary surgery with intraoperative magnetic resonance imaging. Neurosurgery. 2006;58(Suppl. 1):44—51.

[51] Theodosopoulos PV, Leach J, Kerr RG, Zimmer L a, Denny AM, Guthikonda B, et al. Maximizing the extent of tumor resection during transsphenoidal surgery for pituitary macroadenomas: can endoscopy replace intraoperative magnetic resonance imaging? J Neurosurg [Internet] 2010;112(4):736—43. Available from: http://www.ncbi.nlm.nih.gov/pubmed/19835472.

[52] Bellut D, Hlavica M, Schmid C, Bernays RL. Intraoperative magnetic resonance imaging-assisted transsphenoidal pituitary surgery in patients with acromegaly. Neurosurg Focus 2010;29(4):E9. Available from: https://doi.org/10.3171/2010.7.FOCUS10164.

[53] Fahlbusch R, Ganslandt O, Buchfelder M, Schott W, Nimsky C. Intraoperative magnetic resonance imaging during transsphenoidal surgery. J Neurosurg [Internet] 2001;95(3):381—90. Available from: http://www.ncbi.nlm.nih.gov/pubmed/11565857.

[54] Fahlbusch R, v Keller B, Ganslandt O, Kreutzer J, Nimsky C. Transsphenoidal surgery in acromegaly investigated by intraoperative high-field magnetic resonance imaging. Eur J Endocrinol 2005.

[55] Laws ER, Vance ML, Thapar K. Pituitary surgery for the management of acromegaly. Horm Res Paediatr [Internet] 2000;53(3):71—5. Available from: https://www.karger.com/Article/FullText/23538.

[56] Barker 2nd FG, Klibanski A, Swearingen B. Transsphenoidal surgery for pituitary tumors in the United States, 1996-2000: mortality, morbidity, and the effects of hospital and surgeon volume J Clin Endocrinol Metab [Internet] 2003;88(10):4709—192003/10/15. Available from: https://www.ncbi.nlm. nih.gov/pubmed/14557445.

[57] Cappabianca P, Cavallo LM, Colao A, de Divitiis E. Surgical complications associated with the endoscopic endonasal transsphenoidal approach for pituitary adenomas. J Neurosurg [Internet] 2002;97 (2):293—8. Available from: https://www.ncbi.nlm.nih.gov/pubmed/12186456.

[58] Dehdashti AR, Ganna A, Karabatsou K, Gentili F. Pure endoscopic endonasal approach for pituitary adenomas: early surgical results in 200 patients and comparison with previous microsurgical series Neurosurg [Internet] 2008;62(5):1006—72008/06/27. Available from: https://www.ncbi.nlm.nih.gov/pubmed/18580798.

[59] Frank G, Pasquini E, Farneti G, Mazzatenta D, Sciarretta V, Grasso V, et al. The endoscopic versus the traditional approach in pituitary surgery. Neuroendocrinology [Internet] 2006;83(3—4):240—8. Available from: https://www.ncbi.nlm.nih.gov/pubmed/17047389.

[60] Gondim JA, Schops M, de Almeida JP, de Albuquerque LA, Gomes E, Ferraz T, et al. Endoscopic endonasal transsphenoidal surgery: surgical results of 228 pituitary adenomas treated in a pituitary center Pituit [Internet] 2010;13(1):68−772009/08/22. Available from: https://www.ncbi.nlm.nih.gov/pubmed/19697135.

[61] Jho HD. Endoscopic transsphenoidal surgery J Neurooncol [Internet] 2001;54(2):187−952002/01/05. Available from: https://www.ncbi.nlm.nih.gov/pubmed/11761435.

[62] Holdaway IM, Bolland MJ, Gamble GD. A meta-analysis of the effect of lowering serum levels of GH and IGF-I on mortality in acromegaly Eur J Endocrinol [Internet] 2008;159(2):89−952008/06/06. Available from: https://www.ncbi.nlm.nih.gov/pubmed/18524797.

[63] Del Porto LA, Liubinas SV, Kaye AH. Treatment of persistent and recurrent acromegaly J Clin Neurosci [Internet] 2011;18(2):181−902010/12/21. Available from: https://www.ncbi.nlm.nih.gov/pubmed/21167718.

[64] Giustina A, Barkan A, Casanueva FF, Cavagnini F, Frohman L, Ho K, et al. Criteria for cure of acromegaly: a consensus statement. J Clin Endocrinol Metab [Internet] 2000;85(2):526−9. Available from: https://www.ncbi.nlm.nih.gov/pubmed/10690849.

[65] Giustina a, Chanson P, Bronstein MD, Klibanski a, Lamberts S, Casanueva FF, et al. A consensus on criteria for cure of acromegaly. J Clin Endocrinol Metab [Internet] 2010;95(7):3141−8. Available from: http://www.ncbi.nlm.nih.gov/pubmed/20410227.

[66] Mathioudakis N, Salvatori R. Management options for persistent postoperative acromegaly. Neurosurg Clin N Am [Internet] 2012;23(4):621−38. Available from: https://linkinghub.elsevier.com/retrieve/pii/S1042368012000800.

[67] Espinosa E, Ramirez C, Mercado M. The multimodal treatment of acromegaly: current status and future perspectives. Endocr Metab Immune Disord Targets 2014;14(3):169−81.

[68] Espinosa-de-Los-Monteros AL, Sosa E, Cheng S, Ochoa R, Sandoval C, Guinto G, et al. Biochemical evaluation of disease activity after pituitary surgery in acromegaly: a critical analysis of patients who spontaneously change disease status Clin Endocrinol [Internet] 2006;64(3):245−92006/02/21. Available from: https://www.ncbi.nlm.nih.gov/pubmed/16487431.

[69] Nomikos P, Buchfelder M, Fahlbusch R. The outcome of surgery in 668 patients with acromegaly using current criteria of biochemical "cure" Eur J Endocrinol [Internet] 2005;152(3):379−872005/03/11. Available from: https://www.ncbi.nlm.nih.gov/pubmed/15757854.

[70] Tindall GT, Oyesiku NM, Watts NB, Clark RV, Christy JH, Adams DA. Transsphenoidal adenomectomy for growth hormone-secreting pituitary adenomas in acromegaly: outcome analysis and determinants of failure J Neurosurg [Internet] 1993;78(2):205−151993/02/01. Available from: https://www.ncbi.nlm.nih.gov/pubmed/8421204.

[71] Kurosaki M, Luedecke DK, Abe T. Effectiveness of secondary transnasal surgery in GH-secreting pituitary macroadenomas Endocr J [Internet] 2003;50(5):635−422003/11/14. Available from: https://www.ncbi.nlm.nih.gov/pubmed/14614221.

[72] Rudnik A, Kos-Kudla B, Larysz D, Zawadzki T, Bazowski P. Endoscopic transsphenoidal treatment of hormonally active pituitary adenomas. Neuro Endocrinol Lett [Internet] 2007;28(4):438−44. Available from: https://www.ncbi.nlm.nih.gov/pubmed/17693972.

[73] Arafah BM, Rosenzweig JL, Fenstermaker R, Salazar R, McBride CE, Selman W. Value of growth hormone dynamics and somatomedin C (insulin-like growth factor I) levels in predicting the long-term benefit after transsphenoidal surgery for acromegaly J Lab Clin Med [Internet] 1987;109(3):346−541987/03/01. Available from: https://www.ncbi.nlm.nih.gov/pubmed/3102658.

[74] Valdemarsson S, Bramnert M, Cronquist S, Elner A, Eneroth CM, Hedner P, et al. Early postoperative basal serum GH level and the GH response to TRH in relation to the long-term outcome of surgical treatment for acromegaly: a report on 39 patients. J Intern Med [Internet] 1991;230(1):49−54. Available from: https://www.ncbi.nlm.nih.gov/pubmed/1906090.

[75] Ronchi CL, Varca V, Giavoli C, Epaminonda P, Beck-Peccoz P, Spada A, et al. Long-term evaluation of postoperative acromegalic patients in remission with previous and newly proposed criteria J Clin Endocrinol Metab [Internet] 2005;90(3):1377−822004/12/09. Available from: https://www.ncbi.nlm.nih.gov/pubmed/15585548.

[76] Biermasz NR, van Dulken H, Roelfsema F. Ten-year follow-up results of transsphenoidal microsurgery in acromegaly J Clin Endocrinol Metab [Internet] 2000;85(12):4596−6022001/01/03. Available from: https://www.ncbi.nlm.nih.gov/pubmed/11134114.

[77] Biermasz NR, Smit JW, van Dulken H, Roelfsema F. Postoperative persistent thyrotrophin releasing hormone-induced growth hormone release predicts recurrence in patients with acromegaly Clin Endocrinol [Internet] 2002;56(3):313−192002/04/10. Available from: https://www.ncbi.nlm.nih.gov/pubmed/11940042.

[78] Freda PU, Nuruzzaman AT, Reyes CM, Sundeen RE, Post KD. Significance of "abnormal" nadir growth hormone levels after oral glucose in postoperative patients with acromegaly in remission with normal insulin-like growth factor-I levels J Clin Endocrinol Metab [Internet] 2004;89(2):495−5002004/02/07. Available from: https://www.ncbi.nlm.nih.gov/pubmed/14764751.

[79] Freda PU, Post KD, Powell JS, Wardlaw SL. Evaluation of disease status with sensitive measures of growth hormone secretion in 60 postoperative patients with acromegaly J Clin Endocrinol Metab [Internet] 1998;83(11):3808−161998/11/14. Available from: https://www.ncbi.nlm.nih.gov/pubmed/9814451.

[80] Jane Jr. JA, Starke RM, Elzoghby MA, Reames DL, Payne SC, Thorner MO, et al. Endoscopic transsphenoidal surgery for acromegaly: remission using modern criteria, complications, and predictors of outcome J Clin Endocrinol Metab [Internet] 2011;96(9):2732−402011/07/01. Available from: https://www.ncbi.nlm.nih.gov/pubmed/21715544.

[81] Colao A, Auriemma RS, Galdiero M, Lombardi G, Pivonello R. Effects of initial therapy for five years with somatostatin analogs for acromegaly on growth hormone and insulin-like growth factor-I levels, tumor shrinkage, and cardiovascular disease: a prospective study. J Clin Endocrinol Metab [Internet] 2009;94(10):3746−56. Available from: https://www.ncbi.nlm.nih.gov/pubmed/19622615.

[82] Maiza JC, Vezzosi D, Matta M, Donadille F, Loubes-Lacroix F, Cournot M, et al. Long-term (up to 18 years) effects on GH/IGF-1 hypersecretion and tumour size of primary somatostatin analogue (SSTa) therapy in patients with GH-secreting pituitary adenoma responsive to SSTa Clin Endocrinol [Internet] 2007;67(2):282−92007/05/26. Available from: https://www.ncbi.nlm.nih.gov/pubmed/17524029.

[83] Mercado M, Borges F, Bouterfa H, Chang TC, Chervin A, Farrall AJ, et al. A prospective, multicentre study to investigate the efficacy, safety and tolerability of octreotide LAR (long-acting repeatable octreotide) in the primary therapy of patients with acromegaly. Clin Endocrinol [Internet] 2007;66(6):859−68. Available from: https://www.ncbi.nlm.nih.gov/pubmed/17465997.

[84] Colao A, Auriemma RS, Rebora A, Galdiero M, Resmini E, Minuto F, et al. Significant tumour shrinkage after 12 months of lanreotide Autogel-120 mg treatment given first-line in acromegaly Clin Endocrinol [Internet] 2009;71(2):237−452008/12/20. Available from: https://www.ncbi.nlm.nih.gov/pubmed/19094074.

[85] Fleseriu M. Clinical efficacy and safety results for dose escalation of somatostatin receptor ligands in patients with acromegaly: a literature review Pituit [Internet] 2011;14(2):184−932010/12/17. Available from: https://www.ncbi.nlm.nih.gov/pubmed/21161602.

[86] Giustina A, Mazziotti G, Torri V, Spinello M, Floriani I, Melmed S. Meta-analysis on the effects of octreotide on tumor mass in acromegaly PLoS One [Internet] 2012;7(5):e364112012/05/11. Available from: https://www.ncbi.nlm.nih.gov/pubmed/22574156.

[87] Tutuncu Y, Berker D, Isik S, Ozuguz U, Akbaba G, Kucukler FK, et al. Comparison of octreotide LAR and lanreotide autogel as post-operative medical treatment in acromegaly Pituit [Internet] 2012;15(3):398−4042011/08/25. Available from: https://www.ncbi.nlm.nih.gov/pubmed/21863263.

[88] Auriemma RS, Pivonello R, Galdiero M, De Martino MC, De Leo M, Vitale G, et al. Octreotide-LAR vs lanreotide-SR as first-line therapy for acromegaly: a retrospective, comparative, head-to-head study J Endocrinol Invest [Internet] 2008;31(11):956−652009/01/27. Available from: https://www.ncbi.nlm.nih.gov/pubmed/19169050.

[89] Nunes VS, Correa JMS, Puga MES, Silva EMK, Boguszewski CL. Preoperative somatostatin analogues versus direct transsphenoidal surgery for newly-diagnosed acromegaly patients: a systematic review and meta-analysis using the GRADE system. Pituit [Internet] 2015;18(4):500−8. Available from: http://doi.org/10.1007/s11102-014-0602-9.

[90] Barkan AL, Lloyd RV, Chandler WF, Hatfield MK, Gebarski SS, Kelch RP, et al. Preoperative treatment of acromegaly with long-acting somatostatin analog SMS 201-995: shrinkage of invasive pituitary macroadenomas and improved surgical remission rate J Clin Endocrinol Metab [Internet] 1988;67(5):1040−81988/11/01. Available from: https://www.ncbi.nlm.nih.gov/pubmed/2903168.

[91] Colao A, Ferone D, Cappabianca P, del Basso De Caro ML, Marzullo P, Monticelli A, et al. Effect of octreotide pretreatment on surgical outcome in acromegaly J Clin Endocrinol Metab [Internet] 1997;82(10):3308−141997/11/05. Available from: https://www.ncbi.nlm.nih.gov/pubmed/9329359.

[92] Stevenaert A, Harris AG, Kovacs K, Beckers A. Presurgical octreotide treatment in acromegaly Metab [Internet] 1992;41(9 Suppl. 2):51−81992/09/01. Available from: https://www.ncbi.nlm.nih.gov/pubmed/1518434.

[93] Biermasz NR, van Dulken H, Roelfsema F. Direct postoperative and follow-up results of transsphenoidal surgery in 19 acromegalic patients pretreated with octreotide compared to those in untreated matched controls J Clin Endocrinol Metab [Internet] 1999;84(10):3551−51999/10/16. Available from: https://www.ncbi.nlm.nih.gov/pubmed/10522994.

[94] Kristof RA, Stoffel-Wagner B, Klingmuller D, Schramm J. Does octreotide treatment improve the surgical results of macro-adenomas in acromegaly? A randomized study Acta Neurochir [Internet] 1999;141(4):399−4051999/06/03. Available from: https://www.ncbi.nlm.nih.gov/pubmed/10352750.

[95] Sandret L, Maison P, Chanson P. Place of cabergoline in acromegaly: a meta-analysis. J Clin Endocrinol Metab [Internet] 2011;96(5):1327−35. Available from: https://www.ncbi.nlm.nih.gov/pubmed/21325455.

[96] Galoiu S, Poiana C. Current therapies and mortality in acromegaly. J Med Life [Internet] 2015;8(4):411−15. Available from: https://www.ncbi.nlm.nih.gov/pubmed/26664461.

[97] Trainer PJ. ACROSTUDY: the first 5 years Eur J Endocrinol [Internet] 2009;161(Suppl):S19−242009/08/18. Available from: https://www.ncbi.nlm.nih.gov/pubmed/19684052.

[98] Trainer PJ, Ezzat S, D'Souza GA, Layton G, Strasburger CJ. A randomized, controlled, multicentre trial comparing pegvisomant alone with combination therapy of pegvisomant and long-acting octreotide in patients with acromegaly Clin Endocrinol [Internet] 2009;71(4):549−572009/05/15. Available from: https://www.ncbi.nlm.nih.gov/pubmed/19438906.

[99] Wang YY, Higham C, Kearney T, Davis JR, Trainer P, Gnanalingham KK. Acromegaly surgery in Manchester revisited--the impact of reducing surgeon numbers and the 2010 consensus guidelines for disease remission Clin Endocrinol [Internet] 2012;76(3):399−4062011/08/10. Available from: https://www.ncbi.nlm.nih.gov/pubmed/21824170.

[100] Espinosa de los Monteros AL, Gonzalez B, Vargas G, Sosa E, Guinto G, Mercado M. Surgical reintervention in acromegaly: is it still worth trying? Endocr Pr [Internet] 2009;15(5):431−72009/06/06. Available from: https://www.ncbi.nlm.nih.gov/pubmed/19491070.

[101] Melmed S, Colao A, Barkan A, Molitch M, Grossman AB, Kleinberg D, et al. Guidelines for acromegaly management: an update J Clin Endocrinol Metab [Internet] 2009;94(5):1509−172009/02/12. Available from: https://www.ncbi.nlm.nih.gov/pubmed/19208732.

[102] Minniti G, Clarke E, Scaringi C, Enrici RM. Stereotactic radiotherapy and radiosurgery for nonfunctioning and secreting pituitary adenomas. Rep Pr Oncol Radiother [Internet] 2016;21(4):370−8. Available from: https://www.ncbi.nlm.nih.gov/pubmed/27330422.

[103] Barrande G, Pittino-Lungo M, Coste J, Ponvert D, Bertagna X, Luton JP, et al. Hormonal and metabolic effects of radiotherapy in acromegaly: long-term results in 128 patients followed in a single center J Clin Endocrinol Metab [Internet] 2000;85(10):3779−852000/11/04. Available from: https://www.ncbi.nlm.nih.gov/pubmed/11061538.

[104] Brada M, Rajan B, Traish D, Ashley S, Holmes-Sellors PJ, Nussey S, et al. The long-term efficacy of conservative surgery and radiotherapy in the control of pituitary adenomas. Clin Endocrinol [Internet] 1993;38(6):571−8. Available from: https://www.ncbi.nlm.nih.gov/pubmed/8334743.

[105] Epaminonda P, Porretti S, Cappiello V, Beck-Peccoz P, Faglia G, Arosio M. Efficacy of radiotherapy in normalizing serum IGF-I, acid-labile subunit (ALS) and IGFBP-3 levels in acromegaly. Clin Endocrinol [Internet] 2001;55(2):183−9. Available from: https://www.ncbi.nlm.nih.gov/pubmed/11531924.

[106] Jenkins PJ, Bates P, Carson MN, Stewart PM, Wass JA. Conventional pituitary irradiation is effective in lowering serum growth hormone and insulin-like growth factor-I in patients with acromegaly. J Clin Endocrinol Metab [Internet] 2006;91(4):1239−45. Available from: https://www.ncbi.nlm.nih.gov/pubmed/16403824.

[107] Powell JS, Wardlaw SL, Post KD, Freda PU. Outcome of radiotherapy for acromegaly using normalization of insulin-like growth factor I to define cure J Clin Endocrinol Metab [Internet] 2000;85(5):2068−712000/06/08. Available from: https://www.ncbi.nlm.nih.gov/pubmed/10843197.

[108] Biermasz NR, Dulken HV, Roelfsema F. Postoperative radiotherapy in acromegaly is effective in reducing GH concentration to safe levels Clin Endocrinol [Internet] 2000;53(3):321−72000/09/06. Available from: https://www.ncbi.nlm.nih.gov/pubmed/10971449.

[109] Cozzi R, Barausse M, Asnaghi D, Dallabonzana D, Lodrini S, Attanasio R. Failure of radiotherapy in acromegaly Eur J Endocrinol [Internet] 2001;145(6):717−262001/11/27. Available from: https://www.ncbi.nlm.nih.gov/pubmed/11720896.

[110] Gonzalez B, Vargas G, Espinosa-de-los-Monteros AL, Sosa E, Mercado M. Efficacy and safety of radiotherapy in acromegaly Arch Med Res [Internet] 2011;42(1):48−522011/03/08. Available from. Available from: https://www.ncbi.nlm.nih.gov/pubmed/21376263.

[111] Minniti G, Jaffrain-Rea ML, Osti M, Esposito V, Santoro A, Solda F, et al. The long-term efficacy of conventional radiotherapy in patients with GH-secreting pituitary adenomas Clin Endocrinol [Internet] 2005;62(2):210−162005/01/27. Available from: https://www.ncbi.nlm.nih.gov/pubmed/15670198.

[112] Lecumberri B, Estrada J, García-Uría J, Millán I, Pallardo LF, Caballero L, et al. Neurocognitive long-term impact of two-field conventional radiotherapy in adult patients with operated pituitary adenomas. Pituitary 2015.

[113] Bex M, Abs R, T'Sjoen G, Mockel J, Velkeniers B, Muermans K, et al. AcroBel--the Belgian registry on acromegaly: a survey of the "real-life" outcome in 418 acromegalic subjects Eur J Endocrinol [Internet] 2007;157(4):399−4092007/09/26. Available from: https://www.ncbi.nlm.nih.gov/pubmed/17893253.

[114] Colao A, Vandeva S, Pivonello R, Grasso LF, Nachev E, Auriemma RS, et al. Could different treatment approaches in acromegaly influence life expectancy? A comparative study between Bulgaria and Campania (Italy) Eur J Endocrinol [Internet] 2014;171(2):263−732014/06/01. Available from: https://www.ncbi.nlm.nih.gov/pubmed/24878680.

[115] Mestron A, Webb SM, Astorga R, Benito P, Catala M, Gaztambide S, et al. Epidemiology, clinical characteristics, outcome, morbidity and mortality in acromegaly based on the Spanish Acromegaly Registry (Registro Espanol de Acromegalia, REA) Eur J Endocrinol [Internet] 2004;151(4):439−462004/10/13. Available from: https://www.ncbi.nlm.nih.gov/pubmed/15476442.

[116] Minniti G, Scaringi C, Amelio D, Maurizi Enrici R. Stereotactic irradiation of GH-secreting pituitary adenomas Int J Endocrinol [Internet] 2012;2012:4828612012/04/21. Available from: https://www.ncbi.nlm.nih.gov/pubmed/22518123.

[117] Minniti G, Osti MF, Niyazi M. Target delineation and optimal radiosurgical dose for pituitary tumors. Radiat Oncol [Internet] 2016;11(1):135. Available from: https://www.ncbi.nlm.nih.gov/pubmed/27729088.

[118] Petrovich Z, Jozsef G, Yu C, Apuzzo ML. Radiotherapy and stereotactic radiosurgery for pituitary tumors Neurosurg Clin N Am [Internet] 2003;14(1):147–662003/04/15. Available from: https://www.ncbi.nlm.nih.gov/pubmed/12690986.

[119] Roberts BK, Ouyang DL, Lad SP, Chang SD, Harsh 4th GR, Adler Jr. JR, et al. Efficacy and safety of CyberKnife radiosurgery for acromegaly. Pituit [Internet] 2007;10(1):19–25. Available from: https://www.ncbi.nlm.nih.gov/pubmed/17273921.

[120] Losa M, Gioia L, Picozzi P, Franzin A, Valle M, Giovanelli M, et al. The role of stereotactic radiotherapy in patients with growth hormone-secreting pituitary adenoma. J Clin Endocrinol Metab [Internet] 2008;93(7):2546–52. Available from: https://www.ncbi.nlm.nih.gov/pubmed/18413424.

[121] Ronchi CL, Attanasio R, Verrua E, Cozzi R, Ferrante E, Loli P, et al. Efficacy and tolerability of gamma knife radiosurgery in acromegaly: a 10-year follow-up study Clin Endocrinol [Internet] 2009;71 (6):846–522009/06/11. Available from: https://www.ncbi.nlm.nih.gov/pubmed/19508606.

[122] Yang I, Kim W, De Salles A, Bergsneider M. A systematic analysis of disease control in acromegaly treated with radiosurgery Neurosurg Focus [Internet] 2010;29(4):E132010/10/05. Available from: https://www.ncbi.nlm.nih.gov/pubmed/20887123.

[123] Chittiboina P, Lalith Talagala S, Merkle H, Sarlls JE, Montgomery BK, Piazza MG, et al. Endosphenoidal coil for intraoperative magnetic resonance imaging of the pituitary gland during transsphenoidal surgery. J Neurosurg [Internet] 2016;1–9. Available from: http://www.ncbi.nlm.nih.gov/pubmed/26991390.

Medical management of pituitary gigantism and acromegaly

12

Adrian F. Daly[1] and Albert Beckers[2]

[1]*Department of Endocrinology, Centre Hospitalier Universitaire de Liège, Liège Université, Liège, Belgium*
[2]*Service d'Endocrinologie, Domaine Universitaire du Sart Tilman, Université de Liège, Liège, Belgium*

12.1 Introduction

Acromegaly is a rare disease caused by chronically raised growth hormone (GH) and insulin-like growth factor 1 (IGF-1) secretion; typically, this is due to a GH-secreting pituitary adenoma [1]. Acromegaly affects slightly more females than males and has a range of clinical signs/symptoms, the most classically recognized being enlargement of the extremities, soft tissue swelling, and mandibular enlargement affecting the face and skin and voice changes [2]. At a systemic level, acromegaly has pathological effects on the cardiovascular, metabolic and musculoskeletal systems that lead to a high burden of disease and increased mortality if GH and IGF-1 are not controlled [3]. While acromegaly usually presents in early middle age, there is significant phenotypic variation, and it can occur in children/adolescents and in the elderly. Pituitary gigantism represents one of the most severe clinical presentations of acromegaly, affecting as it does children and adolescents [4]. The young age at presentation reflects aggressive underlying molecular genetic processes leading to somatotrope tumor formation. Unlike in sporadic acromegaly where inheritable genetic causes are rare, nearly half of all cases of pediatric gigantism are caused by a pathological genetic or genomic change in a known gene [5]. These genetic abnormalities lead to dysregulation of somatotrope proliferation and secretion that result in tumorigenesis, by affecting molecular signals such as the aryl hydrocarbon receptor interacting protein (AIP), GPR101, $G_s$alpha, Protein kinase A, among others [6]. Individually, these dysregulated cellular processes can lead to the growth of large pituitary adenomas, hyperplasia, and high secreted levels of GH. The young age of patients with pituitary gigantism raises significant obstacles to optimal management as compared with sporadic acromegaly in adults. The presence of a macroadenoma in children of a very young age, such as those with X-linked acrogigantism (X-LAG), makes surgery more challenging. Similarly, the craniofacial fibrous dysplasia that can accompany McCune−Albright syndrome (MAS) can complicate the surgical approach.

An extensive series of consensus guidelines exist for the diagnosis and management of acromegaly, including the use of medical therapies [7−9]. Medical therapy in acromegaly consists of somatostatin analogs, the GH receptor antagonist pegvisomant, and dopamine agonists such as cabergoline. Among the somatostatin analogs, two general classes exist: long-acting depot versions of octreotide and lanreotide, which target somatostatin receptor subtype 2 and the more recently introduced multi-somatostatin receptor ligand, pasireotide, that preferentially binds receptor

Gigantism and Acromegaly. DOI: https://doi.org/10.1016/B978-0-12-814537-1.00002-6

subtypes 5, 2, and 3. Pegvisomant acts by blocking the GH receptor and preventing the stimulation of IGF-1 release, while dopamine agonists have an adjunctive role in suppressing GH secretion from the pituitary. These medical therapies have been subject to large clinical trials, mainly placebo-controlled, and are integrated into consensus guidelines that recommend the order in which they should be used [7,8]. According to the 2014 Guidelines from the Endocrine Society, the main biochemical goals for treatment of acromegaly are an age-normalized serum IGF-1 value (disease control), a random $GH < 1.0 \, \mu g/L$ (correlates with disease control), and to use the same hormonal assays throughout management [7]. For medical therapy, the guidelines note the following options:

- Medical therapy recommended in patients with persistent disease following surgery.
  - Somatostatin analogs or pegvisomant recommended as initial adjuvant medical therapy in patients with significant disease, namely moderate to severe signs and symptoms of acromegaly but without tumor mass effects.
  - Those with modestly elevated serum IGF-1 and mild clinical signs/symptoms of acromegaly, could receive a dopamine agonist, usually cabergoline, as initial adjuvant medical therapy.
  - Addition of pegvisomant or cabergoline is recommended in patients who respond inadequately to a somatostatin analog.
- Primary medical therapy with a somatostatin analog is recommended in those who cannot be cured by surgery, or have extensive cavernous sinus invasion, and do not have chiasmal compression, or are poor surgical candidates.

Medical therapy also requires vigilant follow-up for safety including, but not limited to assessments for cholelithiasis in symptomatic patients treated with somatostatin analogs, while hepatic function tests should be measured on a 6-monthly basis for patients receiving pegvisomant. Due to rare cases of tumor expansion during pegvisomant therapy, a plan should be in place for serial monitoring of tumor status using MRI [7]. While pasireotide has been licensed for use in acromegaly in many jurisdictions, it has not yet been integrated into all consensus guidelines. However, based on its specific labeling, pasireotide is generally used as second- or third-line therapy in acromegaly when surgery has failed or is not possible or where octreotide/lanreotide treatment has not led to control of acromegaly. In a recent update on treatment outcomes, an Expert Group provided some indications as to how pasireotide could be integrated into the treatment algorithm as shown in Fig. 12.1. Management of patients receiving pasireotide for the treatment of acromegaly requires specific assessment of glucose control as pasireotide is associated with impairment of glucose tolerance and diabetes mellitus [10].

While the goals for the treatment of acromegaly and pituitary gigantism are generally similar, some clinically relevant differences exist between pediatric and adult patients that can impact upon the expected efficacy of medical treatment and the timing of changes in treatment modalities. These factors include the importance of effective treatment to limit final adult height, the challenges of dose selection for medical treatment in the pediatric/adolescent population, the anatomical differences between the adult and pediatric patients, and the high incidence of molecular genetic diseases among the pituitary gigantism population.

Pituitary gigantism is a very rare condition and most data have come from case reports and small series. To address this relative lack of information, we implemented a large international collaborative study from 2011–15 that recruited 208 pituitary gigantism patients [5]. Focusing on the management of these patients, we noted that the therapeutic journey was often complex. Multimodal therapy is usual

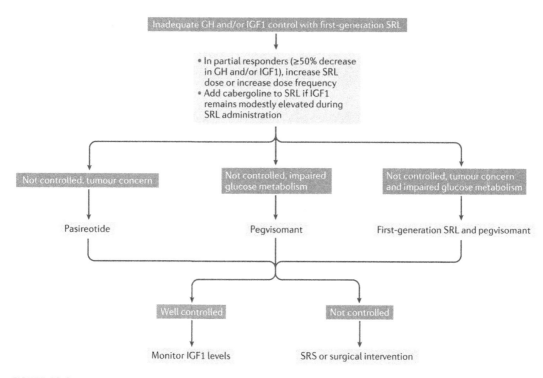

FIGURE 12.1

Suggested Expert Group algorithm for the medical management of acromegaly in patients with inadequate responses to octreotide and lanreotide. *IGF-I*, Insulin-like growth factor I; *SRL*, somatostatin receptor ligand/somatostatin analog; *SRS*, stereotactic radiosurgery.

Reproduced from Melmed S, Bronstein MD, Chanson P, et al. A Consensus Statement on acromegaly therapeutic outcomes. Nat Rev Endocrinol 2018; 14(9):552– 61. doi: 10.1038/s41574-018-0058-5 [9] under Creative Commons 4.0 Attribution International License.

in pituitary gigantism, and following primary medical or surgical therapy, patients often have recourse to reoperation or a wide variety of different combinations of medical therapy. As shown in Fig. 12.2, of 208 patients, only 7% and 15% of cases were controlled by primary medical therapy or surgery, respectively. Irrespective of whether surgery or medical therapy was the primary treatment option, pituitary gigantism patients frequently end up going through surgery on multiple occasions with adjuvant use of medical therapy. Multimodal therapy ($\geq$3 separate surgeries or medical modalities) occurred in nearly a third of pituitary gigantism cases, and despite the high treatment burden only 39% of patients had long-term hormonal control [5].

12.1.1 *AIP* mutations

Germline mutations in the *AIP* gene are the main known cause of pituitary gigantism. In the first statistically controlled series of somatotropinomas, gigantism occurred in 32% of the *AIP*-mutated group as compared with 6.5% wild-type controls ($P < .000001$) [11]. Among pituitary gigantism

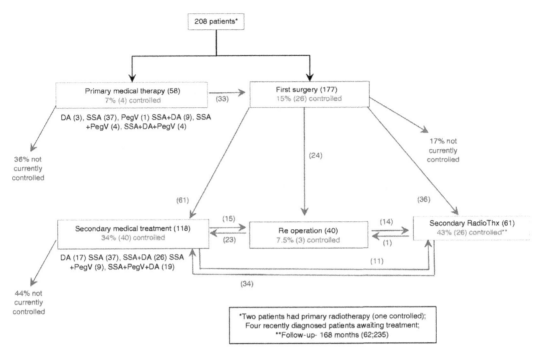

FIGURE 12.2

Schematic representation of treatments used in the management of patients with pituitary gigantism. Numbers in parentheses indicate the number of patients. *DA*, Dopamine agonists; *PegV*, pegvisomant; *RadioThx*, radiotherapy; *SSA*, somatostatin analog.

Reproduced by the authors from their work in Rostomyan L, Daly AF, Petrossians P, et al. Clinical and genetic characterization of pituitary gigantism: an international collaborative study in 208 patients. Endocr Relat Cancer 2015;22(5):745–57. doi: 10.1530/ERC-15-0320 [5] with permission.

patients themselves, 29% have an *AIP* mutation. The phenotype of *AIP*-mutated somatotropinomas is one that leads to large, extensive, and invasive tumors that predominantly secrete GH or cosecrete GH and prolactin. Patients with *AIP* mutations frequently present with familial isolated pituitary adenomas (FIPA), and family members with acromegaly or gigantism can be identified [11,12]. The median onset of disease is significantly younger among *AIP*-mutated somatotropinomas patients versus nonmutated acromegaly controls (17.5 vs 38.0 years; $P < .000001$). This 20-year difference is disease onset means that somatotropinomas due to *AIP* mutations typically overlap the years of maximum vertical growth, leading to a high risk of pituitary gigantism if hormonal control is not achieved.

The aggressive disease features of *AIP* mutation–related somatotropinomas also extend to significantly higher GH secretion at baseline than control acromegaly cases ($P = .00068$), while prolactin cosecretion is 1.9 times more frequent with *AIP* mutations than controls ($P = .00023$).

The young age at disease onset and the large, extensive tumor phenotype in *AIP*-related somatotropinomas imply that the tumor has either undergone years of occult growth or alternatively that tumor grows and invades rapidly. Evidence to date suggests the latter possibility is the more likely. During the identical median latency period from first symptoms to diagnosis (5 years) in *AIP* mutation—related and acromegaly controls, tumors grow to a significantly larger size in the *AIP*-mutated group. This was illustrated recently in one *AIP* mutation—related somatotropinomas in a young female patient with pituitary gigantism, where dramatic progression of the tumor occurred over the 15 months preceding her diagnosis [13].

In *AIP*-mutated cases of pituitary gigantism, the large tumor size and the presence of invasion may reduce greatly the possibility of primary surgical control and reoperation in these patients is frequent [11]. As *AIP*-mutated tumors have high GH secretory potential, reoperation to debulk the tumor could be clinically helpful in an effort to facilitate hormonal decreases with subsequent somatostatin analog use, as has been shown in adult acromegaly [14]. A further problem associated with medical treatment of *AIP*-mutated cases of pituitary gigantism relates to their relative resistance to first-generation somatostatin analogs. As compared with nonmutated cases, significant impairment of GH ($P = .0004$) and IGF-1 inhibition ($P = .028$) is seen in *AIP* mutation—related somatotropinomas [11].

Resistance to first-generation somatostatin analogs like octreotide and lanreotide can be countered by use of the GH receptor antagonist, pegvisomant, in the setting of acromegaly and gigantism with or without *AIP* mutations [15—18]. Treatment with pegvisomant can take the form of monotherapy or it can be used in combination with somatostatin analogs. Either alone or in combination with somatostatin analogs, pegvisomant is an important and useful option for controlling IGF-1 excess and decreasing height gain in *AIP*-mutated pituitary gigantism cases that are resistant to somatostatin analogs alone [5,19]. The combination including a somatostatin analog could be useful to help control tumor growth, which can complicate pegvisomant therapy in a minority of cases [20]. As noted above, patients with *AIP* mutations frequently have tumoural cosecretion of prolactin and GH. Combination treatment with pegvisomant and the dopamine agonist cabergoline has also been shown to be successful in somatostatin analog resistant/intolerant pediatric cases of *AIP*-mutated pituitary gigantism [15].

Recently we reported two cases of *AIP* mutation—related somatotropinomas that were resistant to octreotide/lanreotide during extensive multiyear follow-up [13]. In one of the cases who had incipient pituitary gigantism, somatostatin receptor subtype 2 staining was very low/absent, but subtype 5 receptors were present. On switching to treatment with pasireotide, a multi-somatostatin receptor ligand, both patients experienced control of their GH and IGF-1 hypersecretion. With continued therapy over some years, significant tumor regression occurred (Fig. 12.3), and one patient stopped pasireotide without a return of elevated GH and IGF-1 and still remains controlled off treatment at this time. These data suggest that in some *AIP*-mutated patients with large and resistant tumors, pasireotide therapy might be a useful option, although the diabetes mellitus that can accompany pasireotide therapy adds to the therapeutic burden.

12.1.2 X-linked acrogigantism

Duplications on chromosome Xq26.3 involving the gene *GPR101* are responsible for X-LAG, the earliest onset form of pituitary gigantism described [21—23]. Transgenic pituitary over-expression of *Gpr101* in mice leads to chronic GH hypersecretion and overgrowth [24]. Patients with X-LAG typically develop signs of overgrowth in the first 12—36 months of life, with increased height and

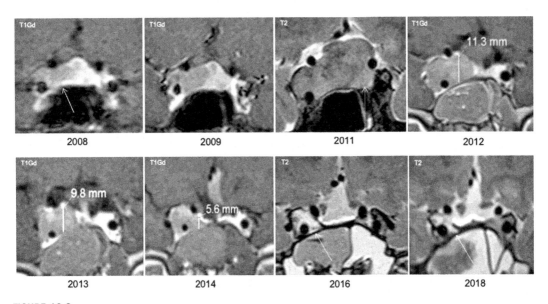

FIGURE 12.3

Evolution of pituitary adenoma size on MRI in a patient with incipient gigantism and a germline *AIP* mutation from diagnosis as a microadenoma to a large invasive pituitary macroadenoma. A sizable postoperative residue in 2012 shrank progressively on pasireotide LAR therapy and the follow-up image from 2018 shows only a small residue remained. *T1Gd*, T1-weighted gadolinium-enhanced image; *T2*, T2-weighted MRI image.

Reproduced by the authors from their work in Daly A, Rostomyan L, Betea D, et al. AIP-mutated acromegaly resistant to first-generation somatostatin analogs: long-term control with pasireotide LAR in two patients. Endocr Connect 2019; 8(4):367–77. doi: 10.1530/EC-19-0004 [13] under Creative Commons 4.0 Attribution-NonCommercial-NoDerivatives 4.0 International (CC BY-NC-ND 4.0).

weight becoming increasingly prominent over time [21]. Along with increased stature, these young children also frequently have signs that are more typical of adult acromegaly, including enlarged hands and feet, increased interdental spacing, and coarsening of the facial features [21,22]. Children with X-LAG also may have increased appetite and signs of insulin resistance [22]. GH and IGF-1 levels (and almost always prolactin) are markedly elevated at diagnosis. In X-LAG pituitary macroadenoma occur either alone or in combination with hyperplasia; rare cases of hyperplasia alone can occur. In some patients in whom it was measured preoperatively, GH releasing hormone was increased in conjunction with GH, IGF-1, and prolactin, suggesting a hypothalamic element in the pathophysiology of X-LAG [25]. Without treatment ,patients with X-LAG suffer increasing tumor size and hormonal excess that can lead to severe gigantism [26,27].

Initial management of X-LAG usually involves the use of surgery or first-generation somatostatin analogs [5,21–23,28,29]. To control X-LAG surgically, a complete resection of the tumor and/or hyperplasia is required, as very small tumor remnants can maintain excess GH and IGF-1 secretion, in some cases for decades [21–23,30,31]. The benefits of extensive resection on GH secretion come at the cost of high rates of hypopituitarism, which is a problem in pituitary gigantism generally [5]. With extensive hyperplasia, discrete boundaries between pathological and normal tissue are not evident, which further complicates primary surgical cure and can necessitate anterior hypophysectomy in some cases [21]. Despite expressing somatostatin subtype 2 receptors, X-LAG patients have a poor response to first-generation somatostatin analogs, even at the full adult dose [22]. Data on the efficacy of the multi-somatostatin receptor ligand pasireotide in X-LAG are not available. In cases where extensive surgical resection (debulking) has taken place, second line therapy with somatostatin analogs can reduce or control GH and IGF-1 in some cases [21,22]. In the growing patient, suboptimal control of excess GH/IGF-1 control should be avoided so as to prevent excessive final adult stature. Given the extended duration of disease in X-LAG due to its early onset and resistant phenotype, the risk of excessive height is very prominent. In those X-LAG patients, the addition of pegvisomant can provide rapid and prolonged IGF-1 control and blunting of height gain [21–23,25]. To optimize hormonal and growth parameters, when surgery and somatostatin analogs have proven inadequate, early introduction of pegvisomant can be considered. Pegvisomant is generally not indicated for treatment of pediatric patients, although case series and anecdotal reports have been widely published [15,17,18,32,33]. In X-LAG cases, addition of pegvisomant at a low dose has proven safe and effective in some individuals [21,22,25].

12.1.3 McCune–Albright syndrome

In MAS, mosaicism for a postzygotic activating mutation of *GNAS* leads to variable and diverse pathology affecting multiple tissues, classically, fibrous dysplasia of the bone, hormonal overactivity, and distinctive café au lait macules [34]. Acromegaly forms a part of the MAS spectrum and rarely can have an early onset in children/adolescents, leading to pituitary gigantism [35]. In the largest series of acromegaly characteristics in MAS, Salenave et al. reported on 112 cases published in the literature [36]. Among these, 40 were diagnosed between the ages of 3 and 16 years and in most cases, hormonal measures were confirmatory of somatotrope overactivity. Whereas excess GH/IGF-1 occurs in 20%–30% of patients with MAS, pituitary gigantism was less frequent than might be expected, possibly because of the counteracting influence of precocious puberty to reduce final height. The interpretation of height and hormonal abnormalities in young children with MAS is, therefore, complex [37]. In our series of 208 pituitary gigantism patients, 5% had MAS [5].

When somatotropinomas occur in the setting of MAS, craniofacial fibrous dysplasia is a significant factor to be considered, as it can complicate surgical access, and progression of cranial bone pathology in parallel with GH excess can worsen outcomes [36,38,39]. In MAS, the *GNAS*-activating mutation leads to diffuse disease affecting much of the anterior pituitary and can involve a mixture of hyperplasia and single or multiple adenomas [40]. The tumor/hyperplasia tissue type is usually mammosomatotrope or somatotrope in nature. Due to the difficult approach through skull base fibrous dysplasia (which can be highly vascular as well as thick) and the diffuse pituitary disease, surgical cures only occur exceptionally and medical therapy is necessary. The response to first-generation somatostatin analogs in terms of hormonal control in the Salenave et al. series was similar to that of unselected acromegaly patients (14%–46%) [36]. While hyperprolactinemia is usually present in MAS patient with pituitary disease, the use of dopamine agonists has relatively low efficacy in terms of hormonal control. Pegvisomant is, is contrast, an important treatment option for MAS patients with acromegaly or pituitary gigantism, as control of IGF-1 is usually achieved in compliant patients [36,41]. As with pituitary gigantism in general, early diagnosis and hormonal control in MAS-related acromegaly leads to improved clinical outcomes, particularly in terms of optic nerve impingement in patients with craniofacial fibrous dysplasia [38].

12.1.4 Other genetic forms of pituitary gigantism

Rare conditions like Carney complex and multiple endocrine neoplasia type 1 (MEN1) account for a small minority of all pituitary gigantism cases (1% each) [5]. In Carney complex, *PRKAR1A* germline mutations lead to a syndrome of endocrine and nonendocrine pathologies that include pituitary adenomas/hyperplasia in about 10% of cases [42]. As with other genetic causes of pituitary adenomas, Carney complex is associated with an early onset phenotype that leads to somatotrope or somatomammotrope tumors/hyperplasia, resulting in pituitary gigantism in rare individuals [42–45]. In MEN1, pituitary adenomas form an integral part of the pathological entity, and have a tendency to occur at an earlier age than non-MEN1 cases [46]. Only exceptional cases of pituitary gigantism have been reported in MEN1 [47]. Management of pituitary gigantism in Carney complex and/or MEN1 is multimodal. Cases of overgrowth or gigantism linked to neurofibromatosis type 1 have been linked to optic tract gliomas, but the actual causative mechanism for overgrowth remains speculative [4].

12.2 Summary

Recent advances in pituitary adenoma research mean that nearly 50% of cases of pituitary gigantism now have an identified genetic cause. In general, pituitary gigantism represents the most severe form of acromegaly disease spectrum, as it occurs in the youngest patients and many of the genetic forms have an aggressive phenotype. In addition, certain genetic forms like *AIP* mutations and X-LAG syndrome lead to pharmacological resistance to the main medical form of therapy, somatostatin analogs. Indeed, the young age of patients at diagnosis and the aggressiveness of the pituitary disease mean that treatment is often multimodal in nature and that control of GH/IGF-1 and tumor require approaches that combine surgery with medical therapy. These factors provide a series of unique challenges and specific aims when managing the medical treatment of pituitary gigantism (Box 12.1). Unlike adult acromegaly, pituitary gigantism has a particularly important window of opportunity to modify final adult height. In addition to the consensus approaches to

---

Box 12.1 Aims and challenges in the management of pituitary gigantism.

Aims

- Control hormonal hypersecretion (GH, IGF-1, prolactin)
- Control tumor growth/expansion
- Decrease signs and symptoms of hormonal excess
- Avoid or minimize side effects of treatment (hypopituitarism)
- Minimize or prevent an abnormal final adult height
  - Early control of GH-IGF-1 decreases final height

(Continued)

Box 12.1 (Continued)

Challenges

- Early pediatric age at onset can complicate surgery
- Macroadenomas are frequent even in children/adolescents
- Genetic forms of pituitary gigantism often lead to extensive involvement of the anterior pituitary (hyperplasia plus adenoma)
- Need for extensive surgical resection can elevate the risk for hypopituitarism
- Onset in adolescence overlaps with pubertal growth spurt which can delay diagnosis
- Lack of dosing information or clinical trials of medical therapies in pediatric patients
- In McCune–Albright syndrome craniofacial fibrous dysplasia can complicate surgical access
- *AIP* mutation–related somatotropinomas are associated with decreased hormonal and tumor size control rates with first-generation somatostatin analogs
  - There may be a role for pasireotide in *AIP* mutation related acromegaly/acrogigantism
- In X-LAG, minimal tumor residue can lead to active disease and continued GH-IGF-1 driven overgrowth
  - Pegvisomant therapy is potentially useful to control IGF-1

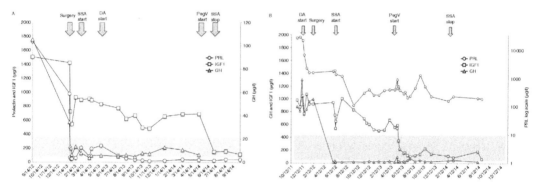

FIGURE 12.4

Treatment of two patients with X-LAG showing GH, IGF-1, and prolactin (PRL) levels over time. Figure A shows a male patient (56 months at diagnosis) who had elevated IGF-1 levels post operatively (IGF-1 normal range shown in gray) that did not normalize on a somatostatin analog and cabergoline. His growth was arrested by the addition of pegvisomant and octreotide was withdrawn. Figure B shows a similar profile in a female patient, who had a gross total resection but remained clinically and hormonally uncontrolled. Addition of octreotide LAR 30 mg/month and cabergoline did not control IGF-1. Pegvisomant rapidly brought IGF-1 levels and growth under control and octreotide was withdrawn.

Reproduced by the authors from their work in Beckers A, Lodish MB, Trivellin G, et al. X-linked acrogigantism syndrome: clinical profile and therapeutic responses. Endocr Relat Cancer 2015; 22(3):353–67. doi: 10.1530/ERC-15-0038 [22] with permission.

acromegaly in adults, management of pituitary gigantism requires early diagnosis and effective control to reduce final height. In pituitary gigantism cases where treatment with surgery and somatostatin analogs is not effective in the short to medium term (3–6 months), IGF-1 control with pegvisomant can often be achieved Fig. 12.4 and Table 12.1.

Table 12.1 Key differences in the characteristics of acromegaly and pituitary gigantism.

Features	Pituitary gigantism ($n = 208$)	Acromegaly ($n = 3173$)
Sex	78% male	54.5% female
Age at diagnosis (median; years)	21.0	45.2
Age at first symptoms (median; years)	14.0	33.5
Delay in diagnosis (median; years)	5.3	9.0
Maximum tumor diameter (median; mm)	22.0	15.0
Macroadenoma (%)	84.3	71.8
Invasion at diagnosis (%)	54.5	47.6
Prolactin cosecretion (%)	34	10

Adapted from Beckers A, Petrossians P, Hanson J, Daly AF. The causes and consequences of pituitary gigantism. Nat Rev Endocrinol 2018. doi: 10.1038/s41574-018-0114-1 [4] with permission.

References

[1] Melmed S. Chapter 15 — acromegaly. The Pituitary 2017;423—66. Available from: https://doi.org/10.1016/B978-0-12-804169-7.00015-5.

[2] Petrossians P, Daly AF, Natchev E, et al. Acromegaly at diagnosis in 3173 patients from the Liège Acromegaly Survey (LAS) Database. Endocr Relat Cancer 2017;24(10):505—18. Available from: https://doi.org/10.1530/ERC-17-0253.

[3] Melmed S, Casanueva FF, Klibanski A, et al. A consensus on the diagnosis and treatment of acromegaly complications. Pituitary 2013;16(3):294—302. Available from: https://doi.org/10.1007/s11102-012-0420-x.

[4] Beckers A, Petrossians P, Hanson J, Daly AF. The causes and consequences of pituitary gigantism. Nat Rev Endocrinol 2018. Available from: https://doi.org/10.1038/s41574-018-0114-1.

[5] Rostomyan L, Daly AF, Petrossians P, et al. Clinical and genetic characterization of pituitary gigantism: an international collaborative study in 208 patients. Endocr Relat Cancer 2015;22(5):745—57. Available from: https://doi.org/10.1530/ERC-15-0320.

[6] Vandeva S, Daly AF, Petrossians P, Zacharieva S, Beckers A. Genetics in endocrinology: somatic and germline mutations in the pathogenesis of pituitary adenomas. Eur J Endocrinol 2019. Available from: https://doi.org/10.1530/EJE-19-0602.

[7] Katznelson L, Laws ER, Melmed S, et al. Acromegaly: an Endocrine Society Clinical Practice Guideline. J Clin Endocrinol Metab 2014;99(11):3933—51. Available from: https://doi.org/10.1210/jc.2014-2700.

[8] Giustina A, Chanson P, Kleinberg D, et al. Expert consensus document: a consensus on the medical treatment of acromegaly. Nat Rev Endocrinol 2014;10(4):243—8. Available from: https://doi.org/10.1038/nrendo.2014.21.

[9] Melmed S, Bronstein MD, Chanson P, et al. A Consensus Statement on acromegaly therapeutic outcomes. Nat Rev Endocrinol 2018;14(9):552—61. Available from: https://doi.org/10.1038/s41574-018-0058-5.

[10] Breitschaft A, Hu K, Hermosillo Reséndiz K, Darstein C, Golor G. Management of hyperglycemia associated with pasireotide (SOM230): healthy volunteer study. Diabet Res Clin Pract 2014;103(3):458—65. Available from: https://doi.org/10.1016/j.diabres.2013.12.011.

[11] Daly AF, Tichomirowa MA, Petrossians P, et al. Clinical characteristics and therapeutic responses in patients with germ-line AIP mutations and pituitary adenomas: an international collaborative study. J Clin Endocrinol Metab 2010;95(11):E373—83. Available from: https://doi.org/10.1210/jc.2009-2556.

[12] Daly AF, Vanbellinghen J-F, Sok KK, et al. Aryl hydrocarbon receptor-interacting protein gene mutations in familial isolated pituitary adenomas: analysis in 73 families. J Clin Endocrinol Metab 2007;92 (5):1891−6. Available from: https://doi.org/10.1210/jc.2006-2513.

[13] Daly A, Rostomyan L, Betea D, et al. AIP-mutated acromegaly resistant to first-generation somatostatin analogs: long-term control with pasireotide LAR in two patients. Endocr Connect 2019;8(4):367−77. Available from: https://doi.org/10.1530/EC-19-0004.

[14] Petrossians P, Borges-Martins L, Espinoza C, et al. Gross total resection or debulking of pituitary adenomas improves hormonal control of acromegaly by somatostatin analogs. Eur J Endocrinol 2005;152 (1):61−6. Available from: https://doi.org/10.1530/eje.1.01824.

[15] Joshi K, Daly AF, Beckers A, Zacharin M. Resistant paediatric somatotropinomas due to AIP mutations: role of pegvisomant. Horm Res Paediatr 2018. Available from: https://doi.org/10.1159/000488856.

[16] Herman-Bonert VS, Zib K, Scarlett JA, Melmed S. Growth hormone receptor antagonist therapy in acromegalic patients resistant to somatostatin analogs. J Clin Endocrinol Metab 2000;85(8):2958−61. < http://www.ncbi.nlm.nih.gov/pubmed/10946911 > .

[17] Goldenberg N, Racine MS, Thomas P, Degnan B, Chandler W, Barkan A. Treatment of pituitary gigantism with the growth hormone receptor antagonist pegvisomant. J Clin Endocrinol Metab 2008;93 (8):2953−6. Available from: https://doi.org/10.1210/jc.2007-2283.

[18] Rix M, Laurberg P, Hoejberg AS, Brock-Jacobsen B. Pegvisomant therapy in pituitary gigantism: successful treatment in a 12-year-old girl. Eur J Endocrinol 2005;153(2):195−201. Available from: https://doi.org/10.1530/eje.1.01956.

[19] Mangupli R, Rostomyan L, Castermans E, et al. Combined treatment with octreotide LAR and pegvisomant in patients with pituitary gigantism: clinical evaluation and genetic screening. Pituitary 2016;19 (5):507−14. Available from: https://doi.org/10.1007/s11102-016-0732-3.

[20] Besser GM, Burman P, Daly AF. Predictors and rates of treatment-resistant tumor growth in acromegaly. Eur J Endocrinol 2005;153(2):187−93. Available from: https://doi.org/10.1530/eje.1.01968.

[21] Trivellin G, Daly AF, Faucz FR, et al. Gigantism and acromegaly due to Xq26 microduplications and GPR101 mutation. N Engl J Med 2014;371(25):2363−74. Available from: https://doi.org/10.1056/NEJMoa1408028.

[22] Beckers A, Lodish MB, Trivellin G, et al. X-linked acrogigantism syndrome: clinical profile and therapeutic responses. Endocr Relat Cancer 2015;22(3):353−67. Available from: https://doi.org/10.1530/ERC-15-0038.

[23] Daly AF, Yuan B, Fina F, et al. Somatic mosaicism underlies X-linked acrogigantism syndrome in sporadic male subjects. Endocr Relat Cancer 2016;23(4):221−33. Available from: https://doi.org/10.1530/ERC-16-0082.

[24] Abboud D, Daly AF, Dupuis N, Bahri MA, Inoue A, Chevigné A, et al. GPR101 drives growth hormone hypersecretion and gigantism in mice via constitutive activation of G_s and $G_{q/11}$. Nat Commun 2020; 11(1):4752. Available from: https://doi.org/10.1038/s41467-020-18500-x.

[25] Daly AF, Lysy PA, Desfilles C, et al. GHRH excess and blockade in X-LAG syndrome. Endocr Relat Cancer 2016;23(3):161−70. Available from: https://doi.org/10.1530/ERC-15-0478.

[26] Naves LA, Daly AF, Dias LA, et al. Aggressive tumor growth and clinical evolution in a patient with X-linked acro-gigantism syndrome. Endocrine 2016;51(2):236−44. Available from: https://doi.org/10.1007/s12020-015-0804-6.

[27] Beckers A, Fernandes D, Fina F, et al. Paleogenetic study of ancient DNA suggestive of X-linked acrogigantism. Endocr Relat Cancer 2017;24(2). Available from: https://doi.org/10.1530/ERC-16-0558.

[28] Iacovazzo D, Caswell R, Bunce B, et al. Germline or somatic GPR101 duplication leads to X-linked acrogigantism: a clinico-pathological and genetic study. Acta Neuropathol Commun 2016;4(1):56. Available from: https://doi.org/10.1186/s40478-016-0328-1.

[29] Wise-Oringer BK, Zanazzi GJ, Gordon RJ, et al. Familial X-linked acrogigantism: postnatal outcomes and tumor pathology in a prenatally diagnosed infant and his mother. J Clin Endocrinol Metab 2019;104 (10):4667–75. Available from: https://doi.org/10.1210/jc.2019-00817.

[30] Daly A, Cuny T, Rabl W, et al. X-Linked acro-gigantism (X-LAG) syndrome : two new cases with long-term follow-up. In: *ENEA 2015 Workshop*. Vol Poster 09; 2015. <https://orbi.uliege.be/handle/2268/189337>. [accessed 26.03.18].

[31] Bergamaschi S, Ronchi CL, Giavoli C, et al. Eight-year follow-up of a child with a GH/prolactin-secreting adenoma: efficacy of pegvisomant therapy. Horm Res Paediatr 2010;73(1):74–9. Available from: https://doi.org/10.1159/000271919.

[32] Maheshwari HG, Prezant TR, Herman-Bonert V, Shahinian H, Kovacs K, Melmed S. Long-acting peptidomimergic control of gigantism caused by pituitary acidophilic stem cell adenoma. J Clin Endocrinol Metab 2000;85(9):3409–16. Available from: https://doi.org/10.1210/jcem.85.9.6824.

[33] Main KM, Sehested A, Feldt-Rasmussen U. Pegvisomant treatment in a 4-year-old girl with neurofibromatosis type 1. Horm Res 2006;65(1):1–5. Available from: https://doi.org/10.1159/000089486.

[34] Dumitrescu CE, Collins MT. McCune-Albright syndrome. Orphanet J Rare Dis 2008;3:12. Available from: https://doi.org/10.1186/1750-1172-3-12.

[35] Vasilev V, Daly AF, Thiry A, et al. McCune-Albright syndrome: a detailed pathological and genetic analysis of disease effects in an adult patient. J Clin Endocrinol Metab 2014;99(10):E2029–38. Available from: https://doi.org/10.1210/jc.2014-1291.

[36] Salenave S, Boyce AM, Collins MT, Chanson P. Acromegaly and McCune-Albright syndrome. J Clin Endocrinol Metab 2014;99(6):1955–69. Available from: https://doi.org/10.1210/jc.2013-3826.

[37] Javaid MK, Boyce A, Appelman-Dijkstra N, et al. Best practice management guidelines for fibrous dysplasia/McCune-Albright syndrome: a consensus statement from the FD/MAS international consortium. Orphanet J Rare Dis 2019;14(1):139. Available from: https://doi.org/10.1186/s13023-019-1102-9.

[38] Boyce AM, Glover M, Kelly MH, et al. Optic neuropathy in McCune-Albright syndrome: effects of early diagnosis and treatment of growth hormone excess. J Clin Endocrinol Metab 2013;98(1):E126–34. Available from: https://doi.org/10.1210/jc.2012-2111.

[39] Boyce AM, Brewer C, DeKlotz TR, et al. Association of hearing loss and otologic outcomes with fibrous dysplasia. JAMA Otolaryngol Neck Surg 2018;144(2):102. Available from: https://doi.org/10.1001/jamaoto.2017.2407.

[40] Vortmeyer AO, Glasker S, Mehta GU, et al. Somatic GNAS mutation causes widespread and diffuse pituitary disease in acromegalic patients with McCune-Albright syndrome. J Clin Endocrinol Metab 2012;97(7):2404–13. Available from: https://doi.org/10.1210/jc.2012-1274.

[41] Akintoye SO, Kelly MH, Brillante B, et al. Pegvisomant for the treatment of gsp-mediated growth hormone excess in patients with McCune-Albright syndrome. J Clin Endocrinol Metab 2006;91(8):2960–6. Available from: https://doi.org/10.1210/jc.2005-2661.

[42] Kamilaris CDC, Faucz FR, Voutetakis A, Stratakis CA. Carney complex. Exp Clin Endocrinol Diabetes 2019;127(2-3):156–64. Available from: https://doi.org/10.1055/a-0753-4943.

[43] Watson JC, Stratakis CA, Bryant-Greenwood PK, et al. Neurosurgical implications of Carney complex. J Neurosurg 2000;92(3):413–18. Available from: https://doi.org/10.3171/jns.2000.92.3.0413.

[44] Carney J Aidan, Gordon H, Carpenter PC, Vittal Shenoy B, Go VLW. The complex of myxomas, spotty pigmentation, and endocrine overactivity. Med (US) 1985;64(4):270–83. Available from: https://doi.org/10.1097/00005792-198507000-00007.

[45] Kurtkaya-Yapicier O, Scheithauer BW, Carney JA, et al. Pituitary adenoma in Carney complex: an immunohistochemical, ultrastructural, and immunoelectron microscopic study. Ultrastruct Pathol 2002;26(6):345–53. Available from: https://doi.org/10.1080/01913120290104656.

[46] Vergès B, Boureille F, Goudet P, et al. Pituitary disease in MEN type 1 (MEN1): data from the France-Belgium MEN1 multicenter study. J Clin Endocrinol Metab 2002;87(2):457−65. <http://www.ncbi.nlm.nih.gov/pubmed/11836268>.

[47] Stratakis CA, Schussheim DH, Freedman SM, et al. Pituitary macroadenoma in a 5-year-old: an early expression of multiple endocrine neoplasia type 1. J Clin Endocrinol Metab 2000;85(12):4776−80. Available from: https://doi.org/10.1210/jcem.85.12.7064.

GHRH-producing tumors and other neuroendocrine neoplasms associated with acromegaly and/or gigantism

13

Sara Pakbaz[1], Anjelica Hodgson[2] and Ozgur Mete[1,2]

[1]*Department of Pathology, University Health Network, University of Toronto, Toronto, ON, Canada*
[2]*Department of Laboratory Medicine and Pathobiology, University of Toronto, Toronto, ON, Canada*

13.1 Introduction

Acromegaly and gigantism are disease states associated with soft tissue and bony overgrowth of the face and/or extremities, with or without tall stature, as a result of increased activity in the hypothalamic−pituitary−growth axis that includes growth hormone (GH)−releasing hormone (GHRH) from the hypothalamus, GH from the anterior pituitary gland and insulin-like growth factor 1 (IGF-1) (also known as somatomedin C) from the liver. Although they are pathogenetically related, acromegaly and gigantism differ in their clinicopathologic phenotypes with regards to the status of growth plate fusion at the time of disease onset. Other disease manifestations include hypertension, diabetes mellitus, respiratory failure, cardiac failure, and other visceral organ dysfunction as a result of enlargement [1,2].

In the majority of patients, acromegaly and gigantism are caused by eutopic GH released by adenohypophyseal pituitary neuroendocrine tumors (PitNETs). Less frequently, other intrasellar GH- or GHRH-producing neuronal/paraneuronal tumors (i.e., hypothalamic gangliocytomas, ganglioneuromas, and paragangliomas) may also cause identical disease manifestations. In addition, extrasellar (i.e., ectopic) overproduction of GHRH and less commonly ectopic GH production may also be the cause of the phenotypic manifestations of acromegaly or gigantism. These interesting cases of neoplastic ectopic hormone secretion have been described in a wide spectrum of tumor types including neuroendocrine neoplasms of variable organs such as thymus, lung, pancreas, luminal gastrointestinal tract, thyroid (medullary thyroid carcinoma), paraganglia and adrenal medulla, and those arising within a mature teratoma of ovary [2−4], as well as a number of exocrine malignancies including endometrial carcinoma, non-endocrine lung cancer, multiple myeloma, and lymphoma [2,5,6]. By definition, paraneoplastic endocrine syndromes cause metabolic changes, signs, and symptoms as a result of production of active hormone peptides by tumors arising in tissues that normally do not produce those peptides. This phenomenon may be the result of tumor transdifferentiation or more commonly as a result of dedifferentiation within those tumors [7]. Of note, a recent report described acromegalic features in a patient with multiple different neoplasms and, as an underlying germline cause was not identified, it was suggested by the authors that GH excess may incite an increased risk for the development of other tumors [8].

Importantly, it should be noted that acromegaly-like manifestations may occur in disease processes that lack GHRH−GH−IGF-1 axis disturbances, and such scenarios fall into the "pseudoacromegaly" realm. A number of heterogeneous disease processes and states may be considered as causes of pseudoacromegaly including Beckwith−Wiedemann syndrome, microduplications in 15q26, Fragile X syndrome, Simpson−Golabi−Behmel syndrome, Sotos syndrome, Tatton−Brown−Rahman syndrome, Weaver syndrome, Berardinelli−Seip congenital lipodystrophy, Malan syndrome, Marshall−Smith syndrome, SETD2-related syndrome, and even maternal diabetes [9]. Given the overlap in clinical features, distinguishing between true cases of acromegaly/gigantism and pseudoacromegaly can be quite challenging; however, accurate distinction is critical due to the vastly different underlying pathologies and other associated phenotypic presentations which commonly exist in pseudoacromegaly conditions.

In this chapter, we discuss the intrasellar (i.e., hypothalamic) GHRH-producing tumors as well as other neuroendocrine neoplasms which demonstrate ectopic production of GHRH and GH. The distinction between these and the more common adenohypophyseal GH-secreting tumors is critically important, particularly due to the different surgical management strategies that may be utilized to treat these different groups of lesions [10,11].

13.2 Epidemiology and background

Acromegaly/gigantism is a rare condition with an incidence of approximately 3−4 cases per million people per year and an estimated prevalence of approximately 40−70 cases per million people worldwide [12−14]. The majority of acromegaly and gigantism cases are caused by eutopic excess in GH production by PitNETs originating from adenohypophyseal cells. In contrast, ectopic production of GH is exceptionally uncommon and has been reported in less than 1% of patients with acromegaly [3,5,7,15,16]. Neoplastic ectopic secretion of GHRH is also rare but is well documented and has been shown to occur in approximately 0.5%−5% of all cases of acromegaly [10,15,17], more frequent in females (about 55%−60%) and in patients in the 4th decade of life [17,18]. In general, ectopic GHRH production most commonly occurs in the context of well-differentiated neuroendocrine neoplasms (NETs) of the lung, luminal gastrointestinal tract, and pancreas [18,19]. Importantly, some of these have been reported in association with germline *MEN1* mutations and are therefore associated with multiple endocrine neoplasia type 1 syndrome [2,14,19−21].

Besides pure PitNETs, other intracranial neoplasms that may cause acromegaly/gigantism include hypothalamic gangliocytomas [22]. Intrasellar gangliocytomas are a heterogeneous group of neuronal and paraneuronal tumors of which the majority appears to manifest as mixed gangliocytoma and PitNET [23]. Overall, pure gangliocytomas and mixed gangliocytoma-adenohypophyseal tumors account for 0.25%−1.25% of all sellar tumors [22,24]. While the presence of isolated gangliocytic tumors is exceptionally rare, mixed gangliocytoma-pituitary tumors producing GHRH and/or GH have been associated with 0.15%−0.5% of all acromegaly cases [23,25]. These tumors also show significant female predominance [26−28]. On the other hand, only about 20% of isolated sellar gangliocytomas (i.e., pure gangliocytomas) have been associated with acromegaly. Besides GHRH secretion, cases of gangliocytoma have been seen in association with corticotropin-releasing hormone (CRH) excess, enkephalin-induced prolactin (PRL) excess, and vasopressin (antidiuretic hormone) secretion [29–32]. Over half of these lesions produce no endocrine symptoms in the setting of pure gangliocytomas [33,34]. Nevertheless, careful clinical, biochemical, and immunohistochemical correlation must be utilized when this diagnosis is being considered.

The first case of acromegaly diagnosed in the absence of a sellar tumor was documented in a French adult male in 1959 who had undergone a previous resection of a well-differentiated pulmonary NET (carcinoid tumor of the bronchus) [35]. More than two decades following this, the GHRH amino acid sequence was discovered in human pancreatic NETs by two separate groups of investigators in 1982 [36,37]. These groups isolated the active peptide that stimulates the pituitary somatotroph cells and results in inappropriate GH release leading to the manifestations of acromegaly. Guillemin et al. identified three peptides in a pancreatic neuroendocrine neoplasm; one main peptide with 44 amino acids (later called "GHRH 1–44") and two minor peptides with 40 and 37 amino acids produced as a result of proteolysis of GHRH 1–44 [36]. They also showed that the metabolically active part of the peptide is within the first 29 amino acid residues from the N-terminal of the molecule. Rivier et al. demonstrated another biologically active peptide (later called "GHRH 1–40") at the same time in a pancreatic NET from an acromegalic patient who did not have a sellar tumor [37]. After these original descriptions, several studies demonstrated GHRH 1–44 and GHRH 1–40 in normal human hypothalamic tissue with the same amino acid sequences [17,38]. Subsequently, other studies have focused on ectopic secretion of GH or GHRH from variable endocrine and exocrine neoplasms associated with paraneoplastic acromegalic manifestations.

Ectopic secretion of bioactive compounds [39–43] has also been reported from pheochromocytomas (intraadrenal paragangliomas arising from adrenal medulla) and paragangliomas (extraadrenal tumors of the paraganglia) such as GHRH, adrenocorticotropic hormone (ACTH), CRH, parathyroid hormone (PTH), PTH-related peptide, bombesin, calcitonin, calcitonin gene–related peptide, vasoactive intestinal peptide, vasopressin (antidiuretic hormone), insulin, IGFs, somatostatin, aldosterone, renin, neuropeptide-Y, and interleukin-6 [43,44]. Among those peptides, ACTH secretion leading to Cushing syndrome is the most common [seen in approximately 1.3% of all pheochromocytomas and paragangliomas (PPGLs)] whereas GH, GHRH, and IGF-1 secreting PPGLs even more rare [43,45]. To the best of our knowledge, only one case of ectopic acromegaly secondary to GHRH-producing mediastinal paraganglioma [33] and six cases of GH/GHRH-secreting pheochromocytoma have been documented in the literature [33,39–43,46].

13.3 Clinical, radiological, and biochemical findings

The hypothalamic–pituitary–liver axis that governs the release of GHRH, GH, and IGF-1 behaves in a very predictable fashion. Excesses in this axis will cause the multisystem manifestations of acromegaly or gigantism which include general somatic overgrowth and soft tissue, coarsening of facial features, frontal bossing, mandibular prognathism, visceral organ enlargement and dysfunction, hypertension, and diabetes mellitus [17,19,33,47]. In general, the features of both of these diseases may be subtle or overt and it is generally recognized that the diagnosis is often delayed and not made until years after disease onset [48].

When a patient is found to exhibit a phenotype suspicious or consistent with a diagnosis of acromegaly or gigantism, the workup to clinch the diagnosis will include extensive clinical as well as radiological and biochemical evaluation. As indicated before, PitNETs resulting in excess GH are the most frequent cause [2]; however, the clinical phenotype of GHRH-producing tumors (both eutopic and ectopic) and ectopic GH-producing tumors is essentially indistinguishable. Although these groups share some similarities, they also differ in terms of other clinicopathologic, radiological, and biochemical features (Table 13.1).

Because the overwhelming majority of cases are due to intracranial and particularly sellar neoplasms, cranial imaging (most commonly magnetic resonance imaging) [49] is employed as a component of the initial diagnostic investigation. It should be noted that the radiological evaluation of sellar lesions is often

Table 13.1 Correlates of ectopic GH- and ectopic/eutopic GHRH-producing neoplasms.

	GHRH-producing tumors		**Ectopic GH-producing neuroendocrine neoplasms**[a]
	Eutopic production	**Ectopic production**	
Clinical findings	Acromegaly/gigantism	Acromegaly/gigantism Possible local tumor effects in the affected organ Possible clinical manifestations due to co-secretion of other active peptides	Acromegaly/gigantism Local tumor effects in the affected organ Clinical manifestations due to co-secretion of other active peptides
Tumor location	Hypothalamus/pituitary region	Variable; reports of ectopic GHRH-producing tumors include NETs of lung, pancreas, and gastrointestinal tract, in addition to pheochromocytomas/paragangliomas	Variable; reports of ectopic GH-producing tumors include NETs in bronchi and pancreas as well as lymphomas and others
Radiological findings	Pituitary hyperplasia/enlarged sella	Pituitary hyperplasia presents as an enlarged sella with no distinct primary pituitary tumor on MRI Variable extrasellar primary tumors may be identified	Pituitary imaging by MRI may be normal or show an empty sella Variable nonpituitary primary tumors may be identified
Biochemical Findings	Elevated GH and IGF-1 (IGFBP3) with or without elevated prolactin Plasma GHRH <100 ng/L	Elevated GH and IGF-1 (IGFBP3) with or without elevated prolactin Elevated plasma GHRH (>300 ng/L) and rarely in small/early tumors can be between 100 and 300 ng/L	Elevated GH and IGF-1 (IGFBP3) with or without elevated prolactin Plasma GHRH <100 ng/L
Dynamic testing	Lack of GH suppression with OGTT	Lack of GH suppression with OGTT Lack of GH response to GHRH 1−44 stimulation test Elevation of serum GH after TRH stimulation test	Lack of GH suppression with OGTT

GH, Growth hormone; *GHRH*, growth hormone−releasing hormone; *NETs*, neuroendocrine tumors; *MRI*, magnetic resonance imaging; *IGF-1*, insulin-like growth factor 1; *IGFBP3*, insulin-like growth factor-binding protein 3; *OGTT*, oral glucose tolerance test; *TRH*, thyrotropin-releasing hormone.
[a]*No clinical or biochemical features are useful to distinguish tumors producing ectopic GH from an eutopic GH-secreting tumors (i.e., those arising in the sella discussed elsewhere); therefore, radiological findings are critical in this context.*

very challenging due to the complex anatomy in the area, extensive differential diagnosis, and radiological mimicry [50] that may occur and, thus, the involvement of a radiologist with expertise and experience with such lesions is typically advised. Unlike extracranial neoplasms, the distinction of a hypothalamic neoplasm with sellar extension from a PitNET with extrasellar growth may be a challenge. Diffuse pituitary enlargement as a result of adenohypophyseal cell hyperplasia (pituitary hyperplasia) without any specific pattern of enhancement is usually seen [2]; therefore such a finding should be a clue to search for GHRH-producing tumor either intracranially or elsewhere.

If a tumor is not subsequently identified in the hypothalamic region, additional imaging of the thorax, abdomen, and pelvis is required to identify an extracranial ectopic hormone−producing source. To do this, various techniques can be used including computed tomography, octreotide scintigraphy, or Gallium-68 DOTATATE-positron emission tomography/computer tomography [2,51,52].

From a biochemical perspective, the plasma GHRH levels can provide additional information for the differential diagnosis of GHRH-producing tumors and ectopic GH-producing tumors. For instance, in patients with acromegaly due to GHRH-secreting hypothalamic gangliocytoma, the plasma GHRH usually remains within normal range. This may reflect the secretion of GHRH directly into the hypophyseal portal circulation system rather than systemic body circulation [53]. Elevated plasma GHRH levels >300 pg/mL has been shown to be a threshold with high specificity for the diagnosis of an ectopic GHRH-secreting neoplasm; in reality, most cases will produce GHRH far more than this [17,19]. In contrast, plasma GHRH levels are typically <100 pg/mL in the context of a GH-secreting neoplasm (due to negative feedback).

It should be noted that GH levels may naturally vary given the variability in its secretion in addition to molecular heterogeneity of the hormone [54]. Because of this, evaluation of GH excess includes the evaluation of serum IGF-1 levels (which are more stable and more reflective of GH activity over a 24-hour period), in addition to dynamic testing to assess GH response by an oral glucose tolerance test. An oral glucose tolerance test involves a challenge of 75 g of glucose, followed by the assessment of GH plasma levels over a number of time intervals up to 2 hours. A normal response to an oral glucose tolerance test is GH suppression to a level less than 1 ng/mL; if suppression is less than this (in conjunction with other findings), the diagnosis of acromegaly/gigantism can be made. Other dynamic tests exist to assess GH response including administration of exogenous GHRH 1−44 (lack of GH response in the cases of ectopic acromegaly) [17] as well as a thyrotropin-releasing hormone (TRH) stimulation test (elevated GH levels after TRH stimulation in cases of ectopic acromegaly) [33]. Prolactin (PRL) may also be elevated in some causes of GHRH excess and this is attributed to mammosomatotroph hyperplasia [17,19,55,56].

It should be noted that tumors are not limited to only the ectopic production of GHRH or GH and may produce multiple hormones (e.g., gastrin, calcitonin, and ACTH) simultaneously [43,57,58]. As such, additional biochemical evaluation may be required in instances where a concomitant clinical syndrome is also suspected in patients with acromegaly/gigantism [21].

13.4 **Pathological diagnosis**

Although clinical, biochemical, and radiological findings are very useful in the diagnostic workup of lesions causing acromegaly/gigantism, pathological examination, including detailed morphological and immunohistochemical biomarker evaluation, is also critical.

In the more common scenario of a pure PitNET, the reticulin network of the adenohypophysis is disrupted. Immunohistochemical evaluation with adenohypophyseal cell lineage transcription factors and hormones shows PIT1-positive PitNETs of which densely granulated somatotroph tumors are the most common histologic subtype in adults.

In the presence of sustained GHRH excess, the anterior pituitary will show hyperplastic changes that are microscopically characterized by expanded adenohypophyseal acini that are composed of a polymorphic population of normal anterior pituitary cell types with an enrichment of somatotrophs and/or mammosomatotrophs, all within a preserved reticulin framework [59]. It should be noted that a "hyperplasia-to-neoplasia" progression with the development of unifocal or multifocal GH-producing PitNETs has also been described [2,60]. Among these, the progression of pituitary hyperplasia to PitNET can occur in the background of GHRH excess. The prototypical example is seen in association with a GHRH-producing gangliocytoma admixed with a somatotroph tumor arising in the background of somatotroph hyperplasia [61]. Somatotroph/mammosomatotroph cell hyperplasia is mostly a secondary phenomenon, and this diagnosis in the setting of acromegaly with biochemically confirmed GHRH overproduction should prompt a comprehensive clinical and imaging search for the origin of GHRH production. For this reason, the diagnosis of "primary" pituitary hyperplasia should be limited to scenarios where an excess of stimulatory peptides (i.e., GHRH), either of intrasellar or extrasellar origin, has been completely excluded. On the other hand, since the coexistence of somatotroph and/or mammosomatotroph hyperplasia and PitNETs have been seen in the context of several familial syndromes (e.g., Multiple Endocrine Neoplasia type 1 and type 4, Carney Complex, McCune Albright Syndrome, and X-linked acrogigantism syndrome), germline testing should also be considered, particularly in acromegaly/gigantism patients who are young, have severe clinical manifestations, or who are resistant to routine treatments [60,62,63].

13.4.1 Neoplasms causing eutopic acromegaly

Anterior pituitary tumors leading to hypersecretion of GH include pure somatotroph tumors (densely granulated or sparsely granulated) and plurihormonal tumors such as mammosomatotroph tumors, mixed somatotroph–lactotroph tumors, acidophil stem cell tumors, and poorly differentiated PIT1 lineage tumors (formerly known as silent subtype III pituitary adenoma) [2,64]. While these are common neoplastic correlates of eutopic acromegaly, neuronal and paraneuronal tumors of the hypothalamus, such as gangliocytomas and ganglioneuromas [65,66], may also cause eutopic acromegaly by way of excess GHRH secretion.

Macroscopically, these neoplasms are usually heterogeneous solid/cystic tumors that do not typically show evidence of hemorrhage or necrosis. Microscopically, isolated gangliocytomas are often composed of evenly distributed and variable-sized ganglion cells with two or more large nuclei and prominent nucleoli, in the background of a finely fibrillary neuropil-rich stroma with scattered dysmorphic nerve cell bodies. Microcalcifications may also be identified [67]. More common than the pure tumors are the mixed gangliocytoma-PitNETs that by definition contain a mixture of an adenohypophyseal NET component (most commonly sparsely granulated somatotroph tumor) composed of epithelioid cells with eosinophilic to amphophilic cytoplasm resembling hormone-producing anterior pituitary cells (Fig. 13.1). Disruption of the reticulin framework (as seen by histochemical staining) will be identified. Mitotic figures and necrosis are not usually identified. As with other NETs, the Ki-67 labeling index should be reported [22,26].

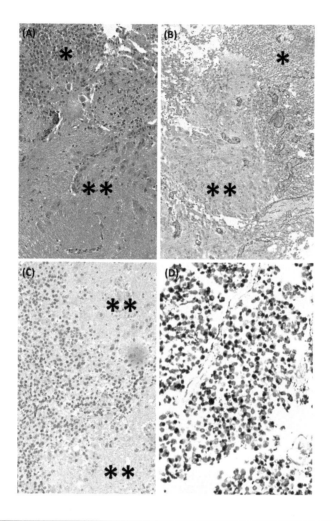

FIGURE 13.1

Eutopic acromegaly due to mixed hypothalamic gangliocytoma and sparsely granulated somatotroph tumor. This photomicrograph illustrates a hypothalamic gangliocytoma in association with a somatotroph tumor (A; single asterisk indicates somatotroph tumor, double asterisks indicate gangliocytoma). Reticulin disruption helps to distinguish somatotroph tumor from somatotroph hyperplasia (B; single asterisk indicates somatotroph tumor, double asterisks indicate gangliocytoma). The somatotroph tumor shows nuclear PIT1 expression (C; single asterisk indicates somatotroph tumor, double asterisks indicate gangliocytoma, the remaining cellular component represents PIT1-positive somatotroph tumor) and juxtanuclear low molecular weight cytokeratin (D) expression, the latter of which is a characteristic of a sparsely granulated somatotroph tumor.

GHRH secretion typically originates from the gangliocytic component, raising the possibility that the GHRH excess leads to autonomous overstimulation of the adenohypophyseal cells and development of a GH-producing PitNET in the background of gangliocytoma; however, there have been some reported cases where coexpression of GH and GHRH is limited to the adenohypophyseal component and is not identified in the gangliocytic component, refuting this argument [26,65,68].

By immunohistochemistry, the adenohypophyseal tumor component is often immunoreactive for PIT1 and GH and, depending on tumor subtype, may show additional scattered hormonal staining such as focal PRL or even ACTH reactivity. The staining pattern of low molecular weight keratins (i.e., CAM5.2) may further subclassify this component into the sparsely granulated subtype with abundant fibrous bodies versus densely granulated subtype with diffuse perinuclear cytoplasmic reactivity [61,64]. GHRH expression highlighted by immunohistochemistry is usually identified in the gangliocytoma component; however, it may be absent in exceptional circumstances. Immunohistochemically, the gangliocytoma component is highlighted by synaptophysin, neurofilament, and microtubule-associated protein 2 (MAP2) in the ganglion cells. Expression of NeuN, S100, and GFAP (in the presence of glial component) has also been reported [22].

Interestingly, extrahypothalamic gangliocytomas are usually unassociated with endocrine syndromes [69]. The occurrence of endocrine abnormalities in association with hypothalamic gangliocytomas makes these neoplasms of interest. A common origin from embryonal pituitary cell rests or the presence of hypothalamic precursor cells with features between neurons and anterior pituitary cells has also been investigated as a potential origin of these tumors [34]. Interestingly, the fact that GH is the first biochemically and immunohistochemically detected pituitary hormone during embryological development may support this theory [70]. Furthermore, the possibility of transdifferentiation between neural and adenohypophyseal cells has been suggested by several studies [26,27,34,71]. A recent study provided additional support to the possibility of transdifferentiation of pituitary neuroendocrine cells into neuronal elements by demonstrating double immunostaining characteristics of a fraction of tumor cells with the coexpression of PIT1 [developmental transcription factor regulating GH, PRL, and thyroid stimulating hormone (TSH)] and neuronal-associated cytoskeletal proteins in the gangliocytoma component of mixed gangliocytomas and somatotroph tumors [65].

The gangliocytic component may also express other hypothalamic-releasing hormones such as TRH, gonadotropin-releasing hormone, and corticotropin releasing hormone (CRH) in addition to other biologically active peptides (e.g., enkephalin, serotonin, somatostatin, vasoactive intestinal peptide, galanin, and alpha-subunit) [26,27]. In cases where there is secretion of other hypothalamic-releasing hormones, hyperplasia of the associated adenohypophyseal cell types is expected [22,23,29,31,67,72].

13.4.2 Neoplasms causing ectopic acromegaly

Neuroendocrine neoplasms secreting ectopic GHRH are most commonly well-differentiated NETs and typically present as relatively large masses (mean size of 5.5 cm) [17]. In general, there are no primary morphological correlates that can distinguish a tumor which is producing ectopic GHRH or GH from one which is not. GHRH-secreting pulmonary NETs (Fig. 13.2) are usually centrally located (bronchial or segmental) and the majority have been classified as low-grade or as "typical carcinoid of the lung (WHO 2015)" [73]. In the setting of ectopic acromegaly, pancreatic NETs with GHRH or GH secretion are usually large tumors and are most commonly located within the caudal portion of the pancreas; because of this location, these tumors are often undetectable until

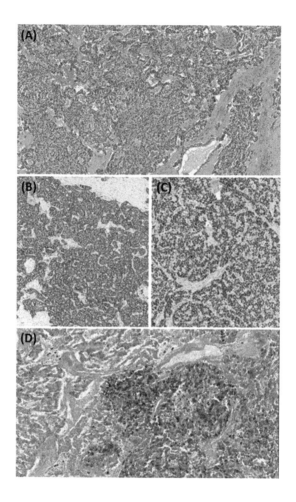

FIGURE 13.2

Ectopic acromegaly due to a GHRH-secreting well-differentiated pulmonary neuroendocrine tumor. This photomicrograph illustrates a central type well-differentiated neuroendocrine tumor of the lung (A). The tumor is positive for AE1/AE3, chromogranin-A, synaptophysin (B), and TTF-1 (C) and is negative for monoclonal carcinoembryonic antigen (CEA). These findings confirm the pulmonary origin of the tumor. The tumor also shows variable GHRH expression (D). This finding explains ectopic acromegaly. *GHRH*, Growth hormone—releasing hormone.

they become significantly enlarged and/or are metastatic. Unfortunately, the risk of metastasis for these tumors is as high as 50% [16,18,21,48]. Interestingly, about half of these pancreatic NETs not only produce GHRH but also other hormones and around 75% of them have been found to occur in the setting of multiple endocrine neoplasia types 1 and 4 [17]. Because of this, germline testing should be considered in all GHRH-producing pancreatic NETs. To facilitate the screening process, immunohistochemical evaluation for menin expression (protein encoded by *MEN1* gene,

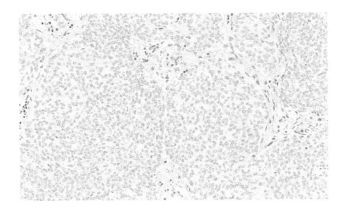

FIGURE 13.3

The role of menin immunohistochemistry in the screening for *MEN1*-driven pathogenesis in patients with ectopic acromegaly. This photomicrograph illustrates global loss of nuclear menin expression in the neuroendocrine tumor, while non-neoplastic elements (e.g., endothelial cells) remain positive for menin. While epigenetic/somatic inactivation of *MEN1* can also result in loss of nuclear staining in various neuroendocrine neoplasms, germline testing confirmed the presence of pathogenic *MEN1* mutation in this patient.

responsible for multiple endocrine neoplasia type 1) (Fig. 13.3) and p27 (protein encoded by *CDKN1B* gene, responsible for multiple endocrine neoplasia type 4) can be utilized for screening of both of these syndromes [2,19,21]. Other gastrointestinal tract NETs originating from foregut (i.e., duodenum), mid-gut, and appendix may also occur in the context of ectopic GHRH production.

The pathological evaluation of well-differentiated NETs producing ectopic hormones is identical to the assessment of any NET. It includes immunohistochemical evaluation to confirm the epithelial and neuroendocrine differentiation in these neoplasms (i.e., chromogranins, synaptophysin, INSM1, and keratins) as well as the proliferative activity by the Ki-67 labeling index [74]. Although immunoreactivity for GHRH is the diagnostic marker for detection of these ectopic sources, site-specific immunobiomarkers should also be used in order to better characterize the cell of origin for each neoplasm; biomarkers which may be used include insulin, glucagon, somatostatin, pancreatic polypeptide, peptide-YY, gastrin, glucagon-like peptide (GLP), serotonin, vasoactive intestinal peptide, PDX-1, islet-1, SATB2 (often weak) and CDX-2 for pancreatic tumors, and TTF-1, bombesin, serotonin, calcitonin, calcitonin gene-related peptide, and alpha-subunit for pulmonary tumors. NETs arising in the gastrointestinal tract may express CDX-2, SATB2, prostatic acid phosphatase (PSAP), PDX-1, VMAT2, and variable hormones including serotonin, gastrin, glucagon, glucagon-like peptide, substance P, somatostatin, pancreatic polypeptide, and vasoactive intestinal peptide [75].

In addition to well-differentiated NETs, PPGLs (pheochromocytomas and paragangliomas) may be associated with ectopic hormone production [43]. PPGLs are nonepithelial neuroendocrine neoplasms arising from paraganglia distributed in a wide spectrum of locations throughout the body [76]. Macroscopically, these tumors are rubbery to firm masses that have a brown cut surface and which may have a pseudocapsule or true capsule. From a microscopic perspective, these tumors usually show a nested (zellballen) or trabecular architecture, often with high vascularity. The tumor cells may show some variability in cytological features and may be epithelioid to oval or spindle-shaped with abundant granular eosinophilic or basophilic cytoplasm. While nuclear atypia can be identified, mitotic activity is often low and necrosis is uncommon in most cases. Immunohistochemically, these tumors express chromogranin-A and synaptophysin but lack expression of keratins [67]. Expression of GATA-3 and tyrosine hydroxylase (rate-limiting enzyme in the biosynthesis of catecholamines) depending on the functional status of the tumor is also useful when making the diagnosis of PPGL [74]. Although extremely rare, immunostaining for GH and GHRH should be considered when acromegaly-related manifestations are present in these patients [43].

Poorly differentiated neuroendocrine carcinomas (often small-cell neuroendocrine carcinoma of the lung) and rare examples of adrenocortical neoplasms and medullary thyroid carcinoma can also lead to acromegaly due to production of GHRH [2,8,77−79].

In general, GHRH expression is very useful in confirming ectopic secretion from a particular neoplasm [2,60]. One should also recognize that positive staining does not always correlate with acromegaly-related manifestations. It should be emphasized that despite rarely leading to clinical acromegaly, immunoexpression for GHRH can be identified in up to 25% of neuroendocrine neoplasms [80] such as PPGLs, gastroenteropancreatic NETs, and pulmonary NETs and neuroendocrine carcinomas (small-cell carcinoma) as well as rare examples of nonendocrine neoplasms [7]. This discrepancy is potentially explained by a number of mechanisms including inability of the tumor cells to secrete the ectopic hormone which they express, secretion of a hormone with severely reduced activity, or secretion of the hormone which is promptly inactivated or otherwise rendered ineffective by the cosecretion of inhibitory hormones [77,80,81].

13.5 Treatment

In general, the primary treatment modality for GHRH-producing tumors (both eutopic and ectopic) as well as ectopic GH-producing tumors will be complete surgical resection of the responsible tumor [73]. In instances where complete resection is not possible (e.g., sensitive tumor location, poor surgical candidate, widespread disease, or metastases), medical therapeutic modalities may be employed. For example, octreotide, a long-acting somatostatin analog, is the most commonly used agent in this instance and it functions to directly inhibit GHRH secretion from tumor cells, thereby leading to decreased levels of GHRH and subsequently GH and IGF-1 [48]. Dopamine agonists may also have some utility [2]. Chemotherapy or radiation may be used, depending on the clinical scenario, particularly in the management of metastatic disease.

13.6 Prognosis

The overall prognosis for patients with ectopic acromegaly/gigantism is favorable with the 5-year survival rate being estimated to be 85% [19,82]. After resection of the responsible tumor and with a median follow-up of 2 years, the cure rate has been reported to be up to 87% [48]. Delayed diagnosis and secondary hypopituitarism as a result of unnecessary pituitary surgery are significant causes of comorbidity in these patients [2,62,82]. Reversibility of acromegalic features after the successful removal of the ectopic hormone−producing tumor usually occurs; however, in hereditary cases, a worse prognosis and higher morbidity rate are expected [20,62,83]. In patients with mixed gangliocytoma-PitNETs, the outcome is driven by the course of the adenohypophyseal component.

Generally, the risk for recurrence after resection of original GHRH-producing tumor resection is very low (less than 5%). Effective response to somatostatin analogue therapy with or without combined dopamine agonists has been documented in the treatment of residual or recurrent GHRH-producing tumors [2,14,19,82]. In terms of surveillance following resection, evaluation of plasma GHRH is an accurate monitoring tool for the detection of recurrence [19]. In cases where somatostatin analogues are used as a component of the treatment strategy, monitoring of GHRH levels may not be useful as these therapeutics function primarily at the pituitary level (GH production); therefore, GHRH may not be significantly decreased in these circumstances and monitoring of GH and IGF-1 will be a better alternative [19,84].

Skeletal abnormalities, degenerative arthropathies, cardiac problems including biventricular hypertrophy valvular insufficiency, hypertension, atherosclerosis, and arrhythmia in addition to respiratory dysfunction are the most common complications and causes of increased morbidity and mortality in patients with acromegaly [7,85]. Overall, avoiding unnecessary interventions involving the pituitary gland, careful follow-up after treatment, and controlling GH and IGF-1 hypersecretion can decrease morbidity and improve outcome for these patients.

References

[1] Katznelson L, Laws ER, Melmed S, Molitch ME, Murad MH, Utz A, et al. Acromegaly: an endocrine society clinical practice guideline. J Clin Endocrinol Metab 2014;99(11):3933−51.

[2] Akirov A, Asa SL, Amer L, Shimon I, Ezzat S. The clinicopathological spectrum of acromegaly. J Clin Med 2019;8(11).

[3] Ozkaya M, Sayiner ZA, Kiran G, Gul K, Erkutlu I, Elboga U. Ectopic acromegaly due to a growth hormone-secreting neuroendocrine-differentiated tumor developed from ovarian mature cystic teratoma. Wien Klin Wochenschr 2015;127(11−12):491−3.

[4] Biermasz NR, Smit JWA, Pereira AM, Frölich M, Romijn JA, Roelfsema F. Acromegaly caused by growth hormone-releasing hormone-producing tumors: long-term observational studies in three patients. Pituitary 2007;10(3):237−49.

[5] Beuschlein F, Strasburger CJ, Siegerstetter V, Moradpour D, Lichter P, Bidlingmaier M, et al. Acromegaly caused by secretion of growth hormone by a non-Hodgkin's lymphoma. N Engl J Med 2000;342(25):1871−6.

[6] Guilmette J, Nosé V. Paraneoplastic syndromes and other systemic disorders associated with neuroendocrine neoplasms. Semin Diagn Pathol 2019;36(4):229−39.

[7] Krug S, Boch M, Rexin P, Pfestroff A, Gress T, Michl P, et al. Acromegaly in a patient with a pulmonary neuroendocrine tumor: case report and review of current literature. BMC Res Notes 2016;9:326.

[8] Jawiarczyk-Przybyłowska A, Wojtczak B, Whitworth J, Sutkowski K, Bidlingmaier M, Korbonits M, et al. Acromegaly associated with GIST, non-small cell lung carcinoma, clear cell renal carcinoma, multiple myeloma, medulla oblongata tumour, adrenal adenoma, and follicular thyroid nodules. Endokrynol Pol 2019;70(2):213−17.

[9] Marques P, Korbonits M. Pseudoacromegaly. Front Neuroendocrinol 2019;52:113−43.

[10] Butler PW, Cochran CS, Merino MJ, Nguyen DM, Schrump DS, Gorden P. Ectopic growth hormone-releasing hormone secretion by a bronchial carcinoid tumor: clinical experience following tumor resection and long-acting octreotide therapy. Pituitary 2012;15(2):260−5.

[11] van Hoek M, Hofland LJ, de Rijke YB, van Nederveen FH, de Krijger RR, van Koetsveld PM, et al. Effects of somatostatin analogs on a growth hormone-releasing hormone secreting bronchial carcinoid, in vivo and in vitro studies. J Clin Endocrinol Metab 2009;94(2):428−33.

[12] Holdaway IM, Rajasoorya C. Epidemiology of acromegaly. Pituitary 1999;2(1):29−41.

[13] Scacchi M, Cavagnini F. Acromegaly. Pituitary 2006;9(4):297−303.

[14] Ribeiro-Oliveira A, Barkan A. The changing face of acromegaly−advances in diagnosis and treatment. Nat Rev Endocrinol 2012;8(10):605−11.

[15] Melmed S, Ezrin C, Kovacs K, Goodman RS, Frohman LA. Acromegaly due to secretion of growth hormone by an ectopic pancreatic islet-cell tumor. N Engl J Med 1985;312(1):9−17.

[16] Ezzat S, Ezrin C, Yamashita S, Melmed S. Recurrent acromegaly resulting from ectopic growth hormone gene expression by a metastatic pancreatic tumor. Cancer 1993;71(1):66−70.

[17] Borson-Chazot F, Garby L, Raverot G, Claustrat F, Raverot V, Sassolas G, et al. Acromegaly induced by ectopic secretion of GHRH: a review 30 years after GHRH discovery. Ann Endocrinol (Paris) 2012;73(6):497−502.

[18] Van den Bruel A, Fevery J, Van Dorpe J, Hofland L, Bouillon R. Hormonal and volumetric long term control of a growth hormone-releasing hormone-producing carcinoid tumor. J Clin Endocrinol Metab 1999;84(9):3162−9.

[19] Garby L, Caron P, Claustrat F, Chanson P, Tabarin A, Rohmer V, et al. Clinical characteristics and outcome of acromegaly induced by ectopic secretion of growth hormone-releasing hormone (GHRH): a French nationwide series of 21 cases. J Clin Endocrinol Metab 2012;97(6):2093−104.

[20] Weiss DE, Vogel H, Lopes MBS, Chang SD, Katznelson L. Ectopic acromegaly due to a pancreatic neuroendocrine tumor producing growth hormone-releasing hormone. Endocr Pract 2011;17(1):79−84.

[21] Duan K, Ezzat S, Asa SL, Mete O. Pancreatic neuroendocrine tumors producing GHRH, GH, Ghrelin, PTH, or PTHrP. In: La Rosa S, Sessa F, editors. Pancreatic neuroendocrine neoplasms. Cham: Springer; 2015.

[22] Asa SL, Scheithauer BW, Bilbao JM, Horvath E, Ryan N, Kovacs K, et al. A case for hypothalamic acromegaly: a clinicopathological study of six patients with hypothalamic gangliocytomas producing growth hormone-releasing factor. J Clin Endocrinol Metab 1984;58(5):796−803.

[23] Kurosaki M, Saeger W, Lüdecke DK. Intrasellar gangliocytomas associated with acromegaly. Brain Tumor Pathol 2002;19(2):63−7.

[24] Crowley RK, Al-Derazi Y, Lynch K, Rawluk D, Thompson CJ, Farrell M, et al. Acromegaly associated with gangliocytoma. Ir J Med Sci 2012;181(3):353−5.

[25] Koutourousiou M, Kontogeorgos G, Wesseling P, Grotenhuis AJ, Seretis A. Collision sellar lesions: experience with eight cases and review of the literature. Pituitary 2010;13(1):8−17.

[26] Teramoto S, Tange Y, Ishii H, Goto H, Ogino I, Arai H. Mixed gangliocytoma-pituitary adenoma containing GH and GHRH co-secreting adenoma cells. Endocrinol Diabetes Metab Case Rep 2019;2019.

[27] Puchner MJ, Lüdecke DK, Saeger W, Riedel M, Asa SL. Gangliocytomas of the sellar region−a review. Exp Clin Endocrinol Diabetes 1995;103(3):129−49.

[28] Kontogeorgos G, Mourouti G, Kyrodimou E, Liapi-Avgeri G, Parasi E. Ganglion cell containing pituitary adenomas: signs of neuronal differentiation in adenoma cells. Acta Neuropathol 2006;112(1):21−8.

[29] Saeger W, Puchner MJ, Lüdecke DK. Combined sellar gangliocytoma and pituitary adenoma in acromegaly or Cushing's disease. a report of 3 cases. Virchows Arch 1994;425(1):93−9.

[30] Asa SL. The role of hypothalamic hormones in the pathogenesis of pituitary adenomas. Pathol Res Pract 1991;187(5):581−3.

[31] Serri O, Berthelet F, Bélair M, Vallette S, Asa SL. An unusual association of a sellar gangliocytoma with a prolactinoma. Pituitary 2008;11(1):85−7.

[32] Asa SL, Ezzat S, Kelly DF, Cohan P, Takasumi Y, Barkhoudarian G, et al. Hypothalamic vasopressin-producing tumors: often inappropriate diuresis but occasionally Cushing disease. Am J Surg Pathol 2019;43(2):251−60.

[33] Ghazi AA, Amirbaigloo A, Dezfooli AA, Saadat N, Ghazi S, Pourafkari M, et al. Ectopic acromegaly due to growth hormone releasing hormone. Endocrine 2013;43(2):293−302.

[34] Geddes JF, Jansen GH, Robinson SF, Gömöri E, Holton JL, Monson JP, et al. "Gangliocytomas" of the pituitary: a heterogeneous group of lesions with differing histogenesis. Am J Surg Pathol 2000;24(4):607−13.

[35] Altmann HW, Schuetz W. [On a bronchus carcinoid containing bone (morphological and clinical observations in an acromegalic patient)]. Beitr Pathol Anat 1959;120:455−73.

[36] Guillemin R, Brazeau P, Böhlen P, Esch F, Ling N, Wehrenberg WB. Growth hormone-releasing factor from a human pancreatic tumor that caused acromegaly. Science 1982;218(4572):585−7.

[37] Rivier J, Spiess J, Thorner M, Vale W. Characterization of a growth hormone-releasing factor from a human pancreatic islet tumour. Nature 1982;300(5889):276−8.

[38] Ling N, Esch F, Böhlen P, Brazeau P, Wehrenberg WB, Guillemin R. Isolation, primary structure, and synthesis of human hypothalamic somatocrinin: growth hormone-releasing factor. Proc Natl Acad Sci USA 1984;81(14):4302−6.

[39] Sasaki A, Yumita S, Kimura S, Miura Y, Yoshinaga K. Immunoreactive corticotropin-releasing hormone, growth hormone-releasing hormone, somatostatin, and peptide histidine methionine are present in adrenal pheochromocytomas, but not in extra-adrenal pheochromocytoma. J Clin Endocrinol Metab 1990;70(4):996−9.

[40] Sano T, Saito H, Yamazaki R, Kameyama K, Ikeda M, Hosoi E, et al. Production of growth hormone-releasing factor in pheochromocytoma. N Engl J Med 1984;311(23):1520.

[41] Saito H, Sano T, Yamasaki R, Mitsuhashi S, Hosoi E, Saito S. Demonstration of biological activity of a growth hormone-releasing hormone-like substance produced by a pheochromocytoma. Acta Endocrinol 1993;129(3):246−50.

[42] Roth KA, Wilson DM, Eberwine J, Dorin RI, Kovacs K, Bensch KG, et al. Acromegaly and pheochromocytoma: a multiple endocrine syndrome caused by a plurihormonal adrenal medullary tumor. J Clin Endocrinol Metab 1986;63(6):1421−6.

[43] Angelousi A, Peppa M, Chrisoulidou A, Alexandraki K, Berthon A, Faucz FR, et al. Malignant pheochromocytomas/paragangliomas and ectopic hormonal secretion: a case series and review of the literature Cancers (Basel) 2019;11(5) [cited 2020 Jan 23] [Internet]. Available from: https://www.ncbi.nlm.nih.gov/pmc/articles/PMC6563134/.

[44] Chejfec G, Lee I, Warren WH, Gould VE. Bombesin in human neuroendocrine (NE) neoplasms. Peptides. 1985;6(Suppl. 3):107−12.

[45] Ballav C, Naziat A, Mihai R, Karavitaki N, Ansorge O, Grossman AB. Mini-review: pheochromocytomas causing the ectopic ACTH syndrome. Endocrine 2012;42(1):69−73.

[46] Vieira Neto L, Taboada GF, Corrêa LL, Polo J, Nascimento AF, Chimelli L, et al. Acromegaly secondary to growth hormone-releasing hormone secreted by an incidentally discovered pheochromocytoma. Endocr Pathol 2007;18(1):46−52.

[47] Melmed S. Medical progress: acromegaly. N Engl J Med 2006;355(24):2558−73.

[48] Losa M, Schopohl J, von Werder K. Ectopic secretion of growth hormone-releasing hormone in man. J Endocrinol Invest 1993;16(1):69−81.

[49] Chaudhary V, Bano S. Imaging of the pituitary: recent advances. Indian J Endocrinol Metab 2011;15 (Suppl. 3):S216−23.

[50] Zamora C, Castillo M. Sellar and parasellar imaging. Neurosurgery 2017;80(1):17−38.

[51] Tirosh A, Kebebew E. The utility of 68Ga-DOTATATE positron-emission tomography/computed tomography in the diagnosis, management, follow-up and prognosis of neuroendocrine tumors. Future Oncol 2018;14(2):111−22.

[52] Tierney JF, Kosche C, Schadde E, Ali A, Virmani S, Pappas SG, et al. 68Gallium-DOTATATE positron emission tomography-computed tomography (PET CT) changes management in a majority of patients with neuroendocrine tumors. Surgery 2019;165(1):178−85.

[53] Ilie I, Korbonits M. Diagnosis of acromegaly. In: Huhtaniemi I, Martini L, editors. Encyclopedia of endocrine disease. 2nd ed. 2018. p. 223−9.

[54] Cook DM, Ezzat S, Katznelson L, Kleinberg DL, Laws ER, Nippoldt TB, et al. AACE Medical Guidelines for Clinical Practice for the diagnosis and treatment of acromegaly. Endocr Pract 2004;10(3):213−25.

[55] Asa SL, Kovacs K, Stefaneanu L, Horvath E, Billestrup N, Gonzalez-Manchon C, et al. Pituitary mammosomatotroph adenomas develop in old mice transgenic for growth hormone-releasing hormone. Proc Soc Exp Biol Med 1990;193(3):232−5.

[56] Stefaneanu L, Kovacs K, Horvath E, Asa SL, Losinski NE, Billestrup N, et al. Adenohypophysial changes in mice transgenic for human growth hormone-releasing factor: a histological, immunocytochemical, and electron microscopic investigation. Endocrinology 1989;125(5):2710−18.

[57] Bevan JS, Asa SL, Rossi ML, Esiri MM, Adams CB, Burke CW. Intrasellar gangliocytoma containing gastrin and growth hormone-releasing hormone associated with a growth hormone-secreting pituitary adenoma. Clin Endocrinol (Oxf) 1989;30(3):213−24.

[58] Zornitzki T, Rubinfeld H, Lysyy L, Schiller T, Raverot V, Shimon I, et al. pNET co-secreting GHRH and calcitonin: ex vivo hormonal studies in human pituitary cells. Endocrinol Diabetes Metab Case Rep [Internet] 2016; [cited 2020 Feb 3]. Available from: https://www.ncbi.nlm.nih.gov/pmc/articles/PMC4762224/.

[59] Asa SL. Tumors of the pituitary gland. Washington, DC: American Registry of Pathology in collaboration with the Armed Forces Institute of Pathology; 2011.

[60] Gomez-Hernandez K, Ezzat S, Asa SL, Mete Ö. Clinical implications of accurate subtyping of pituitary adenomas: perspectives from the treating physician. Turk Patoloji Derg 2015;31(Suppl. 1):4−17.

[61] Mete O, Asa SL. Clinicopathological correlations in pituitary adenomas. Brain Pathol 2012;22(4):443−53.

[62] Trouillas J, Labat-Moleur F, Sturm N, Kujas M, Heymann M-F, Figarella-Branger D, et al. Pituitary tumors and hyperplasia in multiple endocrine neoplasia type 1 syndrome (MEN1): a case-control study in a series of 77 patients versus 2509 non-MEN1 patients. Am J Surg Pathol 2008;32(4):534−43.

[63] Schernthaner-Reiter MH, Trivellin G, Stratakis CA. MEN1, MEN4, and Carney complex: pathology and molecular genetics. Neuroendocrinology 2016;103(1):18−31.

[64] Mete O, Lopes MB. Overview of the 2017 WHO classification of pituitary tumors. Endocr Pathol 2017;28(3):228−43.

[65] Lopes MBS, Sloan E, Polder J. Mixed gangliocytoma-pituitary adenoma: insights on the pathogenesis of a rare sellar tumor. Am J Surg Pathol 2017;41(5):586−95.

[66] Lopes MBS. The 2017 World Health Organization classification of tumors of the pituitary gland: a summary. Acta Neuropathol 2017;134(4):521−35.

[67] Mete O, Asa SL, editors. Endocrine pathology. UK: Cambridge University Press; 2016.

[68] Sabel MC, Hans VH, Reifenberger G. Mixed gangliocytoma/pituitary adenoma. Arch Neurol 2000;57(4):587−8.

[69] Felix I, Bilbao JM, Asa SL, Tyndel F, Kovacs K, Becker LE. Cerebral and cerebellar gangliocytomas: a morphological study of nine cases. Acta Neuropathol 1994;88(3):246−51.

[70] Asa SL, Kovacs K. Functional morphology of the human fetal pituitary. Pathol Annu 1984;19(Pt 1):275−315.

[71] Towfighi J, Salam MM, McLendon RE, Powers S, Page RB. Ganglion cell-containing tumors of the pituitary gland. Arch Pathol Lab Med 1996;120(4):369−77.

[72] Kissiedu JO, Prayson RA. Sellar gangliocytoma with adrenocorticotropic and prolactin adenoma. J Clin Neurosci 2016;24:141−2.

[73] Faglia G, Arosio M, Bazzoni N. Ectopic acromegaly. Endocrinol Metab Clin North Am 1992;21(3):575−95.

[74] Duan K, Mete O. Algorithmic approach to neuroendocrine tumors in targeted biopsies: practical applications of immunohistochemical markers. Cancer Cytopathol 2016;124(12):871−84.

[75] Dayal Y, Lin HD, Tallberg K, Reichlin S, DeLellis RA, Wolfe HJ. Immunocytochemical demonstration of growth hormone-releasing factor in gastrointestinal and pancreatic endocrine tumors. Am J Clin Pathol 1986;85(1):13−20.

[76] Asa SL, Ezzat S, Mete O. The diagnosis and clinical significance of paragangliomas in unusual locations. J Clin Med 2018;7(9).

[77] Gola M, Doga M, Bonadonna S, Mazziotti G, Vescovi PP, Giustina A. Neuroendocrine tumors secreting growth hormone-releasing hormone: pathophysiological and clinical aspects. Pituitary 2006;9(3):221−9.

[78] Weng S-F, Liao K-M, Su D-H, Yu C-J, Tseng F-Y. Small cell lung cancer and acromegaly: a case report. Endocrinologist 2005;15(6):370−3.

[79] Melmed S, Kleinberg D. Pituitary masses and tumors. Williams textbook of endocrinology. 13th ed. Elsevier; 2016. p. 232−99.

[80] Asa SL, Kovacs K, Thorner MO, Leong DA, Rivier J, Vale W. Immunohistological localization of growth hormone-releasing hormone in human tumors. J Clin Endocrinol Metab 1985;60(3):423−7.

[81] Sasaki A, Sato S, Yumita S, Hanew K, Miura Y, Yoshinaga K. Multiple forms of immunoreactive growth hormone-releasing hormone in human plasma, hypothalamus, and tumor tissues. J Clin Endocrinol Metab 1989;68(1):180−5.

[82] Kyriakakis N, Trouillas J, Dang MN, Lynch J, Belchetz P, Korbonits M, et al. Diagnostic challenges and management of a patient with acromegaly due to ectopic growth hormone-releasing hormone secretion from a bronchial carcinoid tumour. Endocrinol Diabetes Metab Case Rep 2017;2017.

[83] Vergès B, Boureille F, Goudet P, Murat A, Beckers A, Sassolas G, et al. Pituitary disease in MEN type 1 (MEN1): data from the France-Belgium MEN1 multicenter study. J Clin Endocrinol Metab 2002;87(2):457−65.

[84] Ch'ng JL, Anderson JV, Williams SJ, Carr DH, Bloom SR. Remission of symptoms during long term treatment of metastatic pancreatic endocrine tumours with long acting somatostatin analogue. Br Med J (Clin Res Ed) 1986;292(6526):981−2.

[85] Colao A, Pivonello R, Marzullo P, Auriemma RS, De Martino MC, Ferone D, et al. Severe systemic complications of acromegaly. J Endocrinol Invest 2005;28(5 Suppl):65−77.

Increased growth hormone secretion from lesions outside the anterior pituitary

14

Christina Tatsi[1] and Constantine A. Stratakis[2]

[1]*Section on Endocrinology and Genetics, Eunice Kennedy Shriver National Institute of Child Health and Human Development (NICHD), National Institutes of Health, Bethesda, MD, United States* [2]*Section on Endocrinology and Genetics, Eunice Kennedy Shriver National Institute of Child Health and Human Development (NICHD), Bethesda, MD, United States*

14.1 Introduction

Acromegaly defines the constellation of signs and symptoms resulting from growth hormone (GH) excess; if it occurs before fusion of the epiphyses the disorder is called gigantism [1]. Excessive GH secretion leads to elevated insulin-like growth factor-1 (IGF-1) levels that mediate most of the adverse effects noted in patients with acromegaly/gigantism, including accelerated growth (children), soft tissue edema and thickness, musculoskeletal problems (arthralgias, prognathism, and others), cardiovascular disease (hypertension, ventricular hypertrophy), sleep apnea, impaired glucose metabolism, hypertriglyceridemia, and others [1].

The most common cause of acromegaly are GH-secreting pituitary adenomas (PAs) that represent approximately 9%−15% of all PAs in adults and 10%−15% in children (Table 14.1) [2−4].

Table 14.1 Causes of acromegaly/gigantism.

Cause	Source	Mechanism
Isolated GH excess	Pituitary	GH-secreting pituitary adenomas or somatotroph hyperplasia due to an underlying genetic defect (Carney complex, McCune−Albright syndrome, and others)
	Ectopic GH secretion	Ectopic secretion of GH by peripheral tumors
GHRH excess	Peripheral tumors (pancreatic, bronchial, medullary thyroid, pheochromocytoma, and others)	Ectopic secretion of GHRH stimulates pituitary somatotroph cells to produce excess GH
	Hypothalamic tumors (gangliocytomas, choristomas, hamartomas, and gliomas)	Ectopic secretion of GHRH stimulates pituitary somatotroph cells to produce excess GH
	Hypothalamic dysregulation	Xq26.3 duplication (including GPR101 gene)
Loss of somatostatin tone	Optic pathway gliomas or other hypothalamic lesions	Loss of the inhibitory somatostatin tone leads to aberrant GH production

Gigantism and Acromegaly. DOI: https://doi.org/10.1016/B978-0-12-814537-1.00007-5

Most of these do not occur in the context of familial/syndromic presentation, but a minority of cases has an identifiable germline genetic defect, such as multiple endocrine neoplasia (MEN) syndromes or familial isolated pituitary adenoma syndrome (FIPA) [5]. Ectopic GH releasing hormone (GHRH) or rarely GH secretion represent less than 1% of all cases of acromegaly and occur in the context of neuroendocrine tumors, such as pancreatic, bronchial, medullary thyroid carcinomas, and others [6−8]. In rare occasions, abnormal hypothalamic signals to the pituitary may result in GH excess. These may originate from excessive hypothalamic GHRH secretion or by disruption of the inhibitory somatostatin tone. This is most likely the mechanism of GH oversecretion in neurofibromatosis type 1 (NF1).

14.2 Physiology of growth hormone secretion

The hypothalamic−pituitary−GH axis involves several hormones that contribute by either stimulating or inhibiting pituitary secretion of GH. In hypothalamus, two major hormones regulate pituitary GH secretion: GHRH, secreted in the hypothalamus, stimulates GH release, while somatostatin inhibits GH production [9,10]. Additionally, ghrelin produced both in the hypothalamus and the stomach and suppressed by food intake, was originally described as a stimulant to GH secretagogue receptor, but has now been recognized to exert several additional actions [11−13]. In contrast, IGF-1 acts through negative feedback at both the hypothalamus and the pituitary, to maintain the homeostasis of GH secretion

Additional factors contribute to stimulate or inhibit overall GH actions. There is a well-described age-, gender- and puberty-related pattern of GH secretion, with older people showing decreasing GH secretion, while females and pubertal adolescents show higher overall GH levels; estrogens increase GH secretion both in vivo and in vitro [14−16]. Obesity, on the other hand, leads to decreased overall GH secretion, potentially through increased insulin and free fatty acid concentrations [15].

The incoming stimulating or inhibiting signals from the hypothalamus and periphery lead to a pulsatile secreting pattern of GH that follows a circadian rhythm [17,18]. GH may be secreted in pulses throughout 24 hours; however, GH's secretion has its maximum frequency and amplitude during sleep and typically at nighttime. The first pulse often occurs few minutes after slow wave sleep in each sleep cycle, with approximately a total of up to eight peaks occurring over 24 hours [19].

14.3 Causes of abnormal pituitary function without pituitary tumors

Several genetic conditions have been currently identified as causes of increased GH secretion due to abnormal somatotroph function, with or without pituitary tumors [2]. Most of them have an identifiable mechanism that leads to dysregulation of GH secretion at the pituitary level.

For example, Carney complex (CNC), caused in more than 70% of cases by inactivating mutations in the gene coding for the regulatory subunit type 1A of protein kinase A (*PRKAR1A*) often presents with GH and/or prolactin excess [20,21]. Although most of the patients have an identifiable PA on MRI, histopathologic evaluation of the resected adenomas and surrounding pituitary tissue shows that the mechanism of GH excess in CNC involves a progressive development of GH hyperplasia that later evolves to distinct PAs [22,23]. Similar conditions involve McCune−Albright syndrome, where mosaic mutation of the cAMP-regulating protein Gsalpha (*GNAS*) leads to somatotroph hyperplasia in up to 21% of cases, and others [24,25].

14.4 Causes of increased growth hormone secretion without abnormal pituitary function

Aberrant GHRH secretion or abnormal somatostatin secretion may contribute to abnormal GH production from the pituitary, without primary pituitary pathology. Although in these cases, pituitary hyperplasia or other pituitary abnormalities may be noted, this is considered to be the response and sequela to abnormal incoming signals from the hypothalamus or the periphery.

Aberrant GHRH secretion has been reported in bronchial carcinoids, pancreatic neuroendocrine tumors, breast tumors, pheochromocytomas, adrenal adenomas, and others [8,26]. Although several tumors express GHRH in histologic specimens, clinical signs of acromegaly are only present in a small percentage [27]. In a review of 99 patients with ectopic acromegaly, most cases were due to lung (51.5%) or pancreatic tumors (34.3%). Of note, the identification of the ectopic cause of acromegaly was delayed by 2−18 years, and several patients have undergone transsphenoidal surgery for exploration of a pituitary source [8]. Furthermore, as previously mentioned, many patients had abnormal enlargement of the pituitary gland (46%), while PAs were reported in 30% of cases [8].

Less commonly, hypothalamic tumors may produce excessive GHRH and lead to acromegaly/gigantism. The most commonly known hypothalamic tumors involved in excess GHRH secretion include choristomas, hamartomas, gliomas, and gangliocytomas [26]. Most commonly, gangliocytomas, which are neuronal tumors that present in the sellar or hypothalamic region, have been associated with excess GHRH secretion and linked in several case reports with clinical acromegaly [28].

More recently, X-linked acrogigantism (X-LAG syndrome) has been reported as another genetic cause of GH excess in childhood [29]. Although the mechanism underlying the pathogenesis of GH excess in X-LAG is not clearly understood, the patients present with Xq26.3 duplication, that involves *GPR101* gene, and pituitary adenomas and/or hyperplasia [30]. It has been postulated that this is the result of excess GHRH levels most probably due to dysregulated hypothalamic function, and thus may represent another cause of GH excess related to GHRH [31].

Abnormal somatostatin tone has been also hypothesized as the cause of GH excess in extrapituitary/hypothalamic brain tumors. This is most commonly described in the context of NF1that is reviewed below.

14.4.1 Neurofibromatosis type 1

Neurofibromatosis type 1 (NF1) is an autosomal dominant syndrome caused by genetic defects in the neurofibromin gene (*NF1*) [32]. It presents in 1:3000−4500 individuals and its main manifestations include several skin findings (café-au-lait macules, axillary/inguinal freckling), eye findings (Lisch nodules), and tumors (neurofibromas as well as susceptibility to other benign or malignant tumors) [32,33]. One of the characteristic tumors of NF1 are optic pathway gliomas (OPGs) that present in 15% of cases by the age of 6 years [34]. OPGs often involve endocrine abnormalities, most commonly precocious puberty but also others, such as delayed puberty, diencephalic syndrome, and GH deficiency [35].

In 5.5%−10.9% of patients with NF1 and OPGs, patients present with GH excess [35,36]. Only one patient with NF1 has been reported with a GH-secreting PA, suggesting that the pathogenesis of the remaining cases results probably from abnormal input signals to the pituitary [37].

The pathophysiologic mechanism of GH excess in patients with NF1 is still not clear. Although the presence of OPG seems to be a mandatory factor for developing GH excess, investigation of OPG tumors themselves did not reveal expression of GH, GHRH, or somatostatin [38,39]. Thus ectopic aberrant secretion of these hormones does not seem to be a likely cause. In contrast, the most prevailing hypothesis is that OPGs disrupt somatostatin secretion by either blocking its availability to the pituitary or invading somatostatin-secreting cells in the hypothalamus [40]. Indeed, patients with NF1 and GH excess show overall elevated basal GH levels and abnormal pulses that may reflect the status of loss of somatostatin inhibition [41].

Interestingly, there are reports of recovery of GH excess in patients with NF1. Josefson et al. and Bruzzi et al. reported nine total patients with OPGs and GH excess who were initially treated with somatostatin analog [42,43]. After stopping treatment, seven of them had normal GH secretion [42,43]. This implies that GH excess in NF1 may not be a life-long diagnosis, but an evolving disorder that may be affected by other treatments and tumor progression.

Treatment of GH excess in patients with NF1 is important not only to avoid the known complications of acromegaly/gigantism, but also to decrease the hypothetical increased risk of tumor growth, since many of the tumors found in NF1 express GH receptors [44]. Most patients with NF1 and GH excess have been managed with a somatostatin analog, octreotide, or lanreotide [35,36]. In most cases, adequate suppression of IGF-1 was achieved. Few case reports of patients whose GH excess was managed with other treatment options (chemotherapy, pegvisomant) have also been published [45,46].

14.5 Conclusions

Acromegaly/gigantism is most commonly caused by PAs. However, rare causes of GH excess should be evaluated especially in the context of unclear pituitary imaging or an underlying genetic condition. These may result from excessive GHRH secretion or disruption of the somatostatin tone, such as in NF1.

References

[1] Melmed S. Medical progress: acromegaly. N Engl J Med 2006;355(24):2558−73.

[2] Tatsi C, Stratakis CA. The genetics of pituitary adenomas. J Clin Med 2019;9(1).

[3] Daly AF, Rixhon M, Adam C, Dempegioti A, Tichomirowa MA, Beckers A. High prevalence of pituitary adenomas: a cross-sectional study in the province of Liege, Belgium. J Clin Endocrinol Metab 2006;91 (12):4769−75.

[4] Perry A, Graffeo CS, Marcellino C, Pollock BE, Wetjen NM, Meyer FB. Pediatric pituitary adenoma: case series, review of the literature, and a skull base treatment paradigm. J Neurol Surg B Skull Base 2018;79(1):91−114.

[5] Marques P, Korbonits M. Genetic aspects of pituitary adenomas. Endocrinol Metab Clin North Am 2017;46(2):335−74.

[6] Doga M, Bonadonna S, Burattin A, Giustina A. Ectopic secretion of growth hormone-releasing hormone (GHRH) in neuroendocrine tumors: relevant clinical aspects. Ann Oncol 2001;12(Suppl. 2):S89−94.

[7] Ozkaya M, Sayiner ZA, Kiran G, Gul K, Erkutlu I, Elboga U. Ectopic acromegaly due to a growth hormone-secreting neuroendocrine-differentiated tumor developed from ovarian mature cystic teratoma. Wien Klin Wochenschr 2015;127(11−12):491−3.

[8] Ghazi AA, Amirbaigloo A, Dezfooli AA, Saadat N, Ghazi S, Pourafkari M, et al. Ectopic acromegaly due to growth hormone releasing hormone. Endocrine 2013;43(2):293−302.

[9] Ling N, Esch F, Bohlen P, Brazeau P, Wehrenberg WB, Guillemin R. Isolation, primary structure, and synthesis of human hypothalamic somatocrinin: growth hormone-releasing factor. Proc Natl Acad Sci USA 1984;81(14):4302−6.

[10] Spoudeas HA, Matthews DR, Brook CG, Hindmarsh PC. The effect of changing somatostatin tone on the pituitary growth hormone and thyroid-stimulating hormone responses to their respective releasing factor stimuli. J Clin Endocrinol Metab 1992;75(2):453−8.

[11] Kojima M, Hosoda H, Date Y, Nakazato M, Matsuo H, Kangawa K. Ghrelin is a growth-hormone-releasing acylated peptide from stomach. Nature 1999;402(6762):656−60.

[12] Al Massadi O, Nogueiras R, Dieguez C, Girault JA. Ghrelin and food reward. Neuropharmacology 2019;148:131−8.

[13] Gray SM, Page LC, Tong J. Ghrelin regulation of glucose metabolism. J Neuroendocrinol 2019;31(7): e12705.

[14] Leung KC, Johannsson G, Leong GM, Ho KK. Estrogen regulation of growth hormone action. Endocr Rev 2004;25(5):693−721.

[15] Rudman D, Kutner MH, Rogers CM, Lubin MF, Fleming GA, Bain RP. Impaired growth hormone secretion in the adult population: relation to age and adiposity. J Clin Invest 1981;67(5):1361−9.

[16] Juul A, Bang P, Hertel NT, Main K, Dalgaard P, Jorgensen K, et al. Serum insulin-like growth factor-I in 1030 healthy children, adolescents, and adults: relation to age, sex, stage of puberty, testicular size, and body mass index. J Clin Endocrinol Metab 1994;78(3):744−52.

[17] Plotnick LP, Thompson RG, Kowarski A, de Lacerda L, Migeon CJ, Blizzard RM. Circadian variation of integrated concentration of growth hormone in children and adults. J Clin Endocrinol Metab 1975;40 (2):240−7.

[18] Olarescu NC, Gunawardane K, Hansen TK, Moller N, Jorgensen JOL. Normal physiology of growth hormone in adults. In: Feingold KR, Anawalt B, Boyce A, Chrousos G, Dungan K, Grossman A, Hershman JM, Kaltsas G, Koch C, Kopp P, et al., editors. Endotext. South Dartmouth (MA); 2000.

[19] Murray PG, Higham CE, Clayton PE. 60 Years of neuroendocrinology: the hypothalamo-GH axis: the past 60 years. J Endocrinol 2015;226(2):T123−40.

[20] Kirschner LS, Carney JA, Pack SD, Taymans SE, Giatzakis C, Cho YS, et al. Mutations of the gene encoding the protein kinase A type I-alpha regulatory subunit in patients with the Carney complex. Nat Genet 2000;26(1):89−92.

[21] Courcoutsakis NA, Tatsi C, Patronas NJ, Lee CC, Prassopoulos PK, Stratakis CA. The complex of myxomas, spotty skin pigmentation and endocrine overactivity (Carney complex): imaging findings with clinical and pathological correlation. Insights Imaging 2013;4(1):119−33.

[22] Pack SD, Kirschner LS, Pak E, Zhuang Z, Carney JA, Stratakis CA. Genetic and histologic studies of somatomammotropic pituitary tumors in patients with the "complex of spotty skin pigmentation,

myxomas, endocrine overactivity and schwannomas" (Carney complex). J Clin Endocrinol Metab 2000;85(10):3860−5.

[23] Boikos SA, Stratakis CA. Pituitary pathology in patients with Carney complex: growth-hormone producing hyperplasia or tumors and their association with other abnormalities. Pituitary 2006;9(3):203−9.

[24] Weinstein LS, Shenker A, Gejman PV, Merino MJ, Friedman E, Spiegel AM. Activating mutations of the stimulatory G protein in the McCune-Albright syndrome. N Engl J Med 1991;325(24):1688−95.

[25] Akintoye SO, Chebli C, Booher S, Feuillan P, Kushner H, Leroith D, et al. Characterization of gsp-mediated growth hormone excess in the context of McCune-Albright syndrome. J Clin Endocrinol Metab 2002;87(11):5104−12.

[26] Gola M, Doga M, Bonadonna S, Mazziotti G, Vescovi PP, Giustina A. Neuroendocrine tumors secreting growth hormone-releasing hormone: pathophysiological and clinical aspects. Pituitary 2006;9(3):221−9.

[27] Dayal Y, Lin HD, Tallberg K, Reichlin S, DeLellis RA, Wolfe HJ. Immunocytochemical demonstration of growth hormone-releasing factor in gastrointestinal and pancreatic endocrine tumors. Am J Clin Pathol 1986;85(1):13−20.

[28] Asa SL, Scheithauer BW, Bilbao JM, Horvath E, Ryan N, Kovacs K, et al. A case for hypothalamic acromegaly: a clinicopathological study of six patients with hypothalamic gangliocytomas producing growth hormone-releasing factor. J Clin Endocrinol Metab 1984;58(5):796−803.

[29] Trivellin G, Daly AF, Faucz FR, Yuan B, Rostomyan L, Larco DO, et al. Gigantism and acromegaly due to Xq26 microduplications and GPR101 mutation. N Engl J Med 2014;371(25):2363−74.

[30] Beckers A, Lodish MB, Trivellin G, Rostomyan L, Lee M, Faucz FR, et al. X-linked acrogigantism syndrome: clinical profile and therapeutic responses. Endocr Relat Cancer 2015;22(3):353−67.

[31] Daly AF, Lysy PA, Desfilles C, Rostomyan L, Mohamed A, Caberg JH, et al. GHRH excess and blockade in X-LAG syndrome. Endocr Relat Cancer 2016;23(3):161−70.

[32] Rosner M, Hanneder M, Siegel N, Valli A, Fuchs C, Hengstschlager M. The mTOR pathway and its role in human genetic diseases. Mutat Res 2008;659(3):284−92.

[33] Evans DG, Howard E, Giblin C, Clancy T, Spencer H, Huson SM, et al. Birth incidence and prevalence of tumor-prone syndromes: estimates from a UK family genetic register service. Am J Med Genet A 2010;152A(2):327−32.

[34] Lewis RA, Gerson LP, Axelson KA, Riccardi VM, Whitford RP. von Recklinghausen neurofibromatosis. II. Incidence of optic gliomata. Ophthalmology 1984;91(8):929−35.

[35] Sani I, Albanese A. Endocrine long-term follow-up of children with neurofibromatosis type 1 and optic pathway glioma. Horm Res Paediatr 2017;87(3):179−88.

[36] Cambiaso P, Galassi S, Palmiero M, Mastronuzzi A, Del Bufalo F, Capolino R, et al. Growth hormone excess in children with neurofibromatosis type-1 and optic glioma. Am J Med Genet A 2017;173 (9):2353−8.

[37] Hozumi K, Fukuoka H, Odake Y, Takeuchi T, Uehara T, Sato T, et al. Acromegaly caused by a somatotroph adenoma in patient with neurofibromatosis type 1. Endocr J 2019;.

[38] Drimmie FM, MacLennan AC, Nicoll JA, Simpson E, McNeill E, Donaldson MD. Gigantism due to growth hormone excess in a boy with optic glioma. Clin Endocrinol (Oxf) 2000;53(4):535−8.

[39] Fuqua JS, Berkovitz GD. Growth hormone excess in a child with neurofibromatosis type 1 and optic pathway tumor: a patient report. Clin Pediatr (Phila) 1998;37(12):749−52.

[40] Manski TJ, Haworth CS, Duval-Arnould BJ, Rushing EJ. Optic pathway glioma infiltrating into somatostatinergic pathways in a young boy with gigantism. Case Rep J Neurosurg 1994;81(4):595−600.

[41] Josefson J, Listernick R, Fangusaro JR, Charrow J, Habiby R. Growth hormone excess in children with neurofibromatosis type 1-associated and sporadic optic pathway tumors. J Pediatr 2011;158(3):433−6.

[42] Bruzzi P, Sani I, Albanese A. Reversible growth hormone excess in two girls with neurofibromatosis type 1 and optic pathway glioma. Horm Res Paediatr 2015;84(6):414−22.

[43] Josefson JL, Listernick R, Charrow J, Habiby RL. Growth hormone excess in children with optic pathway tumors is a transient phenomenon. Horm Res Paediatr 2016;86(1):35−8.

[44] Cunha KS, Barboza EP, Fonseca EC. Identification of growth hormone receptor in plexiform neurofibromas of patients with neurofibromatosis type 1. Clin (Sao Paulo) 2008;63(1):39−42.

[45] Drake AJ, Lowis SP, Bouffet E, Crowne EC. Growth hormone hypersecretion in a girl with neurofibromatosis type 1 and an optic nerve glioma: resolution following chemotherapy. Horm Res 2000;53 (6):305−8.

[46] Main KM, Sehested A, Feldt-Rasmussen U. Pegvisomant treatment in a 4-year-old girl with neurofibromatosis type 1. Horm Res 2006;65(1):1−5.

Index

Note: Page numbers followed by "*f*" and "*t*" refer to figures and tables, respectively.

Printed in the United States
by Baker & Taylor Publisher Services